Oscillation Theory of Partial Differential Equations

Norio Yoshida

University of Toyama, Japan

Oscillation Theory of Partial Differential Equations

NEW JERSEY · LONDON · SINGAPORE · BEIJING · SHANGHAI · HONG KONG · TAIPEI · CHENNAI

Published by

World Scientific Publishing Co. Pte. Ltd.
5 Toh Tuck Link, Singapore 596224
USA office: 27 Warren Street, Suite 401-402, Hackensack, NJ 07601
UK office: 57 Shelton Street, Covent Garden, London WC2H 9HE

British Library Cataloguing-in-Publication Data
A catalogue record for this book is available from the British Library.

OSCILLATION THEORY OF PARTIAL DIFFERENTIAL EQUATIONS

ISBN-13 978-981-283-543-7
ISBN-10 981-283-543-1

Printed in Singapore.

To my family
Chikiko,
Aki, Hikari, Sachi, and Hidemasa

Preface

Oscillation theory of differential equations, originated from the monumental paper of C. Sturm published in 1836, has now been recognized as an important branch of mathematical analysis from both theoretical and practical viewpoints.

Roughly speaking, the objective of oscillation theory is to acquire as much information as possible about the qualitative properties of solutions of differential equations through the analysis of laws governing the distribution of zeros of solutions as well as the asymptotic behavior of solutions of differential equations under consideration.

Oscillation theory has been attracting wide attention of researchers for the past five decades, and as a consequence of their great efforts it has grown up to be a fertile field covering ordinary differential equations, functional differential equations with deviating arguments, partial differential equations with or without functional arguments and difference equations.

It goes without saying that a vast literature, research papers and books, dealing with oscillation theory has been published so far. As regards books and monographs on the subject, most of the existing ones are exclusively concerned with ordinary differential equations and/or difference equations, and there seems to be none which is devoted to the study of oscillation of partial differential equations.

The present book was motivated by this observation and is designed to provide the reader with an exposition of interesting aspects, from rudimentary to advanced, of oscillation theory of partial differential equations which dates back to the publication in 1955 of a paper by Ph. Hartman and A. Wintner.

The main body of this book is divided into eight chapters. In the first three chapters basic oscillation theory is developed for the standard three types of partial differential equations without functional arguments, namely, elliptic equations (Chapter 1), parabolic equations (Chapter 2) and hyperbolic equations (Chapter 3). After discussing oscillation of a class of fourth order Beam equations in the short Chapter 4, we will be concerned with partial differential equations with functional arguments in the following three chapters; more specifically, we present modern oscillation results for functional elliptic equations, functional parabolic equations and functional hyperbolic equations in Chapters 5, 6 and 7, respectively. In the final Chapter 8 we establish Picone type identities which generalize the celebrated identity of M. Picone discovered in 1909, and show that they can be effectively applied to the study of oscillation of a variety of partial differential equations involving half-linear elliptic equations and quasi-linear parabolic equations and systems. The multi-dimensional Riccati method is also investigated in connection with Picone type identities. It should be emphasized that in order to circumvent the difficulties of multi-dimensional analysis the oscillation problem for partial differential equations is often reduced to that of one-dimensional differential equations or inequalities, which is easier to handle, so that knowledge of oscillation theory of ordinary differential equations is indispensable to the investigation of oscillation properties of partial differential equations.

It is hoped that this book will be a standard textbook on oscillation theory of partial differential equations which is useful for both specialists and graduate students working in the field of differential equations, and also that the book will help stimulate further progress in the study of oscillation theory and related subjects.

I am greatly indebted to Professor T. Kusano for drawing my attention to the attractive subject of oscillation theory of partial differential equations and for his continual guidance and encouragement. I wish to express my sincere gratitude to Professor H. Onose for recommending me to write a book on multi-dimensional oscillation theory. Thanks are also due to my friend Professor M. Naito for many stimulating and beneficial discussions on oscillation problems as well as to my collaborators including Professors J. Jaroš, T. Kiguradze, K. Kreith, J. Manojlović, E. S. Noussair, Y. Shoukaku, S. Tanaka, T. Tanigawa, Y.-S. Tao and L.-T. Yan. Finally, I would like to express my heartfelt thanks to Professor K K Phua, Chairman of World Scientific Publishing Company, for his great consideration, and to Editor Ms. Tan Rok Ting for her expert editing of the manuscript.

N. Yoshida

Contents

Chapter 1

Oscillation of Elliptic Equations

1.1 Nodal Oscillation of Linear Elliptic Equations

We consider the linear elliptic equation

$$\ell[u] := \sum_{i,j=1}^{n} \frac{\partial}{\partial x_i}\left(a_{ij}(x)\frac{\partial u}{\partial x_j}\right) + c(x)u = 0, \quad x \in \Omega, \tag{1.1.1}$$

where Ω is an unbounded domain Ω of n-dimensional Euclidean space $\mathbb{R}^n$, $a_{ij}(x) \in C^1(\Omega;\mathbb{R})$ and $c(x) \in C(\Omega;\mathbb{R})$. We assume that the matrix $\left(a_{ij}(x)\right)_{i,j=1}^{n}$ is symmetric and positive definite in Ω, that is, $a_{ij}(x) = a_{ji}(x)$ in Ω $(i,j = 1,2,...,n)$ and $\sum_{i,j=1}^{n} a_{ij}(x)\xi_i\xi_j > 0$ for all $x \in \Omega$ and all $\xi = (\xi_1, \xi_2, ..., \xi_n) \in \mathbb{R}^n \setminus \{0\}$. The notation $|x|$ will be used for the Euclidean length of $x \in \mathbb{R}^n$.

Definition 1.1.1. A bounded domain $G \subset \Omega$ is said to be a *nodal domain* for (1.1.1) if there exists a nontrivial function $u \in C^2(G;\mathbb{R}) \cap C(\overline{G};\mathbb{R})$ such that $\ell[u] = 0$ in G and $u = 0$ on ∂G.

Definition 1.1.2. Equation (1.1.1) is said to be *nodally oscillatory* in Ω if for any $r > 0$ Eq. (1.1.1) has a nodal domain contained in Ω_r, where

$$\Omega_r = \Omega \cap \{x \in \mathbb{R}^n;\ |x| > r\}.$$

We define the functional $M[u;G]$ by

$$M[u;G] = \int_G \left[\sum_{i,j=1}^{n} a_{ij}(x)\frac{\partial u}{\partial x_i}\frac{\partial u}{\partial x_j} - c(x)u^2\right] dx$$

with domain consisting of all real-valued piecewise C^1-functions u on $\overline{G}$.

Theorem 1.1.1. *Equation* (1.1.1) *is nodally oscillatory in* Ω *if* Ω *contains a sequence of domains* G_k $(k = 1,2,...)$ *with the following properties:*

(i) G_k *are nonempty regular bounded domains with piecewise* C^1-*boundaries;*

(ii) *for any* $r > 0$ *there is a natural integer* $k(r) \in \mathbb{N}$ *for which* $G_{k(r)} \subset \Omega_r$;

(iii) *there is a nontrivial piecewise* C^1-*function* u_k *on each* $\overline{G_k}$ *such that* $u_k = 0$ *on* ∂G_k *and* $M[u_k; G_k] \leq 0$ $(k = 1, 2, ...)$.

Proof. For any $r > 0$ there exists a natural integer $k \in \mathbb{N}$ with the properties that $G_k \subset \Omega_r$ and $M[u_k; G_k] \leq 0$ for some nontrivial piecewise C^1-function u_k such that $u_k = 0$ on ∂G_k. Let $\lambda_1(\tilde{G})$ be the smallest eigenvalue of the eigenvalue problem

$$-\left(\sum_{i,j=1}^{n} \frac{\partial}{\partial x_i}\left(a_{ij}(x)\frac{\partial w}{\partial x_j}\right) + c(x)w\right) = \lambda w \quad \text{in } \tilde{G},$$

$$w = 0 \quad \text{on } \partial\tilde{G}$$

for any $\tilde{G} \subset G_k$ with piecewise C^1-boundary. It follows from Courant's minimum principle [54, p.399] that $\lambda_1(G_k) \leq 0$. Since $\lambda_1(\tilde{G})$ varies continuously when $\tilde{G}$ is deformed continuously in some sense and $\lambda_1(\tilde{G})$ increases to infinity as the domain $\tilde{G}$ shrinks to a point (see, for example, Courant and Hilbert [54, p.421], Headley [99], Noussair [220, p.102], Noussair [221, p.1240]). Hence, there exists a domain $\tilde{G}_k \subset G_k$ such that $\lambda_1(\tilde{G}_k) = 0$. This implies that $\tilde{G}_k (\subset \Omega_r)$ is a nodal domain for (1.1.1), and completes the proof. □

Let (r, θ) denote hyperspherical coordinates in $\mathbb{R}^n$, that is,

$$x_1 = r\prod_{j=1}^{n-1} \sin\theta_j,$$

$$x_i = r\cos\theta_{n-i+1}\prod_{j=1}^{n-i} \sin\theta_j \quad (i = 2, 3, ..., n-1),$$

$$x_n = r\cos\theta_1$$

$$\left(0 \leq r < \infty,\ 0 \leq \theta_i \leq \pi\ (i = 1, 2, ..., n-2),\ 0 \leq \theta_{n-1} \leq 2\pi\right).$$

We denote by $\tilde{\Lambda}(r, \theta)$ the largest eigenvalue $\Lambda(x)$ of the matrix $(a_{ij}(x))$ at $x = (r, \theta)$, and let $\tilde{c}(r, \theta) = c(x)$ at $x = (r, \theta)$. We define the spherical means of $\Lambda(x)$ and $c(x)$ by

$$\bar{\Lambda}(r) = \frac{1}{\omega_n r^{n-1}} \int_{S_r} \Lambda(x)\, dS = \frac{1}{\omega_n} \int_{S_1} \tilde{\Lambda}(r,\theta)\, d\omega,$$

$$\bar{c}(r) = \frac{1}{\omega_n r^{n-1}} \int_{S_r} c(x)\, dS = \frac{1}{\omega_n} \int_{S_1} \tilde{c}(r,\theta)\, d\omega,$$

where $S_r = \{x \in \mathbb{R}^n;\ |x| = r\}$ ($(n-1)$-dimensional sphere) and ω_n denotes the surface area of the unit sphere S_1.

Theorem 1.1.2. *Equation* (1.1.1) *is nodally oscillatory in* $\mathbb{R}^n$ *if the ordinary differential equation*

$$\left(r^{n-1}\bar{\Lambda}(r)y'\right)' + r^{n-1}\bar{c}(r)y = 0 \tag{1.1.2}$$

is oscillatory at $r = \infty$.

Proof. Let $\{r_k\}_{k=1,2,\ldots}$ be the zeros of a nontrivial solution $y_0(r)$ of (1.1.2) such that $r_1 < r_2 < \cdots$, $\lim_{k\to\infty} r_k = \infty$. We take G_k in Theorem 1.1.1 to be the annular domains

$$G_k = \{x \in \mathbb{R}^n;\ r_k < |x| < r_{k+1}\}, \quad k = 1, 2, \ldots.$$

Letting $u_k(x) = y_0(|x|)$, we easily see that $u_k(x) = 0$ on ∂G_k and

$$\begin{aligned} M[u_k; G_k] &= \int_{G_k} \left[\sum_{i,j=1}^{n} a_{ij}(x) \frac{\partial u_k}{\partial x_i} \frac{\partial u_k}{\partial x_j} - c(x) u_k^2 \right] dx \\ &\le \int_{r_k}^{r_{k+1}} \int_{S_1} \left[\tilde{\Lambda}(r,\theta)\big(y_0'(r)\big)^2 - \tilde{c}(r,\theta)\big(y_0(r)\big)^2 \right] r^{n-1} dr\, d\omega \\ &= \omega_n \int_{r_k}^{r_{k+1}} \left[\bar{\Lambda}(r)\big(y_0'(r)\big)^2 - \bar{c}(r)\big(y_0(r)\big)^2 \right] r^{n-1} dr \\ &= -\omega_n \int_{r_k}^{r_{k+1}} \left[\big(r^{n-1}\bar{\Lambda}(r)y_0'(r)\big)' + r^{n-1}\bar{c}(r)y_0(r) \right] y_0(r)\, dr \\ &= 0. \end{aligned}$$

The conclusion follows from Theorem 1.1.1. □

Corollary 1.1.1. *Equation*

$$\Delta u + c(x)u = 0$$

is nodally oscillatory in $\mathbb{R}^2$ *if*

$$\int_{\mathbb{R}^2} c(x)\, dx = \infty,$$

where Δ *denotes the Laplacian in* $\mathbb{R}^2$, *that is,* $\Delta = \frac{\partial^2}{\partial x_1^2} + \frac{\partial^2}{\partial x_2^2}$.

Proof. In case $n = 2$ and $a_{ij}(x) = \delta_{ij}$ (Kronecker's delta), (1.1.2) reduces to

$$(ry')' + r\bar{c}(r)y = 0. \tag{1.1.3}$$

It is easy to see that

$$\int_0^\infty r\bar{c}(r)\,dr = \frac{1}{2\pi}\int_0^\infty r\left[\int_0^{2\pi} \tilde{c}(r,\theta)\,d\theta\right] dr = \frac{1}{2\pi}\int_{\mathbb{R}^2} c(x)\,dx = \infty.$$

Leighton oscillation criterion [165] implies that (1.1.3) is oscillatory at $r = \infty$. The conclusion follows from Theorem 1.1.2. □

Remark 1.1.1. Nodal (non)oscillations of the elliptic equation

$$\Delta u + c(x)u = 0 \quad \text{in} \quad \mathbb{R}^n \tag{1.1.4}$$

have been studied by Glazman [88, p.158].

Definition 1.1.3. Elliptic equation (1.1.1) is called *nodally nonoscillatory* in Ω if (1.1.1) is not nodally oscillatory in Ω.

Remark 1.1.2. A necessary and sufficient condition for (1.1.1) to be nodally nonoscillatory was established by Kuks [148, Theorem 3]. For the nodal oscillation of higher order elliptic equations we refer to Headley [100], Noussair [220, 221]. Several authors have studied nodally nonoscillation theorems for higher order elliptic equations (see Allegretto [10, 12, 13], Headley [101, 102], Noussair and Yoshida [226], Yoshida [294]).

1.2 Superlinear Elliptic Equations

We consider the superlinear elliptic equation

$$\Delta u + c(x, u) = 0, \quad x \in \Omega, \tag{1.2.1}$$

where $\Delta = \sum_{i=1}^n \frac{\partial^2}{\partial x_i^2}$ is the Laplacian in $\mathbb{R}^n$ and Ω is an exterior domain of $\mathbb{R}^n$, that is, Ω contains the complement of some n-ball in $\mathbb{R}^n$. Let

$$\Omega \supset \{x \in \mathbb{R}^n;\ |x| \geq r_0\}$$

for some $r_0 > 0$, where $|x| = \left(\sum_{i=1}^n x_i^2\right)^{1/2}$ (the Euclidean length). We use the notation:

$$A(r_1, r_2) = \{x \in \mathbb{R}^n;\ r_1 < |x| < r_2\} \qquad \text{(annular domain)},$$
$$S_r = \{x \in \mathbb{R}^n;\ |x| = r\} \qquad \text{($(n-1)$-dimensional sphere)}.$$

Definition 1.2.1. A real-valued function u on Ω is said to be *oscillatory* in Ω if u has a zero in $\Omega \cap A(r, \infty)$ for any $r > 0$.

Definition 1.2.2. A real-valued function $y(r)$ defined on an interval $[r_y, \infty)$ is called *eventually positive* (*eventually negative*) if $y(r) > 0$ ($y(r) < 0$) on $[\tilde{r}, \infty)$ for some $\tilde{r} > r_y$.

Let (r, θ) denote hyperspherical coordinates in $\mathbb{R}^n$ (see Section 1.1), and let S and ω denote the measures on S_r and S_1, respectively. Then we obtain

$$dx = dx_1 \cdots dx_n = r^{n-1} dr\, d\omega,$$

$$\int_{S_r} dS = \int_{S_1} r^{n-1} d\omega = r^{n-1}\omega_n = r^{n-1}\frac{2\pi^{n/2}}{\Gamma(n/2)},$$

where

$$\Gamma(s) = \int_0^\infty e^{-t} t^{s-1} dt \quad \text{(the gamma function)}.$$

Associated with each function $u \in C(A(r_0, \infty); \mathbb{R})$, we define the function $U(r)$ by

$$U(r) = \frac{1}{\omega_n\, r^{n-1}} \int_{S_r} u\, dS, \quad r > r_0. \tag{1.2.2}$$

Thus, $U(r)$ denotes the spherical mean of u over S_r.

Lemma 1.2.1. *If $u \in C^2(A(r_0, \infty); \mathbb{R})$, then*

$$\frac{1}{\omega_n\, r^{n-1}} \int_{S_r} \Delta u\, dS = r^{1-n}\left(r^{n-1}U'(r)\right)', \quad r > r_0. \tag{1.2.3}$$

Proof. Let $u(x) = \tilde{u}(r, \theta)$ at $x = (r, \theta)$. Then (1.2.2) can be written in the form

$$U(r) = \frac{1}{\omega_n\, r^{n-1}} \int_{S_1} \tilde{u}\, r^{n-1} d\omega = \frac{1}{\omega_n} \int_{S_1} \tilde{u}\, d\omega.$$

The divergence theorem [242, p.81] shows that

$$\int_{A(a,r)} \Delta u\, dx = \int_{\partial A(a,r)} \sum_{i=1}^{n} \frac{\partial u}{\partial x_i} \nu_i\, dS = \int_{\partial A(a,r)} \frac{\partial u}{\partial \nu} dS, \tag{1.2.4}$$

where $r > a > r_0$ and $\nu = (\nu_1, \nu_2, ..., \nu_n)$ denotes the unit exterior normal vector to $\partial A(a, r)$. It is easy to see that (1.2.4) is equivalent to

$$\int_a^r \left(\int_{S_1} \widetilde{\Delta u}(\rho, \theta) d\omega \right) \rho^{n-1} d\rho = \int_{S_r} \frac{\partial u}{\partial \nu}\, dS + \int_{S_a} \frac{\partial u}{\partial \nu}\, dS.$$

Differentiating the above identity with respect to r yields

$$\int_{S_1} \widetilde{\Delta u}(r, \theta) r^{n-1} d\omega = \frac{d}{dr} \int_{S_r} \frac{\partial u}{\partial \nu}\, dS. \tag{1.2.5}$$

Since

$$\frac{\partial u}{\partial \nu} = \sum_{i=1}^{n} \frac{\partial u}{\partial x_i}\frac{x_i}{r} = \sum_{i=1}^{n} \frac{\partial u}{\partial x_i}\frac{\partial x_i}{\partial r} = \frac{\partial \tilde{u}}{\partial r} \quad \text{on } S_r,$$

we see from (1.2.5) that

$$\int_{S_1} \widetilde{\Delta u}(r,\theta) r^{n-1} d\omega = \frac{d}{dr}\left(\int_{S_1} \frac{\partial \tilde{u}}{\partial r} r^{n-1} d\omega\right).$$

Hence

$$\begin{aligned}\int_{S_r} \Delta u \, dS &= \int_{S_1} \widetilde{\Delta u}(r,\theta) r^{n-1} d\omega \\ &= \frac{d}{dr}\left(\int_{S_1} \frac{\partial \tilde{u}}{\partial r} r^{n-1} d\omega\right) \\ &= \omega_n \frac{d}{dr}\left(r^{n-1}\frac{d}{dr}\left(\frac{1}{\omega_n}\int_{S_1} \tilde{u}\, d\omega\right)\right) \\ &= \omega_n \left(r^{n-1}U'(r)\right)', \quad r > r_0. \qquad (1.2.6)\end{aligned}$$

Dividing (1.2.6) by $\omega_n r^{n-1}$, we obtain the desired identity (1.2.3). □

Lemma 1.2.2. *Assume that:*

(i) $c(x,\xi) \in C(A(r_0,\infty) \times \mathbb{R}; \mathbb{R})$;

(ii) $c(x,\xi) \geq p(|x|)\varphi(\xi)$ *for* $(x,\xi) \in A(r_0,\infty) \times (0,\infty)$, *where* $p(r)$ *is a function of class* $C([r_0,\infty);[0,\infty))$ *and* $\varphi(\xi)$ *is a convex function of class* $C((0,\infty);(0,\infty))$;

(iii) $u \in C^2(\Omega;\mathbb{R})$ *is a solution of* (1.2.1) *such that* $u > 0$ *in* $A(r_1,\infty)$ *for some* $r_1 > r_0$.

Then we obtain

$$\left(r^{n-1}U'(r)\right)' + r^{n-1}p(r)\varphi(U(r)) \leq 0 \qquad (1.2.7)$$

for $r > r_1$.

Proof. Since u is a solution of (1.2.1), Lemma 1.2.1 implies

$$\begin{aligned} r^{1-n}\left(r^{n-1}U'(r)\right)' &= \frac{1}{\omega_n r^{n-1}}\int_{S_r} \Delta u \, dS \\ &= -\frac{1}{\omega_n r^{n-1}}\int_{S_r} c(x,u)\, dS, \quad r > r_0. \qquad (1.2.8)\end{aligned}$$

By the hypotheses (ii) and (iii) we have

$$\begin{aligned}\frac{1}{\omega_n r^{n-1}}\int_{S_r} c(x,u)\, dS &\geq \frac{1}{\omega_n r^{n-1}}\int_{S_r} p(|x|)\varphi(u)\, dS \\ &= p(r)\frac{1}{\omega_n r^{n-1}}\int_{S_r} \varphi(u)\, dS, \quad r > r_1. \qquad (1.2.9)\end{aligned}$$

Application of Jensen's inequality [228, p.160] yields

$$\frac{1}{\omega_n r^{n-1}} \int_{S_r} \varphi(u)\, dS \geq \varphi\left(\frac{1}{\omega_n r^{n-1}} \int_{S_r} u\, dS\right)$$
$$= \varphi\big(U(r)\big), \quad r > r_1. \tag{1.2.10}$$

Combining (1.2.8)–(1.2.10), we obtain

$$r^{1-n}\left(r^{n-1}U'(r)\right)' + p(r)\varphi\big(U(r)\big) \leq 0, \quad r > r_1$$

which is the desired inequality (1.2.7). □

Theorem 1.2.1. *Assume that:*

(i) $c(x,\xi) \in C(\Omega \times \mathbb{R};\mathbb{R})$ *and* $c(x,-\xi) = -c(x,\xi)$ *for* $(x,\xi) \in \Omega \times (0,\infty)$;

(ii) *the hypothesis* (ii) *of Lemma* 1.2.2;

(iii) *the ordinary differential inequality* (1.2.7) *has no eventually positive solution.*

Then every solution $u \in C^2(\Omega;\mathbb{R})$ *of* (1.2.1) *is oscillatory in* Ω.

Proof. Suppose to the contrary that there are a number $r_1 > r_0$ and a solution $u \in C^2(\Omega;\mathbb{R})$ of (1.2.1) such that u has no zero in $A(r_1,\infty)$. First we suppose that $u > 0$ in $A(r_1,\infty)$. Lemma 1.2.2 implies that the spherical mean $U(r)$ of u over S_r satisfies (1.2.7). It is obvious that $U(r) > 0$ for $r > r_1$. This contradicts the hypothesis (iii). If $u < 0$ in $A(r_1,\infty)$, then $v := -u$ satisfies

$$\Delta v + c(x,v) = 0 \quad \text{in } \Omega,$$

$$v > 0 \quad \text{in } A(r_1,\infty).$$

Proceeding as in the case where $u > 0$, we are also led to a contradiction. The proof is complete. □

An important special case of (1.2.1) is the following equation

$$\Delta u + p(|x|)u^\gamma = 0, \quad x \in A(r_0,\infty), \tag{1.2.11}$$

where $p(r) \in C([r_0,\infty);[0,\infty))$ and $\gamma \geq 1$. Then (1.2.7) reduces to

$$\left(r^{n-1}U'(r)\right)' + r^{n-1}p(r)\big(U(r)\big)^\gamma \leq 0. \tag{1.2.12}$$

Now we consider the ordinary differential inequality

$$(a(r)y')' + c(r)y^\gamma \leq 0, \tag{1.2.13}$$

where $a(r) \in C^1([r_0, \infty); (0, \infty))$, $c(r) \in C([r_0, \infty); [0, \infty))$ and $\gamma \geq 1$.

Proposition 1.2.1. *Assume that*

$$\int_{r_0}^{\infty} \frac{1}{a(r)}\, dr = \infty$$

and let $A(r) = \int_{r_0}^{r} (1/a(s))\, ds$. *Then,* (1.2.13) *with* $\gamma > 1$ *has no eventually positive solution if and only if*

$$\int_{r_1}^{\infty} A(r) c(r)\, dr = \infty$$

for some $r_1 \geq r_0$. *Moreover,* (1.2.13) *with* $\gamma = 1$ *has no eventually positive solution if*

$$\int_{r_1}^{\infty} A(r)^2 \left[\ell_m(A(r))^{\mu} \prod_{i=0}^{m} \ell_i(A(r)) \right]^{-1} c(r)\, dr = \infty$$

for some $r_1 \geq r_0$, *some* $\mu > 0$ *and some integer* $m \geq 0$, *where* $\ell_0(r) = r$ *and* $\ell_i(r) = \log\big(\ell_{i-1}(r) + 1\big)$.

Proposition 1.2.2. *Assume that*

$$\int_{r_0}^{\infty} \frac{1}{a(r)}\, dr < \infty$$

and let $\tilde{A}(r) = \left(\int_{r}^{\infty} (1/a(s))\, ds \right)^{-1}$. *Then,* (1.2.13) *with* $\gamma > 1$ *has no eventually positive solution if and only if*

$$\int_{r_1}^{\infty} \tilde{A}(r)^{-\gamma} c(r)\, dr = \infty$$

for some $r_1 \geq r_0$. *Furthermore,* (1.2.13) *with* $\gamma = 1$ *has no eventually positive solution if*

$$\int_{r_1}^{\infty} \left[\ell_m(\tilde{A}(r))^{\mu} \prod_{i=0}^{m} \ell_i(\tilde{A}(r)) \right]^{-1} c(r)\, dr = \infty$$

for some $r_1 \geq r_0$, *some* $\mu > 0$ *and some integer* $m \geq 0$.

For the proofs of Propositions 1.2.1 and 1.2.2, see Naito and Yoshida [212].

Applying Propositions 1.2.1 and 1.2.2 to (1.2.12), we can obtain the following.

Corollary 1.2.1. *Let $\gamma\,(>1)$ be the quotient of odd integers. If*

$$\int_{r_1}^{\infty} (r \log r) p(r)\, dr = \infty \quad (n = 2),$$
$$\int_{r_1}^{\infty} r^{n-1-\gamma(n-2)} p(r)\, dr = \infty \quad (n \geq 3)$$

for some $r_1 \geq r_0$, then every solution $u \in C^2(A(r_0, \infty); \mathbb{R})$ of (1.2.11) is oscillatory in $A(r_0, \infty)$.

Proof. In case $n = 2$, we find that

$$\int_{r_0}^{\infty} \frac{1}{r}\, dr = \infty,$$
$$\int_{r_0}^{\infty} \left(\int_{r_0}^{r} \frac{1}{s}\, ds \right) r p(r)\, dr = \int_{r_0}^{\infty} (\log r - \log r_0) r p(r)\, dr = \infty.$$

Therefore, Proposition 1.2.1 implies that (1.2.12) with $n = 2$ has no eventually positive solution. In the case where $n \geq 3$, we observe that

$$\int_{r_0}^{\infty} \frac{1}{r^{n-1}}\, dr = \frac{1}{n-2} {r_0}^{2-n} < \infty;$$
$$\int_{r_0}^{\infty} \left(\int_{r}^{\infty} \frac{1}{s^{n-1}}\, ds \right)^{\gamma} r^{n-1} p(r)\, dr = \frac{1}{(n-2)^{\gamma}} \int_{r_0}^{\infty} r^{n-1-\gamma(n-2)} p(r)\, dr$$
$$= \infty.$$

It follows from Proposition 1.2.2 that (1.2.12) with $n \geq 3$ has no eventually positive solution. The conclusion follows from Theorem 1.2.1. □

Remark 1.2.1. Equation (1.2.11) with $\gamma > 1$ is called *superlinear* elliptic equation, and Eq. (1.2.11) with $0 < \gamma < 1$ is called *sublinear* elliptic equation.

Remark 1.2.2. Some open questions about oscillation have been posed by Noussair and Swanson [223]. One of them was partially solved by Allegretto [11], Naito and Yoshida [212] which deal with elliptic equations in unbounded domains which are not exterior domains. Oscillation of singular elliptic equations was considered by Kusano and Yoshida [157].

1.3 Sublinear Elliptic Equations

We consider the sublinear elliptic equation

$$\Delta u + c(x)\varphi(u) = 0, \quad x \in \Omega, \tag{1.3.1}$$

where Δ is the Laplacian in $\mathbb{R}^n$ and Ω is an exterior domain of $\mathbb{R}^n$, that is,

$$\Omega \supset \{x \in \mathbb{R}^n;\ |x| \geq r_0\}$$

for some $r_0 > 0$. It is assumed that $c(x) \in C(\Omega;\mathbb{R})$, $\varphi(\xi) \in C(\mathbb{R};\mathbb{R}) \cap C^1(\mathbb{R}\setminus\{0\};\mathbb{R})$, and $\varphi(\xi)$ satisfies

$$\xi\varphi(\xi) > 0 \quad \text{and} \quad \varphi'(\xi) \geq 0 \quad \text{for } \xi \neq 0, \tag{1.3.2}$$

$$\int_0^{\pm\delta} \frac{1}{\varphi(\xi)}\, d\xi < \infty \quad \text{for any } \delta > 0. \tag{1.3.3}$$

Let $\bar{c}(r)$ be the spherical mean of $c(x)$ over S_r, that is,

$$\bar{c}(r) = \frac{1}{\omega_n r^{n-1}} \int_{S_r} c(x)\, dS, \quad r > r_0.$$

Theorem 1.3.1. *Every solution $u \in C^2(\Omega;\mathbb{R})$ of (1.3.1) is oscillatory in Ω if the ordinary differential inequality*

$$\left(r^{n-1}y'\right)' \leq -r^{n-1}\bar{c}(r) \tag{1.3.4}$$

has no eventually positive solution.

Proof. Suppose to the contrary that there exists a solution $u \in C^2(\Omega;\mathbb{R})$ of (1.3.1) which has no zero in $A(r_1,\infty)$ for some $r_1 > r_0$. We observe that $u > 0$ in $A(r_1,\infty)$ or $u < 0$ in $A(r_1,\infty)$. We let

$$U(r) = \frac{1}{\omega_n r^{n-1}} \int_{S_r} \Phi(u)\, dS,$$

where

$$\Phi(u) = \int_0^u \frac{1}{\varphi(\xi)}\, d\xi.$$

Then $U(r) > 0$ for $r > r_1$, and by Lemma 1.2.1 we obtain

$$r^{1-n}\left(r^{n-1}U'(r)\right)' = \frac{1}{\omega_n r^{n-1}} \int_{S_r} \Delta\Phi(u)\, dS, \quad r > r_1. \tag{1.3.5}$$

A direct calculation shows that

$$\Delta\Phi(u) = \frac{\Delta u}{\varphi(u)} - \frac{\varphi'(u)}{\varphi(u)^2}|\nabla u|^2, \tag{1.3.6}$$

where ∇ denotes the nabla, that is,

$$\nabla = \left(\frac{\partial}{\partial x_1}, \frac{\partial}{\partial x_2}, \cdots, \frac{\partial}{\partial x_n}\right).$$

Combining (1.3.5) with (1.3.6) yields

$$\begin{aligned} r^{1-n}\left(r^{n-1}U'(r)\right)' &\le \frac{1}{\omega_n\, r^{n-1}}\int_{S_r}\frac{\Delta u}{\varphi(u)}\,dS \\ &= -\frac{1}{\omega_n\, r^{n-1}}\int_{S_r} c(x)\,dS \\ &= -\,\bar{c}(r), \quad r > r_1 \end{aligned}$$

which implies that $U(r)$ is an eventually positive solution of (1.3.4). The contradiction establishes the theorem. □

Now we consider the ordinary differential inequality

$$\left(a(r)y'\right)' \le g(r), \tag{1.3.7}$$

where $a(r) \in C^1([r_0,\infty);(0,\infty))$ and $g(r) \in C([r_0,\infty);\mathbb{R})$.

Lemma 1.3.1. *Assume that*

$$\int_{r_0}^{\infty}\frac{1}{a(r)}\,dr = \infty.$$

Then every solution y of (1.3.7) *defined on an interval $[r_y,\infty)$ is eventually negative if*

$$\int_{r_0}^{\infty} g(r)\,dr = -\infty. \tag{1.3.8}$$

Proof. Let y be a solution of (1.3.7) defined on $[r_y,\infty)$, and let K be a sufficiently large number. The hypothesis (1.3.8) implies that there exists a number $\tilde{r} > r_y$ such that

$$\int_{r_y}^{r} g(s)\,ds \le -K \quad \text{for } r \ge \tilde{r}.$$

Integrating (1.3.7) over $[r_y, r]$, we obtain

$$a(r)y'(r) - a(r_y)y'(r_y) \le \int_{r_y}^{r} g(s)\,ds \le -K, \quad r \ge \tilde{r}.$$

Hence

$$y'(r) \le (K_0 - K)\frac{1}{a(r)}, \quad r \ge \tilde{r},$$

where $K_0 = a(r_y)y'(r_y)$. Integration of the above inequality over $[\tilde{r}, r]$ yields

$$y(r) - y(\tilde{r}) \le (K_0 - K)\int_{\tilde{r}}^{r}\frac{1}{a(s)}\,ds. \tag{1.3.9}$$

Letting K be chosen so that $K_0 - K < 0$, we see that the right hand side of (1.3.9) tends to $-\infty$ as $r \to \infty$. Hence, $y(r)$ is eventually negative. □

Lemma 1.3.2. *Assume that*

$$\int_{r_0}^{\infty} \frac{1}{a(r)}\, dr < \infty.$$

Then every solution y of (1.3.7) defined on an interval $[r_y, \infty)$ is eventually negative if

$$\int_{r_0}^{\infty} \frac{\tilde{A}(r)^2}{a(r)}\, dr = \infty,$$

$$\int_{r_0}^{\infty} \frac{g(r)}{\tilde{A}(r)}\, dr = -\infty,$$

where

$$\tilde{A}(r) = \left(\int_r^{\infty} \frac{1}{a(s)}\, ds\right)^{-1}.$$

Proof. Since

$$(a(r)y')' = \tilde{A}(r)\left(\frac{a(r)}{\tilde{A}(r)^2}\left(\tilde{A}(r)y\right)'\right)',$$

(1.3.7) can be rewritten as

$$\left(\frac{a(r)}{\tilde{A}(r)^2}\left(\tilde{A}(r)y\right)'\right)' \le \frac{g(r)}{\tilde{A}(r)}. \tag{1.3.10}$$

The conclusion follows by applying Lemma 1.3.1 to (1.3.10). □

Lemma 1.3.3. *If*

$$\limsup_{r\to\infty} \int_{r_1}^{r} \left[1 - \frac{\psi_n(s)}{\psi_n(r)}\right] s\bar{c}(s)\, ds = \infty$$

for some $r_1 \ge r_0$, then (1.3.4) has no eventually positive solution, where

$$\psi_n(r) = \begin{cases} \log r & (n = 2) \\ r^{n-2} & (n \ge 3). \end{cases}$$

Proof. Let y be an eventually positive solution of (1.3.4) and let $y(r) > 0$ on $[r_1, \infty)$ for some $r_1 \ge r_0$. Since (1.3.4) can be written in the form

$$\left(r^{3-n}\left(r^{n-2}y\right)'\right)' \le -r\bar{c}(r), \tag{1.3.11}$$

y satisfies (1.3.11) for $r \ge r_1$. Integrating (1.3.11) twice over $[r_1, r]$, we have

$$r^{n-2}y(r) - r_1^{n-2}y(r_1) \le \alpha_n(\psi_n(r) - \psi_n(r_1)) - \beta_n\psi_n(r)\int_{r_1}^{r}\left[1 - \frac{\psi_n(s)}{\psi_n(r)}\right] s\bar{c}(s)\, ds \tag{1.3.12}$$

for $r \geq r_1$, where $\beta_n = 1$ $(n = 2)$, $\beta_n = (n-2)^{-1}$ $(n \geq 3)$, and $\alpha_n = \beta_n[r^{3-n}(r^{n-2}y)']|_{r=r_1}$. Dividing (1.3.12) by $\psi_n(r)$ and then taking the inferior limit as $r \to \infty$, we obtain

$$0 \leq \liminf_{r\to\infty} \frac{r^{n-2}y(r)}{\psi_n(r)}$$
$$\leq \alpha_n - \beta_n \limsup_{r\to\infty} \int_{r_1}^{r} \left[1 - \frac{\psi_n(s)}{\psi_n(r)}\right] s\bar{c}(s)\, ds$$

which contradicts the hypothesis. The proof is complete. □

Theorem 1.3.2. *Every solution $u \in C^2(\Omega; \mathbb{R})$ of (1.3.1) is oscillatory in Ω if*

$$\int_{r_1}^{\infty} r\bar{c}(r)\, dr = \infty \tag{1.3.13}$$

for some $r_1 \geq r_0$.

Proof. If $n = 2$, (1.3.4) reduces to

$$(ry')' \leq -r\bar{c}(r)$$

which has no eventually positive solution by Lemma 1.3.1. If $n \geq 3$, we find that $\int_{r_0}^{\infty} (1/r^{n-1})\, dr < \infty$, $\tilde{A}(r) = (n-2)r^{n-2}$, and (1.3.10) has the form

$$\left(r^{3-n}(r^{n-2}y)'\right)' \leq -r\bar{c}(r)$$

which has no eventually positive solution by Lemma 1.3.2. Hence, the conclusion follows from Theorem 1.3.1. □

Theorem 1.3.3. *Every solution $u \in C^2(\Omega; \mathbb{R})$ of (1.3.1) is oscillatory in Ω if*

$$\limsup_{r\to\infty} \int_{r_1}^{r} \left[1 - \frac{\psi_n(s)}{\psi_n(r)}\right] s\bar{c}(s)\, ds = \infty \tag{1.3.14}$$

for some $r_1 \geq r_0$.

Proof. The conclusion follows from Theorem 1.3.1 combined with Lemma 1.3.3. □

Remark 1.3.1. Since

$$\int_{r_1}^{r} \left[1 - \frac{\psi_n(s)}{\psi_n(r)}\right] s\bar{c}(s)\, ds = \frac{1}{\psi_n(r)} \int_{r_1}^{r} [\psi_n(r) - \psi_n(s)]\, s\bar{c}(s)\, ds$$
$$= \frac{1}{\psi_n(r)} \int_{r_1}^{r} \left(\int_{s}^{r} \psi_n'(t)\, dt\right) s\bar{c}(s)\, ds$$
$$= \frac{1}{\psi_n(r)} \int_{r_1}^{r} \left(\psi_n'(t) \int_{r_1}^{t} s\bar{c}(s)\, ds\right) dt,$$

we conclude that (1.3.14) is weaker than (1.3.13).

Remark 1.3.2. Under some additional hypotheses on $c(x)$, Swanson [255] showed that (1.3.13) is necessary and sufficient for every solution u of

$$\Delta u + c(x)|u|^{\gamma}\mathrm{sgn}\ u = 0, \quad x \in \Omega \tag{1.3.15}$$

to be oscillatory in Ω, where $0 < \gamma < 1$ and

$$\mathrm{sgn}\ u = \begin{cases} 1 & \text{if } u > 0 \\ 0 & \text{if } u = 0 \\ -1 & \text{if } u < 0. \end{cases}$$

Moreover, Onose [230] showed that

$$\int_{r_1}^{\infty} r(\log r)^{\gamma}\bar{c}(r)\,dr = \infty \quad \text{for some } r_1 \geq r_0$$

is necessary and sufficient for every bounded solution u of (1.3.15) to be oscillatory in Ω under some hypotheses on $c(x)$.

Remark 1.3.3. Sublinear elliptic equations with variable coefficients are considered in Section 1.5.

1.4 Perturbed Elliptic Equations

We deal with the elliptic equation with a forcing term

$$\Delta u + c(x,u) = f(x), \quad x \in \Omega, \tag{1.4.1}$$

where Δ is the Laplacian in $\mathbb{R}^n$ and Ω is an exterior domain of $\mathbb{R}^n$ which contains $\{x \in \mathbb{R}^n;\ |x| \geq r_0\}$ for some $r_0 > 0$. We use the same notation as in Section 1.2.

Let $\bar{f}(r)$ be the spherical mean of $f(x)$ over S_r, that is,

$$\bar{f}(r) = \frac{1}{\omega_n r^{n-1}} \int_{S_r} f(x)\,dS.$$

Theorem 1.4.1. *Assume that the hypotheses* (i) *and* (ii) *of Theorem* 1.2.1 *are satisfied, and that* $f(x) \in C(\Omega;\mathbb{R})$. *Every solution* $u \in C^2(\Omega;\mathbb{R})$ *of* (1.4.1) *is oscillatory in* Ω *if the ordinary differential inequalities*

$$\left(r^{n-1}y'\right)' + r^{n-1}p(r)\varphi(y) \leq r^{n-1}\bar{f}(r), \tag{1.4.2}$$

$$\left(r^{n-1}y'\right)' + r^{n-1}p(r)\varphi(y) \leq -r^{n-1}\bar{f}(r) \tag{1.4.3}$$

are oscillatory at $r = \infty$ *in the sense that neither* (1.4.2) *nor* (1.4.3) *has a solution which is positive on* $[R,\infty)$ *for any* $R > r_0$.

Proof. Assume, for the sake of contradiction, that (1.4.1) has a solution $u \in C^2(\Omega; \mathbb{R})$ which has no zero in $A(r_1, \infty) = \{x \in \mathbb{R}^n;\ |x| > r_1\}$ for some $r_1 > r_0$. First we suppose that $u > 0$ in $A(r_1, \infty)$. Then, the spherical mean $U(r)$ given by (1.2.2) is positive in (r_1, ∞). From Lemma 1.2.1 and (1.4.1) we see that

$$\begin{aligned} r^{1-n}\left(r^{n-1}U'(r)\right)' &= \frac{1}{\omega_n\, r^{n-1}} \int_{S_r} \Delta u \, dS \\ &= -\frac{1}{\omega_n\, r^{n-1}} \int_{S_r} c(x,u)\, dS + \bar{f}(r). \end{aligned} \tag{1.4.4}$$

As in the proof of Lemma 1.2.2, we obtain

$$\frac{1}{\omega_n\, r^{n-1}} \int_{S_r} c(x,u)\, dS \geq p(r)\varphi(U(r)), \quad r > r_1. \tag{1.4.5}$$

Combining (1.4.4) with (1.4.5), we conclude that $U(r)$ is a positive solution of (1.4.2) in (r_1, ∞). Analogously, if $u < 0$ in $A(r_1, \infty)$, then $-U(r)$ is a positive solution of (1.4.3) in (r_1, ∞). This contradicts the hypothesis and the proof is complete. □

Now we consider the ordinary differential inequality

$$\left(a(r)\left(b(r)y\right)'\right)' + h(r,y) \leq g(r), \tag{1.4.6}$$

where $a(r) \in C^1([r_0, \infty); (0, \infty))$, $b(r) \in C^2([r_0, \infty); (0, \infty))$, $h(r, \xi) \in C([r_0, \infty) \times (0, \infty); (0, \infty))$ and $g(r) \in C([r_0, \infty); \mathbb{R})$.

We suppose that $h(r, \xi)$ is nondecreasing in the second variable ξ, and that

$$\int_{r_0}^{\infty} \frac{1}{a(r)}\, dr = \infty. \tag{1.4.7}$$

We define

$$\alpha(r,t) = \int_t^r \frac{1}{a(s)}\, ds, \quad r, t \in [r_0, \infty).$$

Theorem 1.4.2. *Inequality* (1.4.6) *has no eventually positive solution if*

$$\liminf_{r\to\infty} \frac{1}{\alpha(r,R)} \int_R^r \alpha(r,s) g(s)\, ds = -\infty$$

for all sufficiently large R.

Proof. Let $y(r)$ be an eventually positive solution of (1.4.6). Then $y(r) > 0$ on $[r_1, \infty)$ for some $r_1 > r_0$. From (1.4.6) we see that

$$\left(a(r)\left(b(r)y(r)\right)'\right)' = g(r) - h(r, y(r)) \leq g(r), \quad r \geq r_1.$$

Integrating the above inequality over $[r_1, r]$ twice, we obtain

$$\begin{aligned} b(r)y(r) &\le c_1 + c_2 \int_{r_1}^{r} \frac{1}{a(s)}\,ds + \int_{r_1}^{r} \frac{1}{a(s)} \int_{r_1}^{s} g(t)\,dt\,ds \\ &= c_1 + c_2\alpha(r, r_1) + \int_{r_1}^{r} \alpha(r,t)g(t)\,dt, \quad r \ge r_1 \end{aligned}$$

for some constants c_1 and c_2. Dividing the above inequality by $\alpha(r, r_1)$ and taking the inferior limit as $r \to \infty$, we conclude that

$$\liminf_{r\to\infty} \frac{b(r)y(r)}{\alpha(r, r_1)} = -\infty.$$

This contradicts the positivity of $y(r)$ on $[r_1, \infty)$. □

Theorem 1.4.3. *Suppose that the unperturbed inequality*

$$\big(a(r)\big(b(r)y\big)'\big)' + h(r, y) \le 0 \tag{1.4.8}$$

has no eventually positive solution. Inequality (1.4.6) *has no eventually positive solution if there exists a function* $\theta(r) \in C^2([r_0, \infty); \mathbb{R})$ *which is oscillatory at* $r = \infty$ *and satisfies*

$$\big(a(r)\big(b(r)\theta(r)\big)'\big)' = g(r), \quad r \ge r_0 \tag{1.4.9}$$

and

$$\liminf_{r\to\infty} b(r)\theta(r) = 0. \tag{1.4.10}$$

Proof. Suppose that there exists a solution $y(r)$ of (1.4.6) which is positive on $[r_1, \infty)$ for some $r_1 > r_0$. Letting $z(r) = y(r) - \theta(r)$, we obtain

$$\big(a(r)\big(b(r)z(r)\big)'\big)' \le -h(r, y(r)) < 0, \quad r \ge r_1. \tag{1.4.11}$$

Since $a(r)\big(b(r)z(r)\big)'$ is decreasing, we observe that $\big(b(r)z(r)\big)'$ is positive on $[r_1, \infty)$ or eventually negative on $[r_1, \infty)$. Hence, $b(r)z(r)$ is a strictly monotone function. Consequently, $z(r)$ is eventually of constant sign. If $z(r) < 0$ on $[r_2, \infty)$ for some $r_2 > r_1$, then $0 < y(r) < \theta(r)$ on $[r_2, \infty)$. Since $\theta(r)$ is oscillatory at $r = \infty$, this is impossible. Hence, $z(r) > 0$ on $[r_2, \infty)$ for some $r_2 > r_1$. From (1.4.7) and (1.4.11) we conclude that $\big(b(r)z(r)\big)' > 0$ for $r \ge r_2$. An integration of (1.4.11) yields

$$a(r)\big(b(r)z(r)\big)' \ge \int_{r}^{\infty} h(r, y(r))\,dr, \quad r \ge r_2. \tag{1.4.12}$$

Dividing (1.4.12) by $a(r)$ and then integrating over $[r_1, \infty)$, we obtain

$$b(r)z(r) \ge c_3 + \int_{r_2}^{r} \frac{1}{a(s)} \int_{s}^{\infty} h(t, y(t))\,dt\,ds, \quad r \ge r_2$$

or

$$b(r)y(r) \geq c_3 + b(r)\theta(r) + \int_{r_2}^{r} \frac{1}{a(s)} \int_{s}^{\infty} h(t, y(t))\, dt\, ds, \quad r \geq r_2,$$

where $c_3 = b(r_2)z(r_2) > 0$. Using (1.4.10), we have

$$b(r)y(r) \geq \frac{c_3}{2} + \int_{R}^{r} \frac{1}{a(s)} \int_{s}^{\infty} h(t, y(t))\, dt\, ds, \quad r \geq R \tag{1.4.13}$$

for some $R > r_2$. Then

$$y(r) \geq w(r) := \frac{1}{b(r)} \left(\frac{c_3}{2} + \int_{R}^{r} \frac{1}{a(s)} \int_{s}^{\infty} h(t, y(t))\, dt\, ds \right), \quad r \geq R.$$

It is easy to see that

$$\big(a(r)\big(b(r)w(r)\big)'\big)' + h(r, y(r)) = 0, \quad r \geq R.$$

Since $h(r, w(r)) \leq h(r, y(r))$, $r \geq R$, we find that $w(r)$ is an eventually positive solution of (1.4.8). This contradiction proves the theorem. □

Theorem 1.4.4. *Assume that there exists a function $\theta(r) \in C^2([r_0, \infty); \mathbb{R})$ which is oscillatory at $r = \infty$ and satisfies (1.4.9). Inequality (1.4.6) has no eventually positive solution if*

$$\int_{R}^{\infty} h(s, \theta_+(s))\, ds = \infty$$

for all sufficiently large R, where

$$\theta_+(s) = \max\{\theta(s),\ 0\}.$$

Proof. Let $y(r)$ be a solution of (1.4.6) which is positive on $[r_1, \infty)$ for some $r_1 > r_0$. As in the proof of Theorem 1.4.3, we see that (1.4.11) holds, and that $z(r) > 0$ on $[r_2, \infty)$ for some $r_2 > r_1$, that is, $y(r) > \theta(r)$ for $r \geq r_2$. Since $y(r) > 0$ on $[r_2, \infty)$, we obtain $y(r) \geq \theta_+(r)$. Therefore, $h(r, y(r)) \geq h(r, \theta_+(r))$. From (1.4.11) it follows that

$$\big(a(r)\big(b(r)z(r)\big)'\big)' \leq -h(r, y(r)) \leq -h(r, \theta_+(r)), \quad r \geq r_2. \tag{1.4.14}$$

Integrating (1.4.14) over $[r_2, r]$, we have

$$a(r)\big(b(r)z(r)\big)' \leq d_1 - \int_{r_2}^{r} h(s, \theta_+(s))\, ds, \quad r \geq r_2, \tag{1.4.15}$$

where d_1 is a constant. Proceeding as in the proof of Theorem 1.4.3, we observe that $\big(b(r)z(r)\big)' > 0$ for $r \geq r_2$, and therefore $a(r)\big(b(r)z(r)\big)' > 0$ for $r \geq r_2$. Letting $r \to \infty$, we conclude that the right hand side of (1.4.15) tends to $-\infty$. This contradicts the positivity of the left hand side of (1.4.15). The proof is complete. □

We now turn to the elliptic equation (1.4.1). Applying Theorems 1.4.2–1.4.4 to (1.4.2) and (1.4.3), we can derive the following oscillation results for (1.4.1).

Theorem 1.4.5. *Suppose that $c(x,\xi)$ and $f(x)$ satisfy the hypotheses of Theorem 1.4.1. Suppose that if $n = 2$, then*

$$\liminf_{r\to\infty} \int_R^r \left(1 - \frac{\log s}{\log r}\right) s\bar{f}(s)\, ds = -\infty, \tag{1.4.16}$$

$$\limsup_{r\to\infty} \int_R^r \left(1 - \frac{\log s}{\log r}\right) s\bar{f}(s)\, ds = \infty \tag{1.4.17}$$

for all sufficiently large R, and if $n \geq 3$, then

$$\liminf_{r\to\infty} \int_R^r \left[1 - \left(\frac{s}{r}\right)^{n-2}\right] s\bar{f}(s)\, ds = -\infty, \tag{1.4.18}$$

$$\limsup_{r\to\infty} \int_R^r \left[1 - \left(\frac{s}{r}\right)^{n-2}\right] s\bar{f}(s)\, ds = \infty \tag{1.4.19}$$

for all sufficiently large R. Then every solution $u \in C^2(\Omega;\mathbb{R})$ of (1.4.1) is oscillatory in Ω.

Proof. It is easily seen that

$$\alpha(r,s) = \begin{cases} \log r - \log s & (n=2) \\ \dfrac{1}{n-2}(r^{n-2} - s^{n-2}) & (n \geq 3). \end{cases}$$

In case $n = 2$, (1.4.2) and (1.4.3) reduce to

$$(ry')' + rp(r)\varphi(y) \leq \pm r\bar{f}(r). \tag{1.4.20}$$

Theorem 1.4.2 and the hypotheses imply that (1.4.20) have no eventually positive solutions. If $n \geq 3$, then (1.4.2) and (1.4.3) can be rewritten in the form

$$\left(r^{3-n}(r^{n-2}y)'\right)' + rp(r)\varphi(y) \leq \pm r\bar{f}(r) \tag{1.4.21}$$

in view of the fact that

$$(r^{n-1}y')' = r^{n-2}\left(r^{3-n}(r^{n-2}y)'\right)'.$$

As in the case where $n = 2$, it can be shown that (1.4.21) have no eventually positive solutions. The conclusion follows from Theorem 1.4.1. □

Theorem 1.4.6. *Suppose that $c(x, \xi)$ and $f(x)$ satisfy the hypotheses of Theorem* 1.4.1. *Moreover, suppose that the inequality*

$$\left(r^{3-n}(r^{n-2}y)'\right)' + rp(r)\varphi(y) \leq 0$$

has no eventually positive solution. Every solution $u \in C^2(\Omega; \mathbb{R})$ of (1.4.1) *is oscillatory in Ω if there exists a function $\theta(r) \in C^2([r_0, \infty); \mathbb{R})$ with the properties that:*

(i) *$\theta(r)$ is oscillatory at $r = \infty$;*
(ii) $\left(r^{3-n}\left(r^{n-2}\theta(r)\right)'\right)' = r\bar{f}(r), \quad r \geq r_0;$
(iii) $\lim_{r \to \infty} r^{n-2}\theta(r) = 0.$

Proof. The conclusion follows from Theorem 1.4.1 combined with Theorem 1.4.3. □

Theorem 1.4.7. *Suppose that $c(x, \xi)$ and $f(x)$ satisfy the hypotheses of Theorem* 1.4.1. *Moreover, suppose that there exists a function $\theta(r) \in C^2([r_0, \infty); \mathbb{R})$ which satisfies the conditions* (i) *and* (ii) *of Theorem* 1.4.6. *Every solution $u \in C^2(\Omega; \mathbb{R})$ of* (1.4.1) *is oscillatory in Ω if*

$$\int_R^\infty sp(s)\varphi(\theta_\pm(s))\,ds = \infty$$

for all sufficiently large R, where

$$\theta_\pm(s) = \max\{\pm\theta(s),\ 0\}.$$

Proof. The conclusion follows by combining Theorem 1.4.1 with Theorem 1.4.4. □

Example 1.4.1. We consider the equation

$$\Delta u + \frac{1}{4|x|}\,u^3 = |x| \sin|x|, \quad x \in \Omega, \tag{1.4.22}$$

where $\Omega = \{x \in \mathbb{R}^3;\ |x| \geq 1\}$. It is easy to see that $\bar{f}(r) = r \sin r$, and that

$$\liminf_{r \to \infty} \int_R^r \left(1 - \frac{s}{r}\right) s^2 \sin s\, ds = -\infty,$$

$$\limsup_{r \to \infty} \int_R^r \left(1 - \frac{s}{r}\right) s^2 \sin s\, ds = \infty$$

for all sufficiently large R. Theorem 1.4.5 implies that every solution $u \in C^2(\Omega; \mathbb{R})$ of (1.4.22) is oscillatory in Ω. We note that $u(x) = |x|^{-1/2}$ is a positive solution of the unperturbed equation

$$\Delta u + \frac{1}{4|x|}\,u^3 = 0.$$

Remark 1.4.1. Let $\theta(r)$ be a continuous function defined on some infinite interval $[r_0, \infty)$. The function $\theta(r)$ is said to be *oscillatory* at $r = \infty$ if $\theta(r)$ has arbitrarily large zeros. That is, for every $s > r_0$ there exists a point $t > s$ such that $\theta(t) = 0$. Otherwise $\theta(r)$ is called *nonoscillatory.*

Remark 1.4.2. Some other oscillation results for (1.4.1) were obtained by Chen and Zhang [49]. Forced oscillations were studied by a modification of Sturm's technique (Yoshida [304]). In the case where $f(x) = 0$, oscillation results for (1.4.1) with arbitrary nonlinearity were obtained by Naito, Naito and Usami [211].

1.5 Elliptic Equations with Variable Coefficients

We study the elliptic equation with variable coefficients

$$\sum_{i,j=1}^{n} \frac{\partial}{\partial x_i}\left(a_{ij}(x)\frac{\partial u}{\partial x_j}\right) + c(x,u) = f(x), \quad x \in \Omega, \tag{1.5.1}$$

where Ω is an exterior domain of $\mathbb{R}^n$ which contains $\{x \in \mathbb{R}^n;\ |x| \geq r_0\}$ for some $r_0 > 0$. It is assumed that $c(x,\xi) \in C(\Omega \times \mathbb{R}; \mathbb{R})$, $f(x) \in C(\Omega; \mathbb{R})$, $\frac{\partial^2}{\partial x_k \partial x_\ell} a_{ij}(x)$ are locally Hölder-continuous in Ω $(i, j, k, \ell = 1, 2, ..., n)$, and the matrix $A(x) := \left(a_{ij}(x)\right)_{i,j=1}^{n}$ is symmetric in Ω. Moreover, it is assumed that (1.5.1) is uniformly elliptic, that is, for some positive constant $\mu\,(< 1)$, the inequality

$$\mu|\xi|^2 \leq \xi A(x)\xi^T \leq \mu^{-1}|\xi|^2$$

holds for all $\xi \in \mathbb{R}^n$ and all $x \in \Omega$, where the superscript T denotes the transpose.

We recall that:

$$A(r_0, \infty) = \{x \in \mathbb{R}^n;\ |x| > r_0\}$$

(*cf.* Section 1.2). Let x_0 be a fixed point of $A(r_0,\infty)$. Then we see that there exists a fundamental solution $E(x) \in C^2(A(r_0,\infty) \setminus \{x_0\})$ of the operator $P := \sum_{i,j=1}^{n} \frac{\partial}{\partial x_i} a_{ij}(x) \frac{\partial}{\partial x_j}$ with a singularity at x_0, that is,

$$P[E(x)] = -\delta(x - x_0),$$

where δ is the Dirac function (see Itô [109, p.84]). In agreement with what obtains for the Laplacian Δ, we suppose that

$$\sum_{i,j=1}^{n} a_{ij}(x)\frac{\partial E(x)}{\partial x_i}\frac{\partial E(x)}{\partial x_j} \leq \psi(E(x)), \tag{1.5.2}$$

where

$$\psi(t) = \begin{cases} k^2 \exp(4\pi t), & -\infty < t < \infty \ (n = 2) \\ k^2 t^{2(n-1)/(n-2)}, & 0 < t < \infty \quad (n \ge 3) \end{cases}$$

for some positive constant $k = k(n)$, and that $\{x \in A(a,\infty);\ E(x) = -1/\varepsilon\}$ $(n = 2)$ and $\{x \in A(r_0,\infty);\ E(x) = \varepsilon\}$ $(n \ge 3)$ are compact for every small $\varepsilon > 0$ (see Levine and Payne [166]). We note that the inequality (1.5.2) holds for $|x - x_0| \le r < 1$ in the case where $n \ge 3$, $\Omega = B(x_0, 1)$ (the unit ball with center at x_0) and $E(x)$ is the Green function associated with P in $B(x_0, 1)$ (see Aviles [20, Lemma 2.2]).

We introduce the smooth function

$$\rho(x) = \begin{cases} \exp\left(-2\pi E(x)\right) & (n = 2) \\ \left(\omega_n (n-2) E(x)\right)^{1/(2-n)} & (n \ge 3) \end{cases}$$

and define a "P-sphere" as a set

$$\tilde{S}_r := \{x \in A(r_0,\infty);\ \rho(x) = r\}.$$

We note that $\tilde{S}_r$ is compact for large r, and that there exists a number $r_1 > 0$ such that $\tilde{A}(r_1,\infty) \subset A(r_0,\infty)$, where

$$\tilde{A}(r_1,\infty) = \{x \in \mathbb{R}^n;\ \rho(x) > r_1\}.$$

Let $G(t)$ be a function of class $C^2(\mathbb{R};\mathbb{R})$, and we define

$$M[u](r) = \frac{1}{\omega_n r^{n-1}} \int_{\tilde{S}_r} G(u) \frac{(\nabla\rho(x)) A(x) (\nabla\rho(x))^T}{|\nabla\rho(x)|}\, d\sigma, \quad r > r_1,$$

where $\nabla\rho(x)$ denotes the gradient of $\rho(x)$, σ denotes the measure on $\tilde{S}_r$, and ω_n denotes the surface area of the unit sphere S_1.

Lemma 1.5.1. *If $u \in C^2(\Omega;\mathbb{R})$, then we have*

$$\omega_n \frac{d}{dr}\left(r^{n-1} \frac{d}{dr}(M[u](r))\right)$$
$$= \int_{\tilde{S}_r} \Big[G''(u)(\nabla u)A(x)(\nabla u)^T + G'(u)P[u]\Big] |\nabla\rho|^{-1} d\sigma, \quad r > r_1. \quad (1.5.3)$$

Proof. In view of the fact that $\nabla\rho = -\omega_n \rho^{n-1} \nabla E$, we see that

$$M[u](r) = -\int_{\tilde{S}_r} G(u) \frac{(\nabla E) A(x) (\nabla\rho)^T}{|\nabla\rho|}\, d\sigma.$$

Applying the divergence theorem, we obtain

$$\begin{aligned}
M[u](r) &= -\int_{r_1<\rho(x)<r} G'(u)(\nabla u)A(x)(\nabla E)^T dx \\
&\quad - \int_{\tilde{S}_{r_1}} G(u)\frac{(\nabla E)A(x)(\nabla\rho)^T}{|\nabla\rho|}\, d\sigma \\
&= -\int_{r_1}^{r}\left(\int_{\tilde{S}_\eta} G'(u)(\nabla u)A(x)(\nabla E)^T|\nabla\rho|^{-1}d\sigma\right) d\eta \\
&\quad - \int_{\tilde{S}_{r_1}} G(u)\frac{(\nabla E)A(x)(\nabla\rho)^T}{|\nabla\rho|}\, d\sigma. \qquad (1.5.4)
\end{aligned}$$

Differentiating (1.5.4) yields

$$\begin{aligned}
\frac{d}{dr}M[u](r) &= -\int_{\tilde{S}_r} G'(u)(\nabla u)A(x)(\nabla E)^T|\nabla\rho|^{-1}d\sigma \\
&= \frac{1}{\omega_n r^{n-1}}\int_{\tilde{S}_r} G'(u)(\nabla u)A(x)\left(-\omega_n\rho^{n-1}\nabla E\right)^T|\nabla\rho|^{-1}d\sigma \\
&= \frac{1}{\omega_n r^{n-1}}\int_{\tilde{S}_r} G'(u)(\nabla u)A(x)\frac{(\nabla\rho)^T}{|\nabla\rho|}\, d\sigma. \qquad (1.5.5)
\end{aligned}$$

Using the divergence theorem and (1.5.5), we obtain

$$\begin{aligned}
&\int_{r_1<\rho(x)<r} \nabla\cdot\left(G'(u)(\nabla u)A(x)\right) dx \\
&= \int_{\tilde{S}_r} G'(u)(\nabla u)A(x)\frac{(\nabla\rho)^T}{|\nabla\rho|}\, d\sigma - \int_{\tilde{S}_{r_1}} G'(u)(\nabla u)A(x)\frac{(\nabla\rho)^T}{|\nabla\rho|}\, d\sigma \\
&= \omega_n r^{n-1}\frac{d}{dr}\Big(M[u](r)\Big) - \omega_n {r_1}^{n-1}\frac{d}{dr}\Big(M[u](r)\Big)\Big|_{r=r_1}, \qquad (1.5.6)
\end{aligned}$$

where

$$\nabla\cdot w = \operatorname{div} w = \sum_{i=1}^{n}\frac{\partial w_i}{\partial x_i} \quad \text{(divergence of } w\text{)}$$

for every C^1-function $w = (w_1, w_2, ..., w_n)$. On the other hand, an easy calculation shows that

$$\begin{aligned}
&\int_{r_1<\rho(x)<r} \nabla\cdot\left(G'(u)(\nabla u)A(x)\right) dx \\
&= \int_{r_1<\rho(x)<r} \Big[G''(u)(\nabla u)A(x)(\nabla u)^T + G'(u)P[u]\Big] dx. \qquad (1.5.7)
\end{aligned}$$

Combining (1.5.6) with (1.5.7), we obtain

$$\omega_n r^{n-1}\frac{d}{dr}\Big(M[u](r)\Big) - \omega_n {r_1}^{n-1}\frac{d}{dr}\Big(M[u](r)\Big)\Big|_{r=r_1}$$
$$= \int_{r_1<\rho(x)<r}\Big[G''(u)(\nabla u)A(x)(\nabla u)^T + G'(u)P[u]\Big]\,dx. \tag{1.5.8}$$

Differentiating both sides of (1.5.8) yields the desired identity (1.5.3). □

Remark 1.5.1. In case $P = \Delta$, we choose

$$E(x) = \begin{cases} (2\pi)^{-1}\log\left(|x|^{-1}\right) & (n=2) \\ \left(\omega_n(n-2)\right)^{-1}|x|^{2-n} & (n \geq 3). \end{cases}$$

It can be shown that $\rho(x) = |x|$ and $|\nabla\rho| = 1$. Then $M[u](r)$ reduces to the spherical mean of $G(u)$ over $S_r = \{x \in \mathbb{R}^n;\ |x| = r\}$. In the case where $P = \Delta$ and $G(t) = t$, Lemma 1.5.1 reduces to Lemma 1.2.1.

Lemma 1.5.2. *Letting*

$$K(r) = \int_{\tilde{S}_r}(\nabla\rho)A(x)(\nabla\rho)^T\left(\omega_n r^{n-1}|\nabla\rho|\right)^{-1}d\sigma,$$

we find that $K(r)$ is a positive constant independent of r. Moreover, if $P = \Delta$ or $\Omega = \mathbb{R}^n$, then we obtain $K(r) = 1$.

Proof. Since $\nabla\rho \not\equiv 0$, we see that $K(r) > 0$. Hence, it suffices to show that $K(r)$ is a constant independent of r. A simple computation yields

$$K(r) = \int_{\tilde{S}_r}(-\omega_n\rho^{n-1}\nabla E)A(x)(\nabla\rho)^T\left(\omega_n r^{n-1}|\nabla\rho|\right)^{-1}d\sigma$$
$$= -\int_{\tilde{S}_r}(\nabla E)A(x)\frac{(\nabla\rho)^T}{|\nabla\rho|}\,d\sigma. \tag{1.5.9}$$

From the divergence theorem it follows that

$$\int_{\tilde{S}_r}(\nabla E)A(x)\frac{(\nabla\rho)^T}{|\nabla\rho|}\,d\sigma$$
$$= \int_{A(r_0,\infty)\cap\{\rho(x)<r\}} P[E]\,dx - \int_{S_{r_0}}(\nabla E)A(x)\left(-\frac{x}{r_0}\right)^T dS$$
$$= -1 + \int_{S_{r_0}}(\nabla E)A(x)\left(\frac{x}{r_0}\right)^T dS. \tag{1.5.10}$$

From (1.5.9) and (1.5.10) we observe that

$$K(r) = 1 - \int_{S_{r_0}}(\nabla E)A(x)\left(\frac{x}{r_0}\right)^T dS,$$

where the right hand side is a constant independent of r. If $\Omega = \mathbb{R}^n$, then it is readily seen that

$$\int_{\tilde{S}_r} (\nabla E) A(x) \frac{(\nabla \rho)^T}{|\nabla \rho|} \, d\sigma = \int_{\rho(x)<r} P[E] \, dx = -1.$$

Hence, in view of (1.5.9), we have $K(r) = 1$. In case $P = \Delta$, we obtain $\rho(x) = |x|$ and $|\nabla \rho| = 1$ (see Remark 1.5.1). An easy computation shows that

$$\begin{aligned} K(r) &= \int_{\tilde{S}_r} (\nabla \rho)(\nabla \rho)^T \left(\omega_n r^{n-1} |\nabla \rho|\right)^{-1} d\sigma \\ &= \frac{1}{\omega_n r^{n-1}} \int_{S_r} dS = 1. \end{aligned}$$

□

Remark 1.5.2. In the case where $P = \Delta$, (1.5.2) holds with equality for

$$k = k(n) = \begin{cases} (2\pi)^{-1} & (n = 2) \\ {\omega_n}^{-1} \left(\omega_n (n-2)\right)^{(n-1)/(n-2)} & (n \geq 3). \end{cases}$$

Theorem 1.5.1. *Assume that:*

(i) $c(x, -\xi) = -c(x, \xi)$ *for* $(x, \xi) \in \Omega \times (0, \infty)$;
(ii) $c(x, \xi) \geq p(\rho(x)) \varphi(\xi)$ *for* $(x, \xi) \in \tilde{A}(r_1, \infty) \times (0, \infty)$, *where* $p(r)$ *is a function of class* $C([r_1, \infty); [0, \infty))$ *and* $\varphi(\xi) \in C((0, \infty); (0, \infty))$ *is a convex function.*

Then every solution $u \in C^2(\Omega; \mathbb{R})$ *of* (1.5.1) *is oscillatory in* Ω *if the ordinary differential inequalities*

$$\left(r^{n-1} y'\right)' + \frac{1}{\beta(n)} r^{n-1} p(r) \varphi(y) \leq r^{n-1} \tilde{f}(r), \tag{1.5.11}$$

$$\left(r^{n-1} y'\right)' + \frac{1}{\beta(n)} r^{n-1} p(r) \varphi(y) \leq -r^{n-1} \tilde{f}(r) \tag{1.5.12}$$

are oscillatory at $r = \infty$ *in the sense that neither* (1.5.11) *nor* (1.5.12) *has a solution which is positive on* $[r, \infty)$ *for any* $r > r_1$, *where*

$$\beta(n) = \begin{cases} k^2 (2\pi)^2 & (n = 2) \\ k^2 {\omega_n}^2 \left(\omega_n (n-2)\right)^{2(n-1)/(2-n)} & (n \geq 3) \end{cases}$$

and

$$\tilde{f}(r) = \frac{1}{K \omega_n r^{n-1}} \int_{\tilde{S}_r} f(x) |\nabla \rho|^{-1} d\sigma,$$

where K *is the positive constant defined in Lemma* 1.5.2.

Proof. Suppose to the contrary that there is a solution $u \in C^2(\Omega; \mathbb{R})$ of (1.5.1) which has no zero in $A(r_2, \infty)$ for some $r_2 > r_0$. First we assume that $u > 0$ in $A(r_2, \infty)$. For the number r_2 there exists a number $r_3 > r_1$ such that $\{x \in \mathbb{R}^n;\ \rho(x) > r_3\} \subset A(r_2, \infty)$. Letting $G(t) = t$ in Lemma 1.5.1, we obtain

$$\begin{aligned}\left(r^{n-1}\tilde{U}'(r)\right)' &= \frac{1}{K\omega_n}\int_{\tilde{S}_r} P[u]|\nabla\rho|^{-1}d\sigma \\ &= \frac{1}{K\omega_n}\int_{\tilde{S}_r} (-c(x,u) + f(x))|\nabla\rho|^{-1}d\sigma, \quad r > r_3, \quad (1.5.13)\end{aligned}$$

where

$$\tilde{U}(r) = \frac{1}{K\omega_n r^{n-1}}\int_{\tilde{S}_r} u\,\frac{(\nabla\rho)A(x)(\nabla\rho)^T}{|\nabla\rho|}\,d\sigma.$$

From the hypothesis (ii) it follows that

$$\begin{aligned}\frac{1}{K\omega_n}\int_{\tilde{S}_r} c(x,u)\,d\sigma &\geq (K\omega_n)^{-1}p(r)\int_{\tilde{S}_r}\varphi(u)|\nabla\rho|^{-1}d\sigma \\ &= r^{n-1}p(r)\int_{\tilde{S}_r}\varphi(u)\big(K\omega_n r^{n-1}|\nabla\rho|\big)^{-1}d\sigma. \quad (1.5.14)\end{aligned}$$

Inequality (1.5.2) implies

$$(\nabla\rho)A(x)(\nabla\rho)^T \leq \beta(n).$$

Hence we easily see that

$$\begin{aligned}&\int_{\tilde{S}_r}\varphi(u)\big(K\omega_n r^{n-1}|\nabla\rho|\big)^{-1}d\sigma \\ &\geq \frac{1}{\beta(n)}\int_{\tilde{S}_r}\varphi(u)(\nabla\rho)A(x)(\nabla\rho)^T\big(K\omega_n r^{n-1}|\nabla\rho|\big)^{-1}d\sigma. \quad (1.5.15)\end{aligned}$$

Using Jensen's inequality, we obtain

$$\int_{\tilde{S}_r}\varphi(u)(\nabla\rho)A(x)(\nabla\rho)^T\big(K\omega_n r^{n-1}|\nabla\rho|\big)^{-1}d\sigma \geq \varphi(\tilde{U}(r)). \quad (1.5.16)$$

Combining (1.5.14)–(1.5.16) yields

$$\frac{1}{K\omega_n}\int_{\tilde{S}_r} c(x,u)\,d\sigma \geq \frac{1}{\beta(n)}r^{n-1}p(r)\varphi\big(\tilde{U}(r)\big), \quad r > r_3. \quad (1.5.17)$$

From (1.5.13) and (1.5.17) we see that

$$\left(r^{n-1}\tilde{U}'(r)\right)' \leq -\frac{1}{\beta(n)}r^{n-1}p(r)\varphi\big(\tilde{U}(r)\big) + r^{n-1}\tilde{f}(r), \quad r > r_3$$

which is equivalent to (1.5.11). Hence, $\tilde{U}(r)$ is a positive solution of (1.5.11) in (r_3, ∞). Analogously, if $u < 0$ in $A(r_1, \infty)$, then $-\tilde{U}(r)$ is a positive solution of (1.5.12) in (r_3, ∞). This contradicts the hypothesis and completes the proof. □

Proceeding as in Section 1.4, we can establish various oscillation results for (1.5.1) by applying Theorems 1.4.2–1.4.4 to (1.5.11) and (1.5.12).

Remark 1.5.3. In the case where $P = \Delta$, it is easily verified that $\beta(n) = 1$ and $(\nabla\rho)A(x)(\nabla\rho)^T = 1$.

Next we consider the sublinear elliptic equation

$$\sum_{i,j=1}^{n} \frac{\partial}{\partial x_i}\left(a_{ij}(x)\frac{\partial u}{\partial x_j}\right) + c(x)\varphi(u) = 0, \quad x \in \Omega. \tag{1.5.18}$$

We assume that $c(x)$ and $\varphi(\xi)$ satisfy the same conditions as were stated in Section 1.3.

Theorem 1.5.2. *Every solution $u \in C^2(\Omega;\mathbb{R})$ of (1.5.18) is oscillatory in Ω if the ordinary differential inequality*

$$\left(r^{n-1}y'\right)' \le -Kr^{n-1}\tilde{c}(r) \tag{1.5.19}$$

has no eventually positive solution, where

$$\tilde{c}(r) = \frac{1}{K\omega_n r^{n-1}}\int_{\tilde{S}_r} c(x)|\nabla\rho|^{-1}d\sigma.$$

Proof. Suppose that there exists a solution $u \in C^2(\Omega;\mathbb{R})$ of (1.5.18) which is either positive or negative in $\tilde{A}(r_3,\infty)$ for some $r_3 > r_1$. Defining

$$G(u) = \int_0^u \frac{1}{\varphi(\xi)}\,d\xi,$$

we find that $G(u) > 0$, $G'(u) = \varphi(u)^{-1}$ and $G''(u) = -\varphi'(u)/\varphi(u)^2 \le 0$. Lemma 1.5.1 implies that

$$\begin{aligned}\frac{d}{dr}\left(r^{n-1}\frac{d}{dr}\Big(M[u](r)\Big)\right) &\le \frac{1}{\omega_n}\int_{\tilde{S}_r}\frac{1}{\varphi(u)}P[u]|\nabla\rho|^{-1}d\sigma \\ &= -\frac{1}{\omega_n}\int_{\tilde{S}_r} c(x)|\nabla\rho|^{-1}d\sigma \\ &= -Kr^{n-1}\tilde{c}(r), \quad r \ge r_3.\end{aligned}$$

Hence, $M[u](r)$ is a positive solution of (1.5.19) on $[r_3,\infty)$. This contradicts the hypothesis and the proof is complete. □

Applying Lemma 1.3.3 to (1.5.19), we derive the following theorem which is similar to Theorem 1.3.3.

Theorem 1.5.3. *Every solution $u \in C^2(\Omega;\mathbb{R})$ of (1.5.18) is oscillatory in Ω if*

$$\limsup_{r\to\infty}\int_{r_2}^{r}\left[1-\frac{\psi_n(s)}{\psi_n(r)}\right]s\tilde{c}(s)\,ds = \infty$$

for some $r_2 \ge r_1$.

Example 1.5.1. We consider the sublinear equation

$$\sum_{i,j=1}^{3} \frac{\partial}{\partial x_i}\left(a_{ij}(x)\frac{\partial u}{\partial x_j}\right)+(\nabla\rho)A(x)(\nabla\rho)^T(\log\rho)(\sin\rho)|u|^{\gamma}\operatorname{sgn} u = 0, \quad x \in \Omega, \tag{1.5.20}$$

where $0 < \gamma < 1$. It is easy to check that $\tilde{c}(r) = (\log r)\sin r$. Since

$$\int_{r_2}^{r}\left(1-\frac{s}{r}\right)s(\log s)\sin s\,ds = -(\log r)\sin r + B(r, r_2),$$

where $B(r, r_2)$ is bounded as $r \to \infty$, we see that

$$\limsup_{r\to\infty}\int_{r_2}^{r}\left(1-\frac{s}{r}\right)s(\log s)\sin s\,ds = \infty.$$

Hence, it follows from Theorem 1.5.3 that every solution $u \in C^2(\Omega;\mathbb{R})$ of (1.5.20) is oscillatory in Ω.

Remark 1.5.4. The oscillations of elliptic equations with variable coefficients were also investigated by Bugir and Dobrotvor [39], Noussair and Swanson [225], Usami [274] and Yoshida [296, 297].

1.6 Higher Order Elliptic Equations I

We discuss the higher order elliptic equation with constant coefficients

$$\Delta^m u + a_1\Delta^{m-1}u + \cdots + a_{m-1}\Delta u + a_m u = 0 \quad \text{in } \Omega, \tag{1.6.1}$$

where Δ^k is the kth iterated Laplacian (k-metaharmonic operator) and Ω is an exterior domain in $\mathbb{R}^n$ which contains $\{x \in \mathbb{R}^n;\ |x| \geq r_0\}$ for some $r_0 > 0$. It is assumed that a_i $(i = 1, 2, ..., m)$ are real constants.

The algebraic equation

$$z^m + a_1 z^{m-1} + \cdots + a_{m-1}z + a_m = 0 \tag{1.6.2}$$

is associated with (1.6.1).

Lemma 1.6.1. *If $u \in C^4(\Omega;\mathbb{R})$, then*

$$\frac{1}{\omega_n r^{n-1}}\int_{S_r}\Delta^2 u\,dS = r^{1-n}\left(r^{n-1}\left(r^{1-n}\left(r^{n-1}U'(r)\right)'\right)'\right)', \quad r > r_0, \tag{1.6.3}$$

where ω_n denotes the surface area of the unit sphere S_1, Δ^2 is the biharmonic operator and the spherical mean $U(r)$ is defined by (1.2.2).

Proof. From Lemma 1.2.1 we see that

$$\frac{1}{\omega_n r^{n-1}} \int_{S_r} \Delta^2 u \, dS = r^{1-n} \frac{d}{dr} \left(r^{n-1} \frac{d}{dr} \left(\frac{1}{\omega_n r^{n-1}} \int_{S_r} \Delta u \, dS \right) \right), \quad r > r_0. \tag{1.6.4}$$

Using Lemma 1.2.1 in the right hand side of (1.6.4) again, we obtain the desired identity (1.6.3). □

Theorem 1.6.1. *If the algebraic equation* (1.6.2) *has only simple roots and has no real nonnegative root, then every solution* $u \in C^{2m}(\Omega; \mathbb{R})$ *of* (1.6.1) *is oscillatory in* Ω.

Proof. Assume that there exists a nonoscillatory solution $u \in C^{2m}(\Omega; \mathbb{R})$ of (1.6.1). Without loss of generality we may assume that $u > 0$ in $A(r_1, \infty) = \{x \in \mathbb{R}^n; \ |x| > r_1\}$ for some $r_1 > r_0$. By the hypothesis we have

$$z^m + a_1 z^{m-1} + \cdots + a_{m-1} z + a_m = \prod_{k=1}^{p} (z^2 + 2b_k z + b_k^2 + c_k^2) \prod_{k=2p+1}^{m} (z + d_k^2),$$

where $c_k > 0$ $(k = 1, 2, ..., p)$, $d_k > 0$ $(k = 2p+1, 2p+2, ..., m)$, $-b_i \pm \sqrt{-1}\, c_i \neq -b_j \pm \sqrt{-1}\, c_j$ $(i \neq j)$ and $d_i \neq d_j$ $(i \neq j)$. Hence, (1.6.1) can be rewritten as

$$\left(\prod_{k=1}^{p} (\Delta^2 + 2b_k \Delta + b_k^2 + c_k^2) \prod_{k=2p+1}^{m} (\Delta + d_k^2) \right) u = 0.$$

From a result of Gorbaĭchuk and Dobrotvor [91, Lemma 4] it follows that there exists a unique system $\tilde{u}_k(x)$ $(k = 1, 2, ..., p)$, $u_k(x)$ $(k = 2p+1, 2p+2, ..., m)$ such that

$$(\Delta^2 + 2b_k \Delta + b_k^2 + c_k^2) \tilde{u}_k(x) = 0 \quad (k = 1, 2, ..., p),$$

$$(\Delta + d_k^2)\, u_k(x) = 0 \quad (k = 2p+1, 2p+2, ..., m)$$

and

$$u(x) = \sum_{k=1}^{p} \tilde{u}_k(x) + \sum_{k=2p+1}^{m} u_k(x). \tag{1.6.5}$$

Then we obtain

$$U(r) = \sum_{k=1}^{p} \tilde{U}_k(r) + \sum_{k=2p+1}^{m} U_k(r), \tag{1.6.6}$$

where $\tilde{U}_k(r)$ and $U_k(r)$ denote the spherical means of $\tilde{u}_k(x)$ and $u_k(x)$ over the sphere S_r, respectively. From Lemmas 1.2.1 and 1.6.1 we see that

$$\begin{aligned} r^{1-n}\left(r^{n-1}\left(r^{1-n}\left(r^{n-1}\tilde{U}_k'(r)\right)'\right)'\right)' + 2b_k r^{1-n}\left(r^{n-1}\tilde{U}_k'(r)\right)' \\ + (b_k^2 + c_k^2)\tilde{U}_k(r) = 0, \quad (1.6.7) \\ r^{1-n}\left(r^{n-1}U_k'(r)\right)' + d_k^2 U_k(r) = 0. \end{aligned}$$

The following system

$$y' = \left(A + V(r)\right) y, \quad y = (y_1, y_2, y_3, y_4)^T \tag{1.6.8}$$

is associated with (1.6.7), where the superscript T denotes the transpose,

$$A = \begin{pmatrix} 0 & 1 & 0 & 0 \\ 0 & 0 & 1 & 0 \\ 0 & 0 & 0 & 1 \\ -(b_k^2 + c_k^2) & 0 & -2b_k & 0 \end{pmatrix}$$

and

$$V(r) = \begin{pmatrix} 0 & 0 & 0 & 0 \\ 0 & 0 & 0 & 0 \\ 0 & 0 & 0 & 0 \\ 0 & -\left(2b_k \frac{n-1}{r} - \frac{(n-1)(n-3)}{r^3}\right) & -\frac{(n-1)(n-3)}{r^2} & -\frac{2(n-1)}{r} \end{pmatrix}.$$

Since $\det(A - \lambda I) = \lambda^4 + 2b_k\lambda^2 + b_k^2 + c_k^2$, it can be shown that the characteristic roots of A are $\pm\mu_1 \pm \sqrt{-1}\,\mu_2$, where

$$\begin{aligned} \mu_1 &= 2^{-1/2}\left(-b_k + \left(b_k^2 + c_k^2\right)^{1/2}\right)^{1/2}, \\ \mu_2 &= 2^{-1/2}\left(b_k + \left(b_k^2 + c_k^2\right)^{1/2}\right)^{1/2}. \end{aligned}$$

It is easy to see that the characteristic polynomial for $A + V(r)$ is given by

$$\begin{aligned} \lambda^4 + \frac{2(n-1)}{r}\lambda^3 + \left(2b_k + \frac{(n-1)(n-3)}{r^2}\right)\lambda^2 \\ + \left(2b_k \frac{n-1}{r} - \frac{(n-1)(n-3)}{r^3}\right)\lambda + b_k^2 + c_k^2. \end{aligned}$$

We observe, using Ferrari's formula (see, for example, [276, p.190]), that the characteristic roots $\lambda_i(r)$ of $A + V(r)$ can be written as

$$\begin{aligned} \lambda_i(r) &= -\frac{n-1}{2r} + \mu_1(r) + (-1)^{i+1}\sqrt{-1}\,\mu_2(r) \quad (i = 1, 2), \\ \lambda_i(r) &= -\frac{n-1}{2r} - \mu_1(r) + (-1)^{i+1}\sqrt{-1}\,\mu_2(r) \quad (i = 3, 4), \end{aligned}$$

where $\lim_{r\to\infty} \mu_k(r) = \mu_k$ $(k = 1, 2)$. It is easy to check that

$$\int_{r_0}^{\infty} |V'(r)|\, dr < \infty \quad \text{and} \quad \lim_{r\to\infty} V(r) = 0.$$

Then it can be shown [52, p.92] that there are solutions $Y_i(r)$ $(i = 1, 2, 3, 4)$ of (1.6.8) and r_2 $(r_0 \le r_2 < \infty)$ such that

$$\lim_{r\to\infty} Y_i(r) \exp\left(-\int_{r_2}^{r} \lambda_i(s)\, ds\right) = p_i \quad (i = 1, 2, 3, 4),$$

where each p_i is a characteristic vector of A associated with $\mu_1 + (-1)^{i+1}\sqrt{-1}\,\mu_2$ $(i = 1, 2)$, $-\mu_1 + (-1)^{i+1}\sqrt{-1}\,\mu_2$ $(i = 3, 4)$. Hence we find that

$$\begin{aligned} Y_i(r) r^{\frac{n-1}{2}} &\approx P_i \exp\left(\int_{r_2}^{r} \mu_1(s)\, ds\right) \\ &\quad \times \left(\cos \int_{r_2}^{r} \mu_2(s)\, ds + (-1)^{i+1}\sqrt{-1} \sin \int_{r_2}^{r} \mu_2(s)\, ds\right) \\ &\qquad (r \to \infty;\ i = 1, 2), \end{aligned}$$

$$\begin{aligned} Y_i(r) r^{\frac{n-1}{2}} &\approx P_i \exp\left(-\int_{r_2}^{r} \mu_1(s)\, ds\right) \\ &\quad \times \left(\cos \int_{r_2}^{r} \mu_2(s)\, ds + (-1)^{i+1}\sqrt{-1} \sin \int_{r_2}^{r} \mu_2(s)\, ds\right) \\ &\qquad (r \to \infty;\ i = 3, 4), \end{aligned}$$

where $P_i = K_i p_i$ for some constants $K_i \in \mathbb{R}$ $(i = 1, 2, 3, 4)$. Since $\tilde{U}_k(r)$ is a real-valued function and a linear combination of the first components of $Y_i(r)$ $(i = 1, 2, 3, 4)$, we see that

$$\begin{aligned} r^{\frac{n-1}{2}} \tilde{U}_k(r) &\approx B_k \exp\left(\int_{r_2}^{r} \mu_1(s)\, ds\right) \sin\left(\int_{r_2}^{r} \mu_2(s)\, ds + \sigma_k\right) \\ &\quad + C_k \exp\left(-\int_{r_2}^{r} \mu_1(s)\, ds\right) \sin\left(\int_{r_2}^{r} \mu_2(s)\, ds + \tau_k\right) \quad (r \to \infty) \end{aligned} \tag{1.6.9}$$

for some constants B_k, C_k, σ_k and τ_k $(k = 1, 2, ..., p)$. Using the same arguments as in Dobrotvor [64, p.231], we conclude that

$$r^{\frac{n-1}{2}} U_k(r) \approx A_k \sin\left(\int_{r_2}^{r} \left(d_k^2 - \frac{1 - n^2}{4s^2}\right)^{1/2} ds + \theta_k\right) \quad (r \to \infty) \tag{1.6.10}$$

for some constants A_k and θ_k $(k = 2p+1, 2p+2, ..., m)$. Combining (1.6.6), (1.6.9) and (1.6.10) yields

$$r^{\frac{n-1}{2}}U(r) \approx \sum_{k=2p+1}^{m} A_k \sin\left(\int_{r_2}^{r}\left(d_k^2 - \frac{1-n^2}{4s^2}\right)^{1/2} ds + \theta_k\right)$$
$$+\sum_{k=1}^{p} B_k \exp\left(\int_{r_2}^{r} \mu_1(s)\,ds\right)\sin\left(\int_{r_2}^{r}\mu_2(s)\,ds + \sigma_k\right)$$
$$+\sum_{k=1}^{p} C_k \exp\left(-\int_{r_2}^{r} \mu_1(s)\,ds\right)\sin\left(\int_{r_2}^{r}\mu_2(s)\,ds + \tau_k\right)$$
$$(r \to \infty). \tag{1.6.11}$$

We observe that the left hand side of (1.6.11) is positive for $r > r_1$ in view of the fact that $u > 0$ in $A(r_1, \infty)$. However, the right hand side of (1.6.11) changes sign in an arbitrary interval (r, ∞) (see Gorbaĭchuk and Dobrotvor [91, Lemma 6]). This is a contradiction. In case (1.6.2) has only simple negative or only simple complex roots, then we replace (1.6.5) by

$$u(x) = \sum_{k=1}^{m} u_k(x), \quad (\Delta + d_k^2)u_k(x) = 0,$$

and

$$u(x) = \sum_{k=1}^{m/2} \tilde{u}_k(x), \quad (\Delta^2 + 2b_k\Delta + b_k^2 + c_k^2)\tilde{u}_k(x) = 0,$$

respectively. Arguing as above, we are also led to a contradiction. This completes the proof. □

Remark 1.6.1. The results in this section can be extended to the more general equation

$$L^m u + a_1 L^{m-1} u + \cdots + a_{m-1} L u + a_m u = 0, \tag{1.6.12}$$

where

$$L = \sum_{i,j=1}^{n} a_{ij} \frac{\partial^2}{\partial x_i \partial x_j} \quad (a_{ij} = \text{constant}), \tag{1.6.13}$$

$(a_{ij})_{i,j=1}^n$ being the symmetric and positive definite matrix.

Remark 1.6.2. In the case where $n = 3$, we can extend the results of Naito and Yoshida [213] to

$$L^m u + a_1 L^{m-1} u + \cdots + a_{m-1} L u + a_m u + c(x, u) = f(x),$$

where L is given by (1.6.13).

Remark 1.6.3. If $u \in C^{2m}(\Omega; \mathbb{R})$ and u satisfies (1.6.1), then u is analytic in Ω (see Hörmander [104, p.178]). Therefore, the set of zeros of a nontrivial solution of (1.6.1) does not have interior points.

Remark 1.6.4. Oscillations of the fourth order elliptic equation

$$\Delta^2 u + a_1 \Delta u + a_2 u = 0$$

were studied by Bugir [38], Frydrych [80] and Gorbaĭchuk and Dobrotvor [92]. Equation (1.6.1) with $n = 3$ was treated by Barański [28] and Gorbaĭchuk and Dobrotvor [91]. For Eq. (1.6.1) with $n \geq 2$ we refer to Bugir [36], Dobrotvor [64] and Wachnicki [277]. Equation (1.6.12) was investigated by Bouchekif and Górowski [35], Górowski [93] and Yoshida [307]. We mention in particular the paper of Bouchekif [34] which treats (1.6.1) in the case where the multiplicities of real roots and pure imaginary ones are arbitrary and the multiplicity of complex roots is equal to one.

1.7 Higher Order Elliptic Equations II

In 1964 Śliwiński [247] investigated the problem of finding an oscillatory solution of the higher order elliptic equation

$$\Delta^m u + f(|x|)u = 0 \quad \text{in } \mathbb{R}^3,$$

where Δ^m is the mth iterate of the Laplacian in $\mathbb{R}^3$. Oscillations of the fourth order elliptic equation

$$\Delta^2 u + c(x, u) = 0$$

were studied by Kitamura and Kusano [128] by reducing the multi-dimensional problem to the problem of oscillation of a fourth order ordinary differential inequality. Furthermore, the higher order elliptic equation

$$\Delta^m u + c(x, u) = 0 \tag{1.7.1}$$

was treated by Kitamura and Kusano [130] and Toraev [262], where Δ^m is the mth iterated Laplacian (m-metaharmonic operator). For the forced oscillations of higher order elliptic equations we refer to Bugir [37] and Švaňa [250].

We deal with (1.7.1) in an exterior domain $\Omega \subset \mathbb{R}^n$ which contains $\{x \in \mathbb{R}^n;\ |x| \geq r_0\}$ for some $r_0 > 0$. We recall that: $A(r, \infty) = \{x \in \mathbb{R}^n;\ |x| > r\}$.

Lemma 1.7.1. *If $u \in C^{2m}(\Omega;\mathbb{R})$, then*

$$\frac{1}{\omega_n r^{n-1}} \int_{S_r} \Delta^m u\, dS = \left(r^{1-n} \frac{d}{dr} r^{n-1} \frac{d}{dr} \right)^m U(r), \quad r > r_0, \tag{1.7.2}$$

where $U(r)$ is the spherical mean of u over S_r defined by (1.2.2).

Proof. Arguing as in the proof of Lemma 1.6.1, we obtain the desired identity (1.7.2) with the aid of Lemma 1.2.1. □

Theorem 1.7.1. *Assume that:*

(i) $c(x,\xi) \in C(\Omega \times \mathbb{R};\mathbb{R})$ *and* $c(x,-\xi) = -c(x,\xi)$ *for* $(x,\xi) \in \Omega \times (0,\infty)$;
(ii) $c(x,\xi) \geq p(|x|)\varphi(\xi)$ *for* $(x,\xi) \in A(r_0,\infty) \times (0,\infty)$, *where $p(r)$ is a function of class* $C([r_0,\infty);[0,\infty))$ *and* $\varphi(\xi) \in C((0,\infty);(0,\infty))$ *is a convex function.*

Every solution $u \in C^{2m}(\Omega;\mathbb{R})$ of (1.7.1) is oscillatory in Ω if the ordinary differential inequality

$$\left(r^{1-n} \frac{d}{dr} r^{n-1} \frac{d}{dr} \right)^m y + p(r)\varphi(y) \leq 0 \tag{1.7.3}$$

has no eventually positive solution.

Proof. Suppose to the contrary that (1.7.1) has a solution $u \in C^{2m}(\Omega;\mathbb{R})$ which has no zero in $A(r_1,\infty)$ for some $r_1 > r_0$. Without loss of generality we may assume that $u > 0$ in $A(r_1,\infty)$. Then, Lemma 1.7.1 implies that the spherical mean $U(r)$ of u satisfies

$$\begin{aligned} \left(r^{1-n} \frac{d}{dr} r^{n-1} \frac{d}{dr} \right)^m U(r) &= \frac{1}{\omega_n r^{n-1}} \int_{S_r} \Delta^m u\, dS \\ &= -\frac{1}{\omega_n r^{n-1}} \int_{S_r} c(x,u)\, dS, \quad r > r_1. \end{aligned} \tag{1.7.4}$$

Proceeding as in the proof of Lemma 1.2.2, we see that (1.2.9) and (1.2.10) hold. By (1.2.9) and (1.2.10) we obtain the inequality

$$\frac{1}{\omega_n r^{n-1}} \int_{S_r} c(x,u)\, dS \geq p(r)\varphi(U(r)), \quad r > r_1. \tag{1.7.5}$$

Combining (1.7.4) with (1.7.5), we conclude that $U(r)$ is a positive solution of (1.7.3) in (r_1,∞). This contradicts the hypothesis and completes the proof. □

We shall derive sufficient conditions for no solution of (1.7.3) to be eventually positive as a consequence of the analysis of the more general inequality

$$L_{2m}y + q(r)h(y) \le 0, \tag{1.7.6}$$

where L_{2m} is a $2m$th order differential operator defined by

$$L_{2m}\cdot = \frac{1}{p_{2m}}\frac{d}{dr}\frac{1}{p_{2m-1}}\frac{d}{dr}\cdots\frac{d}{dr}\frac{1}{p_1}\frac{d}{dr}\frac{\cdot}{p_0}.$$

We make the following assumptions:

(a) $p_i(r) \in C([r_0,\infty);(0,\infty))$ $(i = 0,1,...,2m)$ and

$$\int_{r_0}^{\infty} p_i(r)\,dr = \infty \quad (i = 1,2,...,2m-1); \tag{1.7.7}$$

(b) $q(r) \in C([r_0,\infty);(0,\infty))$;

(c) $h(\xi)$ is a nondecreasing function of class $C((0,\infty);(0,\infty))$, and

$$\int_{\delta}^{\infty} \frac{1}{h(\xi)}\,d\xi < \infty \quad \text{for any } \delta > 0;$$

(d) $h(\xi\eta) \ge h(\xi)h(\eta)$ for any ξ, η such that $0 < \xi < 1 < \eta$.

Let $i_k \in \{1,2,...,2m-1\}$ $(k = 1,2,...,2m-1)$ and r, $s \in [r_0,\infty)$. We define $I_0 = 1$ and

$$I_k(r,s;p_{i_k},...,p_{i_1}) = \int_s^r p_{i_k}(\sigma)I_{k-1}(\sigma,s;p_{i_{k-1}},...,p_{i_1})\,d\sigma \ (k = 1,2,...,2m-1).$$

Proposition 1.7.1. *Assume that*

$$\liminf_{r\to\infty} \frac{I_{2m-1}(r,r_0;p_1,...,p_{2m-1})}{I_{2m-1}(r,r_0;p_{2m-1},...,p_1)} > 0 \tag{1.7.8}$$

and that, for each $l \in \{2,3,...,m-1\}$, either

$$\liminf_{r\to\infty} \frac{I_{2m-1}(r,r_0;p_{2m-1},...,p_{2l},p_1,...,p_{2l-1})}{I_{2m-1}(r,r_0;p_{2m-1},...,p_1)} > 0 \tag{1.7.9}$$

or

$$\liminf_{r\to\infty} \frac{I_{2m-1}(r,r_0;p_1,...,p_{2l-2},p_{2m-1},...,p_{2l-1})}{I_{2m-1}(r,r_0;p_{2m-1},...,p_1)} > 0. \tag{1.7.10}$$

(i) *Let $p_0(r) = 1$ and let* (a)–(c) *hold. Then, the condition*

$$\int_{r_0}^{\infty} I_{2m-1}(r,r_0;p_{2m-1},...,p_1)p_{2m}(r)q(r)\,dr = \infty \tag{1.7.11}$$

is a necessary and sufficient condition in order that (1.7.6) *has no eventually positive solution.*

(ii) *Let* $\lim_{r\to\infty} p_0(r) = 0$ *and let* (a)–(d) *hold. Then, the condition*

$$\int_{r_0}^{\infty} I_{2m-1}(r, r_0; p_{2m-1}, \ldots, p_1) p_{2m}(r) h\big(c\, p_0(r)\big) q(r)\, dr = \infty \quad \text{for all } c > 0 \tag{1.7.12}$$

is a necessary and sufficient condition in order that (1.7.6) *has no eventually positive solution.*

For the proof of Proposition 1.7.1, see Kitamura and Kusano [130, Theorem 6].

We are now in a position to state the following oscillation results for (1.7.1).

Theorem 1.7.2. $(n = 2)$ *Assume that:*

(i) *the hypothesis* (i) *of Theorem* 1.7.1 *is satisfied;*
(ii) $c(x, \xi) \geq p(|x|)\varphi(\xi)$ *for* $(x, \xi) \in A(r_0, \infty) \times (0, \infty)$, *where* $p(r)$ *is a function of class* $C([r_0, \infty); [0, \infty))$ *and* $\varphi(\xi)$ *is a convex and non-decreasing function of class* $C((0, \infty); (0, \infty))$;
(iii) $\displaystyle\int_{\delta}^{\infty} \frac{1}{\varphi(\xi)}\, d\xi < \infty$ *for any* $\delta > 0$;
(iv) $\displaystyle\int_{r_0}^{\infty} r^{2m-1}(\log r) p(r)\, dr = \infty.$

Then every solution $u \in C^{2m}(\Omega; \mathbb{R})$ *of* (1.7.1) *is oscillatory in* $\Omega \subset \mathbb{R}^2$.

Proof. In case $n = 2$, the associated inequality (1.7.3) reduces to (1.7.6) with $p_0(r) = 1$, $p_1(r) = p_3(r) = \cdots = p_{2m-1}(r) = r^{-1}$, $p_2(r) = p_4(r) = \cdots = p_{2m}(r) = r$, $q(r) = p(r)$ and $h(y) = \varphi(y)$. It is easily seen that

$$I_{2m-1}(r, r_0; p_{2m-1}, \ldots, p_1) = \frac{r^{2m-2}(\log r)[1 + o(1)]}{[2^{m-1}(m-1)!\,]^2} \quad \text{as} \quad r \to \infty$$

so that the condition (iv) is equivalent to (1.7.11). Therefore, from Proposition 1.7.1 and the Remark 2 of Kitamura and Kusano [130] it follows that (1.7.3) has no eventually positive solution. Hence, the conclusion follows from Theorem 1.7.1. □

Theorem 1.7.3. $(n \geq 3)$ *Assume that:*

(i) *the hypotheses* (i)–(iii) *of Theorem* 1.7.2 *are satisfied;*
(ii) $\varphi(uv) \geq \varphi(u)\varphi(v)$ *for all* u, v *such that* $0 < u < 1 < v$;
(iii) $\displaystyle\int_{r_0}^{\infty} r^{n+2m-3} p(r) \varphi(c\, r^{2-n})\, dr = \infty$ *for all* $c > 0$.

Then every solution $u \in C^{2m}(\Omega; \mathbb{R})$ *of* (1.7.1) *is oscillatory in* $\Omega \subset \mathbb{R}^n$ $(n \geq 3)$.

Proof. It is not possible to apply Theorem 1.7.1 directly to (1.7.3) since

$$\int_{r_0}^{\infty} p_{2i-1}(r)\, dr = \int_{r_0}^{\infty} r^{1-n} dr < \infty \quad (i = 1, 2, ..., m).$$

According to Theorem 1 of Trench [268], the differential operator appearing in (1.7.3) can be rewritten as

$$\left(r^{1-n} \frac{d}{dr} r^{n-1} \frac{d}{dr} \right)^m y = D^{2m}(y; p_0, p_1, ..., p_{2m})$$

so that (1.7.7) holds, where $D^{j+1}(y; p_0, p_1, ..., p_{j+1})$ $(j = 1, 2, ..., 2m-1)$ are defined by

$$D^0(y; p_0) = \frac{y}{p_0}, \quad D^1(y; p_0, p_1) = \frac{1}{p_1} \frac{d}{dr} D^0(y; p_0),$$

$$D^{j+1}(y; p_0, p_1, ..., p_{j+1}) = \frac{1}{p_{j+1}} \frac{d}{dr} D^j(y; p_0, p_1, ..., p_j) \ (j = 1, 2, ..., 2m-1).$$

It can be shown that

$$p_0(r) = r^{2-n}, \ p_{2m}(r) = r \quad \text{and} \quad p_i(r) = p_{2m-i}(r) \ (i = 1, 2, ..., m) \tag{1.7.13}$$

with $p_i(r)$ $(i = 1, 2, ..., m)$ given explicitly by

$$p_i(r) = r \ (i = 1, 2, ..., \nu - 1), \quad p_i(r) = r^{(-1)^{i-\nu}(n-2\nu-1)} \ (i = \nu, \nu + 1, ..., m), \tag{1.7.14}$$

where $\nu = \min\{m, \ [(n-1)/2]\}$. ($[N]$ denotes the largest integer not exceeding N.)

Using (1.7.13) and (1.7.14), we see that either (1.7.8), (1.7.9) or (1.7.8), (1.7.10) are satisfied. In fact, if n is odd or $n \geq 2m+1$, then the situation is as described in the Remark 1 of Kitamura and Kusano [130], while n is even and $n \leq 2m$, then we obtain

$$\liminf_{r \to \infty} \frac{I_{2l-1}(r, r_0; p_1, ..., p_{2l-1})}{I_{2l-1}(r, r_0; p_{2l-1}, ..., p_1)} > 0 \quad (l = 2, 3, ..., (m+1)/2)$$

so that the Remark 2 of Kitamura and Kusano [130] is applicable. Using (1.7.13) and (1.7.14) again, we obtain by direct computation

$$I_{2m-1}(r, r_0; p_{2m-1}, ..., p_1) = \frac{(n-4)!! \ r^{n+2m-4}}{(2m-2)!! \ (2m+n-4)!!} [1 + o(1)]$$

as $r \to \infty$, which shows that the condition (iii) is equivalent to (1.7.12). ($N!! = 1$ if $N \leq 0$, $N!! = 1 \cdot 3 \cdot \cdots \cdot N$ if $N > 0$ is odd, and $N!! = 2 \cdot 4 \cdot \cdots \cdot N$ if $N > 0$ is even.) Therefore, applying (ii) of Proposition 1.7.1, we observe that (1.7.3) has no eventually positive solution. This combined with Theorem 1.7.1 ensures that every solution $u \in C^{2m}(\Omega; \mathbb{R})$ of (1.7.1) is oscillatory in Ω. Thus the proof is complete. □

Applying Theorems 1.7.2 and 1.7.3 to the equation

$$\Delta^m u + c(x)|u|^\gamma \text{sgn}\, u = 0, \quad \gamma > 1, \tag{1.7.15}$$

we have the following result.

Corollary 1.7.1. *Every solution $u \in C^{2m}(\Omega; \mathbb{R})$ of (1.7.15) is oscillatory in Ω if the following conditions are satisfied:*

(i) $c(x) \geq p(|x|)$ *for* $x \in A(r_0, \infty)$, *where* $p(r) \in C([r_0, \infty); [0, \infty))$;

(ii) $\displaystyle\int_{r_0}^{\infty} r^{2m-1}(\log r)p(r)\, dr = \infty \quad (n = 2)$;

(iii) $\displaystyle\int_{r_0}^{\infty} r^{2m-1-(\gamma-1)(n-2)}p(r)\, dr = \infty \quad (n \geq 3)$.

Remark 1.7.1. It is easy to check that the hypotheses (i), (ii) of Theorem 1.7.3 are satisfied by

$$\varphi(\xi) = |\xi|^\alpha \big(\log|\xi| + 1\big)^\beta \text{sgn}\ \xi \quad (\alpha > 1,\ \beta \geq 0 \text{ or } \alpha = 1,\ \beta > 1)$$

but not by

$$\varphi(\xi) = |\xi|^\alpha \exp\big(|\xi|^\beta\big) \text{sgn}\ \xi \quad (\alpha \geq 0,\ \beta > 0).$$

Remark 1.7.2. Corollary 1.7.1 was also established by Toraev [262].

1.8 Notes

Theorems 1.1.1 and 1.1.2 are due to Swanson [252]. Theorem 1.1.2 for (1.1.4) and Corollary 1.1.1 were first established by Kreith and Travis [146].

We mention in particular the paper of Allegretto [9] which deals with nonlinear elliptic equations. The proof of Lemma 1.2.1 is different from that given by Noussair and Swanson [223, Lemma 2]. Theorem 1.2.1 is from Noussair and Swanson [223]. Propositions 1.2.1 and 1.2.2 are taken from Naito and Yoshida [212].

Lemmas 1.3.1 and 1.3.2, Theorems 1.3.1 and 1.3.2 are due to Kitamura and Kusano [129]. Lemma 1.3.3 and Theorem 1.3.3 are taken from Kura [150].

The paper [9] of Allegretto seems to be the first paper dealing with oscillations of perturbed nonlinear elliptic equations. The results in Section 1.4 except Theorems 1.4.4 and 1.4.7 are from Kusano and Naito [152].

Lemma 1.5.1 is due to Suleĭmanov [248]. Theorems 1.5.1 and 1.5.2 are extracted from Yoshida [296].

Theorem 1.6.1 is based on Dobrotvor [64] and Yoshida [307].

All results in Section 1.7 are obtained by Kitamura and Kusano [130].

In 1980 Noussair and Swanson [225] established oscillation criteria for second order elliptic equations by employing Riccati transformations. The Riccati techniques for linear or semilinear elliptic equations were developed by numerous authors, see, for example, Mařík [186], Xu [287,288], Yoshida [297] and the references cited therein.

As was shown in Theorems 1.1.1, 1.2.1, 1.3.1, 1.4.1, 1.5.1, 1.5.2, 1.7.1, there is a close relation between oscillation of ordinary differential equations and that of partial differential equations. We refer to Kreith [138], Swanson [251] for oscillation theory of ordinary and partial differential equations without functional arguments.

Chapter 2

Oscillation of Parabolic Equations

2.1 Boundary Value Problems

We consider the parabolic equation with a forcing term

$$\frac{\partial u}{\partial t} - a(t)\Delta u + c(x,t,u) = f(x,t), \quad (x,t) \in \Omega := G \times (0,\infty), \tag{2.1.1}$$

where Δ is the Laplacian in $\mathbb{R}^n$ and G is a bounded domain of $\mathbb{R}^n$ with piecewise smooth boundary ∂G. It is assumed that $a(t) \in C((0,\infty);[0,\infty))$, $c(x,t,\xi) \in C(\overline{\Omega}\times\mathbb{R};\mathbb{R})$, $f(x,t) \in C(\overline{\Omega};\mathbb{R})$, and moreover that $\xi c(x,t,\xi) \geq 0$ for $(x,t) \in \Omega,\ \xi \in \mathbb{R}$.

Definition 2.1.1. A real-valued function u on Ω is said to be *oscillatory* in Ω if u has a zero in $G \times (t,\infty)$ for any $t > 0$. Otherwise u is called *nonoscillatory* in Ω.

Definition 2.1.2. By a *solution* of Eq. (2.1.1) we mean a function $u \in C^1(\overline{G} \times (0,\infty);\mathbb{R})$ which is twice differentiable in x and satisfies (2.1.1).

It is known that if u satisfies the initial-boundary value problem

$$\frac{\partial u}{\partial t} - \frac{\partial^2 u}{\partial x^2} = 0, \quad (x,t) \in (0,\pi)\times(0,\infty),$$

$$u(0,t) = u(\pi,t) = 0, \quad t \in [0,\infty), \tag{2.1.2}$$

$$u(x,0) = 0, \quad x \in [0,\pi], \tag{2.1.3}$$

then $u \equiv 0$ in $(0,\pi)\times(0,\infty)$ (see, for example, Protter and Weinberger [242, p.160]). However, the parabolic equation with a forcing term

$$\frac{\partial u}{\partial t} - \frac{\partial^2 u}{\partial x^2} = (\sin x)(\cos t + \sin t), \quad (x,t) \in (0,\pi)\times(0,\infty)$$

has a non-trivial oscillatory solution $u = (\sin x)\sin t$ which satisfies (2.1.2) and (2.1.3).

We consider two kinds of boundary conditions:

$$(\mathrm{B}_1) \quad u = \psi \quad \text{on} \quad \partial G \times (0, \infty),$$

$$(\mathrm{B}_2) \quad \frac{\partial u}{\partial \nu} + \mu u = \tilde{\psi} \quad \text{on} \quad \partial G \times (0, \infty),$$

where $\psi,\ \tilde{\psi} \in C(\partial G \times (0, \infty); \mathbb{R})$ and $\mu \in C(\partial G \times (0, \infty); [0, \infty))$.

It is known that the first eigenvalue λ_1 of the eigenvalue problem (EVP)

$$-\Delta w = \lambda w \quad \text{in} \ G, \tag{2.1.4}$$

$$w = 0 \quad \text{on} \ \partial G \tag{2.1.5}$$

is positive and the corresponding eigenfunction $\Phi(x)$ may be chosen so that $\Phi(x) > 0$ in G (see Courant and Hilbert [54]).

Theorem 2.1.1. *Every solution u of the boundary value problem* (2.1.1), (B_1) *is oscillatory in Ω if the ordinary differential inequalities*

$$y' + \lambda_1 a(t) y \leq \pm(-a(t)\Psi(t) + F(t)) \tag{2.1.6}$$

have no eventually positive solutions, where

$$\Psi(t) = \int_{\partial G} \psi \frac{\partial \Phi(x)}{\partial \nu} \, dS,$$

$$F(t) = \int_G f(x, t)\Phi(x) \, dx.$$

Proof. Suppose on the contrary, that there is a solution u of the problem (2.1.1), (B_1) which has no zero in $G \times (t_0, \infty)$ for some $t_0 > 0$. First we assume that $u > 0$ in $G \times (t_0, \infty)$. The hypothesis on $c(x, t, \xi)$ implies that $c(x, t, u) \geq 0$ in $G \times (t_0, \infty)$, and therefore

$$\frac{\partial u}{\partial t} - a(t)\Delta u \leq f(x, t) \quad \text{in} \ G \times (t_0, \infty). \tag{2.1.7}$$

Multiplying (2.1.7) by $\Phi(x)$ and then integrating over G, we obtain

$$\int_G \frac{\partial u}{\partial t} \Phi(x) \, dx - a(t) \int_G (\Delta u)\Phi(x) \, dx \leq \int_G f(x, t)\Phi(x) \, dx, \quad t > t_0. \tag{2.1.8}$$

It follows from Green's formula (see, for example, Protter and Weinberger [242, p.82]) that

$$\begin{aligned} \int_G (\Delta u)\Phi(x) \, dx &= \int_{\partial G} \left[\Phi(x) \frac{\partial u}{\partial \nu} - u \frac{\partial \Phi(x)}{\partial \nu} \right] dS + \int_G u \Delta\Phi(x) \, dx \\ &= -\int_{\partial G} \psi \frac{\partial \Phi(x)}{\partial \nu} \, dS - \lambda_1 \int_G u \, \Phi(x) \, dx \\ &= -\Psi(t) - \lambda_1 U(t), \end{aligned} \tag{2.1.9}$$

where

$$U(t) = \int_G u\,\Phi(x)\,dx.$$

We combine (2.1.8) with (2.1.9) to obtain

$$U'(t) + \lambda_1 a(t)U(t) \le -a(t)\Psi(t) + F(t), \quad t > t_0.$$

Hence, $U(t)$ is a positive solution of

$$y' + \lambda_1 a(t)y \le -a(t)\Psi(t) + F(t)$$

in (t_0, ∞). This contradicts the hypothesis. In case $u < 0$ in $G \times (t_0, \infty)$, we let $v = -u$. Arguing as in the case where $u > 0$, we see that $V(t) := \int_G v\,\Phi(x)\,dx$ is a positive solution of

$$y' + \lambda_1 a(t)y \le -(-a(t)\Psi(t) + F(t))$$

in (t_0, ∞). This also contradicts the hypothesis. The proof is complete. □

Corollary 2.1.1. *Every solution u of the problem* (2.1.1), (B_1) *is oscillatory in Ω if*

$$\liminf_{t\to\infty} \int_T^t \exp(\lambda_1 A(s))\big(-a(s)\Psi(s) + F(s)\big)\,ds = -\infty,$$

$$\limsup_{t\to\infty} \int_T^t \exp(\lambda_1 A(s))\big(-a(s)\Psi(s) + F(s)\big)\,ds = \infty$$

for all large T, where

$$A(s) = \int_0^s a(\tau)\,d\tau.$$

Proof. Let $y_\pm(t)$ be eventually positive solutions of (2.1.6). Then, $y_\pm(t) > 0$ on $[T, \infty)$ for some $T > 0$. Multiplying (2.1.6) by $\exp(\lambda_1 A(t))$ yields

$$\big(y_\pm \exp(\lambda_1 A(t))\big)' \le \pm\exp(\lambda_1 A(t))\big(-a(t)\Psi(t) + F(t)\big). \qquad (2.1.10)$$

Integrating (2.1.10) over $[T, t]$, we have

$$\begin{aligned}&\exp(\lambda_1 A(t))y_\pm(t) - \exp(\lambda_1 A(T))y_\pm(T)\\ &\le \pm\int_T^t \exp(\lambda_1 A(s))\big(-a(s)\Psi(s) + F(s)\big)\,ds, \quad t \ge T. \qquad (2.1.11)\end{aligned}$$

The hypothesis implies that the right hand side of (2.1.11) are not bounded from below, and hence $\exp(\lambda_1 A(t))y_\pm(t)$ cannot be eventually positive. This contradicts the positivity of $\exp(\lambda_1 A(t))y_\pm(t)$ $(t \ge T)$. Hence, (2.1.6) have no eventually positive solutions. The conclusion follows from Theorem 2.1.1. □

Theorem 2.1.2. *Every solution u of the problem* (2.1.1), (B_2) *is oscillatory in Ω if the ordinary differential inequalities*

$$y' \le \pm\big(a(t)\tilde{\Psi}(t) + \tilde{F}(t)\big)$$

have no eventually positive solutions, where

$$\tilde{\Psi}(t) = \frac{1}{|G|}\int_{\partial G} \psi\, dS,$$

$$\tilde{F}(t) = \frac{1}{|G|}\int_{G} f(x,t)\, dx$$

$\Big(|G|$ *denotes the volume of* G, *that is,* $|G| = \displaystyle\int_G dx\Big)$.

Proof. Let u be a solution of the problem (2.1.1), (B_2) such that $u > 0$ in $G \times (t_0, \infty)$ for some $t_0 > 0$. Proceeding as in the proof of Theorem 2.1.1, we see that (2.1.7) holds. Integration of (2.1.7) over G yields

$$\begin{aligned}\frac{d}{dt}\int_G u\, dx &\le a(t)\int_{\partial G} \frac{\partial u}{\partial \nu}\, dS + \int_G f(x,t)\, dx \\ &\le a(t)\int_{\partial G} \psi\, dS + \int_G f(x,t)\, dx, \quad t > t_0\end{aligned}$$

or equivalently

$$\tilde{U}'(t) \le a(t)\tilde{\Psi}(t) + \tilde{F}(t), \quad t > t_0,$$

where

$$\tilde{U}(t) = \frac{1}{|G|}\int_G u\, dx.$$

Hence, $\tilde{U}(t)$ is a positive solution of

$$y' \le a(t)\tilde{\Psi}(t) + \tilde{F}(t)$$

in (t_0, ∞). This contradicts the hypothesis. The case where $u < 0$ in $G \times (t_0, \infty)$ can be treated similarly, and we are also led to a contradiction. The proof is complete. $\square$

We state the analogue of Corollary 2.1.1.

Corollary 2.1.2. *Every solution u of the problem* (2.1.1), (B_2) *is oscillatory in Ω if*

$$\liminf_{t\to\infty} \int_T^t \big(a(s)\tilde{\Psi}(s) + \tilde{F}(s)\big)\, ds = -\infty, \tag{2.1.12}$$

$$\limsup_{t\to\infty} \int_T^t \big(a(s)\tilde{\Psi}(s) + \tilde{F}(s)\big)\, ds = \infty \tag{2.1.13}$$

for all large T.

Example 2.1.1. We consider the problem

$$\frac{\partial u}{\partial t} - \frac{\partial^2 u}{\partial x^2} + u = (\cos x)e^t(3\sin t + \cos t), \quad (x,t) \in (0, \tfrac{\pi}{2}) \times (0,\infty), \tag{2.1.14}$$

$$-\frac{\partial u}{\partial x}(0,t) = 0, \quad \frac{\partial u}{\partial x}(\tfrac{\pi}{2}, t) = -e^t \sin t, \quad t > 0. \tag{2.1.15}$$

Here $n = 1$, $a(t) = 1$, $G = (0, \pi/2)$, $\Omega = (0,\pi/2) \times (0,\infty)$, $c(x,t,\xi) = \xi$, $f(x,t) = (\cos x)e^t(3\sin t + \cos t)$, $\tilde{\Psi}(t) = -(2/\pi)e^t \sin t$ and $\tilde{F}(t) = (2/\pi)e^t(3\sin t + \cos t)$. Since

$$\begin{aligned}
\int_T^t (\tilde{\Psi}(s) + \tilde{F}(s))\, ds &= \int_T^t \frac{2}{\pi} e^s (2\sin s + \cos s)\, ds \\
&= \frac{4}{\pi}\int_T^t e^s \sin s\, ds + \frac{2}{\pi}\int_T^t e^s \cos s\, ds \\
&= \frac{4}{\pi}\left(\frac{1}{\sqrt{2}} e^t \sin(t - \tfrac{\pi}{4}) + \frac{1}{2} e^T(\cos T - \sin T)\right) \\
&\quad + \frac{2}{\pi}\left(\frac{1}{\sqrt{2}} e^t \sin(t + \tfrac{\pi}{4}) - \frac{1}{2} e^T(\cos T - \sin T)\right),
\end{aligned}$$

the conditions (2.1.12), (2.1.13) are satisfied. It follows from Corollary 2.1.2 that every solution u of the problem (2.1.14), (2.1.15) is oscillatory in $(0,\pi/2) \times (0,\infty)$. One such solution is $u = (\cos x)e^t \sin t$.

Example 2.1.2. We consider the problem

$$\begin{aligned}
\frac{\partial u}{\partial t} - \frac{\partial^2 u}{\partial x^2} + u &= (3\cos x + 2)e^t \cos t \\
&\quad -(\cos x + 1)e^t \sin t, \quad (x,t) \in (0,\pi) \times (0,\infty),
\end{aligned} \tag{2.1.16}$$

$$-\frac{\partial u}{\partial x}(0,t) = \frac{\partial u}{\partial x}(\pi,t) = 0, \quad t > 0. \tag{2.1.17}$$

Here $n = 1$, $a(t) = 1$, $G = (0,\pi)$, $\Omega = (0,\pi) \times (0,\infty)$, $c(x,t,\xi) = \xi$ and $f(x,t) = (3\cos x + 2)e^t \cos t - (\cos x + 1)e^t \sin t$. It is easy to check that $\tilde{\Psi}(t) = 0$ and $\tilde{F}(t) = 2e^t \cos t - e^t \sin t$. Since

$$\begin{aligned}
\int_T^t \tilde{F}(s)\, ds &= \int_T^t (2e^s \cos s - e^s \sin s)\, ds \\
&= 2\left(\frac{1}{\sqrt{2}} e^t \sin(t + \tfrac{\pi}{4}) - \frac{1}{2} e^T(\cos T - \sin T)\right) \\
&\quad - \left(\frac{1}{\sqrt{2}} e^t \sin(t - \tfrac{\pi}{4}) + \frac{1}{2} e^T(\cos T - \sin T)\right),
\end{aligned}$$

the conditions (2.1.12), (2.1.13) are satisfied. Corollary 2.1.2 implies that every solution u of the problem (2.1.16), (2.1.17) is oscillatory in $(0,\pi) \times (0,\infty)$. For example, $u = (\cos x + 1)e^t \cos t$ is such a solution.

Remark 2.1.1. The zero set of a solution of a parabolic equation is a subject of interest (see Angenent [19]).

2.2 Initial Value Problems

It is known that the initial value problem

$$\frac{\partial u}{\partial t} - \frac{\partial^2 u}{\partial x^2} = 0, \quad (x,t) \in \mathbb{R} \times (0,\infty),$$

$$u(x,0) = \sin x, \quad x \in \mathbb{R}$$

has a solution $u = (\sin x)e^{-t}$ which vanishes on the lines $\{(m\pi, t);\ t \geq 0, m \in \mathbb{Z}\}$, where $\mathbb{Z}$ denotes the set of all integers. In other words the solution u has arbitrarily large zeros which are far away from the initial line $t = 0$.

We study the initial value problem

$$\frac{\partial u}{\partial t} - \Delta u + p(t)u = 0, \quad (x,t) \in \mathbb{R}^n \times (0,\infty), \tag{2.2.1}$$

$$u(x,0) = \varphi(x), \quad x \in \mathbb{R}^n, \tag{2.2.2}$$

where Δ is the Laplacian in $\mathbb{R}^n$. It is assumed that $p(t) \in C([0,\infty);(0,\infty))$ and $p(t)$ is bounded from above.

Definition 2.2.1. The function space $\mathcal{E}[0,\infty)$ is defined to be the set of all continuous functions $u = u(x,t)$ defined on $\mathbb{R}^n \times [0,\infty)$ satisfying

$$|u(x,t)| \leq L \exp\left(|x|^\beta\right), \quad x \in \mathbb{R}^n,\ t \geq 0$$

with some constants $L > 0$ and $0 < \beta < 2$, where L and β may depend on $u(x,t)$.

Definition 2.2.2. The function space Φ is defined to be the set of all functions $\varphi(x)$ on $\mathbb{R}^n$ such that $\varphi(x)$, $\frac{\partial \varphi}{\partial x_i}(x)$, $\frac{\partial^2 \varphi}{\partial x_i \partial x_j}(x)$ $(i,j = 1,2,...,n)$ are all continuous and bounded on $\mathbb{R}^n$.

Definition 2.2.3. A function $u : \mathbb{R}^n \times (0,\infty) \longrightarrow \mathbb{R}$ is said to be *oscillatory* in $\mathbb{R}^n \times (0,\infty)$ if u has a zero in $\mathbb{R}^n \times (t,\infty)$ for any $t > 0$.

We take $X \in \mathbb{R}^n$ and fix it. Associated with every function $\varphi(x) \in \Phi$ we define

$$\hat{\varphi}_X(t) = \int_{\mathbb{R}^n} H(x,t)\varphi(x+X)\,dx,$$

where

$$H(x,t) = \frac{1}{(4\pi t)^{n/2}} \exp\left(-\frac{|x|^2}{4t}\right)$$

is the Green function of the heat equation $\frac{\partial u}{\partial t} - \Delta u = 0$.

For any $t > 0$, any $\varepsilon > 0$ and each $u \in \mathcal{E}[0, \infty)$ we define

$$U_{t,\varepsilon,X}(s) = \int_{\mathbb{R}^n} H(x, t - s + \varepsilon) u(x + X, s)\, dx, \quad 0 \le s \le t.$$

Lemma 2.2.1. *Let $\varphi(x) \in \Phi$ and let $u \in \mathcal{E}[0, \infty)$ be a classical solution of the initial value problem (2.2.1), (2.2.2) such that $u > 0$ in $\mathbb{R}^n \times (t_0, \infty)$ for some $t_0 > 0$. Then, for any sufficiently large t we see that $U_{t,\varepsilon,X}(s) \in C^1([0,t]; \mathbb{R})$, and that*

$$U_{t,\varepsilon,X}(s) > 0, \quad t_0 < s \le t, \tag{2.2.3}$$

$$\frac{d}{ds} U_{t,\varepsilon,X}(s) < 0, \quad t_0 < s \le t. \tag{2.2.4}$$

Proof. Since $u = u(x,t) \in \mathcal{E}[0, \infty)$, it follows that $u(x + X, t) \in \mathcal{E}[0, \infty)$ for each $X \in \mathbb{R}^n$. In Fujita [86] it was proved that $U_{t,\varepsilon,X}(s)$ exists and is a continuous function of s. Since $p(t)$ is bounded from above, we observe that

$$0 < p(x,t) \le K \quad \text{on } [0, \infty)$$

for some $K > 0$. Hence, it can be shown that

$$|p(t)u(x + X, t)| \le KL \exp(|x|^\beta), \quad (x,t) \in \mathbb{R}^n \times [0, \infty)$$

and consequently $p(t)u(x + X, t) \in \mathcal{E}[0, \infty)$. As was shown in Fujita [86], we observe that $U_{t,\varepsilon,X}(s)$ is continuously differentiable and satisfies the equation

$$\begin{aligned} \frac{d}{ds} U_{t,\varepsilon,X}(s) &= -\int_{\mathbb{R}^n} p(s) H(x, t - s + \varepsilon) u(x + X, s)\, dx \\ &= -p(s) \int_{\mathbb{R}^n} H(x, t - s + \varepsilon) u(x + X, s)\, dx \\ &= -p(s) U_{t,\varepsilon,X}(s), \quad 0 \le s \le t. \end{aligned} \tag{2.2.5}$$

Since $u(x + X, s) > 0$ in $\mathbb{R}^n \times (t_0, t]$, we see that $U_{t,\varepsilon,X}(s) > 0$ in $(t_0, t]$, and (2.2.5) implies

$$\frac{d}{ds} U_{t,\varepsilon,X}(s) < 0, \quad t_0 < s \le t.$$

This completes the proof. □

Theorem 2.2.1. *Let $\varphi(x) \in \Phi$ and let $u \in \mathcal{E}[0,\infty)$ be a classical solution of the initial value problem (2.2.1), (2.2.2). If $\hat{\varphi}_X(t)$ is oscillatory at $t = \infty$ for some $X \in \mathbb{R}^n$, then u is oscillatory in $\mathbb{R}^n \times (0,\infty)$.*

Proof. Suppose to the contrary that u has no zero in $\mathbb{R}^n \times (t_0, \infty)$ for some $t_0 > 0$. First we assume that $u > 0$ in $\mathbb{R}^n \times (t_0, \infty)$. It follows from Lemma 2.2.1 that we can choose t so large that $\hat{\varphi}_X(t) = 0$ and (2.2.3), (2.2.4) hold. From the properties of the Green function $H(x,t)$ we see that

$$\lim_{\varepsilon \to 0} U_{t,\varepsilon,X}(t) = u(X,t) > 0,$$

$$\lim_{\varepsilon \to 0} U_{t,\varepsilon,X}(0) = \hat{\varphi}_X(t) = 0.$$

Hence, for sufficiently small $\varepsilon > 0$, $U_{t,\varepsilon,X}(0)$ exists in a neighborhood of 0. It follows from Lemma 2.2.1 that there exists a number $T \in (0, t_0)$ such that $\frac{d}{ds} U_{t,\varepsilon,X}(s)\big|_{s=T} \geq 0$ and $U_{t,\varepsilon,X}(T) > 0$. We see from (2.2.5) that

$$0 \leq \frac{d}{ds} U_{t,\varepsilon,X}(s)\Big|_{s=T} = -p(T) U_{t,\varepsilon,X}(T) < 0.$$

This is a contradiction. If $u < 0$ in $\mathbb{R}^n \times (t_0, \infty)$, we conclude that $v := -u$ satisfies

$$\frac{\partial v}{\partial t} - \Delta v + p(t) v = 0, \quad (x,t) \in \mathbb{R}^n \times (0,\infty),$$

$$v(x,0) = -\varphi(x), \quad x \in \mathbb{R}^n,$$

and $v > 0$ in $\mathbb{R}^n \times (t_0, \infty)$. Proceeding as in the case where $u > 0$, we are also led to a contradiction. The proof is complete. □

Theorem 2.2.2. *Let $\varphi(x) \in \Phi$ and let $u \in \mathcal{E}[0,\infty)$ be a classical solution of the initial value problem (2.2.1), (2.2.2). If there exists a sequence of zeros $\{t_n\}$ of $\hat{\varphi}_X(t)$ such that $\lim_{n \to \infty} t_n = 0$, then there exists a sequence of zeros $\{(\tilde{x}_n, \tilde{t}_n)\}_{n=1}^{\infty} \subset \mathbb{R}^n \times (0,\infty)$ of the solution u such that $\lim_{n \to \infty} \tilde{t}_n = 0$.*

Proof. Suppose that there exists a positive number $\tilde{t} > 0$ such that u has no zero in $\mathbb{R}^n \times (0, \tilde{t})$. First we assume that $u > 0$ in $\mathbb{R}^n \times (0, \tilde{t})$. Then, $U_{t,\varepsilon,X}(s) > 0$ for $0 < s < \tilde{t}$. By the hypothesis we can choose a positive integer N so that $\hat{\varphi}_X(t_N) = 0$ and $0 < t_N < \tilde{t}$. Arguing as in the proof of Lemma 2.2.1, we find that (2.2.5) with $t = t_N$ holds. Hence we obtain

$$\frac{d}{ds} U_{t_N,\varepsilon,X}(s) = -p(s) U_{t_N,\varepsilon,X}(s) < 0, \quad 0 < s \leq t_N$$

and therefore $U_{t_N,\varepsilon,X}(s)$ is decreasing on $[0, t_N]$. Since $U_{t_N,\varepsilon,X}(s) > 0$ in $(0, t_N]$ and

$$\lim_{\varepsilon\to 0} U_{t_N,\varepsilon,X}(0) = \hat{\varphi}_X(t_N) = 0,$$

we conclude that $\lim_{\varepsilon\to 0} U_{t_N,\varepsilon,X}(t_N) = 0$. However, this contradicts the fact that

$$\lim_{\varepsilon\to 0} U_{t_N,\varepsilon,X}(t_N) = u(X, t_N) > 0.$$

The case where $u < 0$ in $\mathbb{R}^n \times (0, \hat{t})$ can be treated similarly, and we are also led to a contradiction. The proof is complete. □

Remark 2.2.1. We note that $\hat{\varphi}_x(t)$ is a solution of the initial value problem

$$\frac{\partial u}{\partial t} - \Delta u = 0, \quad (x, t) \in \mathbb{R}^n \times (0, \infty),$$

$$u(x, 0) = \varphi(x), \quad x \in \mathbb{R}^n$$

and unique within the class $\mathcal{E}[0, \infty)$ (see, for example, Cannon [42, Theorems 3.5.1 and 3.6.1]).

Example 2.2.1. We consider the initial value problem

$$\frac{\partial u}{\partial t} - \frac{\partial^2 u}{\partial x^2} + u = 0, \quad (x, t) \in \mathbb{R} \times (0, \infty), \tag{2.2.6}$$

$$u(x, 0) = \sin x, \quad x \in \mathbb{R}. \tag{2.2.7}$$

Here $n = 1$, $p(t) = 1$ and $\varphi(x) = \sin x$. It is easy to see that

$$\begin{aligned}\hat{\varphi}_0(t) &= \int_{\mathbb{R}} \frac{1}{2\sqrt{\pi t}} \exp\left(-\frac{|x|^2}{4t}\right) \sin x \, dx \\ &= 0, \quad t > 0\end{aligned}$$

and therefore $\hat{\varphi}_0(t)$ is oscillatory at $t = \infty$. If u is a classical solution of the problem (2.2.6), (2.2.7), then u is oscillatory in $\mathbb{R} \times (0, \infty)$ by virtue of Theorem 2.2.1. In fact, $u = (\sin x)e^{-2t}$ is such a solution.

Example 2.2.2. We consider the initial value problem

$$\frac{\partial u}{\partial t} - \frac{\partial^2 u}{\partial x^2} + 2u = 0, \quad (x, t) \in \mathbb{R} \times (0, \infty), \tag{2.2.8}$$

$$u(x, 0) = \cos x, \quad x \in \mathbb{R}. \tag{2.2.9}$$

Here $n = 1$, $p(t) = 2$ and $\varphi(x) = \cos x$. It is easily verified that

$$\begin{aligned}\hat{\varphi}_{\pi/2}(t) &= \int_{\mathbb{R}} \frac{1}{2\sqrt{\pi t}} \exp\left(-\frac{|x|^2}{4t}\right) \cos\left(x + \frac{\pi}{2}\right) dx \\ &= -\int_{\mathbb{R}} \frac{1}{2\sqrt{\pi t}} \exp\left(-\frac{|x|^2}{4t}\right) \sin x \, dx \\ &= 0, \quad t > 0.\end{aligned}$$

From Theorem 2.2.1 we conclude that the classical solution u of the problem (2.2.8), (2.2.9) is oscillatory in $\mathbb{R} \times (0, \infty)$. Indeed, $u = (\cos x)e^{-3t}$ is such a solution.

The following example shows that there may exist an oscillatory solution $u \notin \mathcal{E}[0, \infty)$ of certain initial value problem in the case where $\varphi(x) \notin \Phi$.

Example 2.2.3. We consider the initial value problem

$$\frac{\partial u}{\partial t} - \frac{\partial^2 u}{\partial x^2} + e^{-t}u = 0, \quad (x, t) \in \mathbb{R} \times (0, \infty), \tag{2.2.10}$$

$$u(x, 0) = e\left(e^{x-k} - e^{-(x-k)}\right), \quad x \in \mathbb{R}, \tag{2.2.11}$$

where k is some number. Here $n = 1$, $p(t) = e^{-t}$ and $\varphi(x) = e\left(e^{x-k} - e^{-(x-k)}\right)$. We easily see that $\varphi(x) \notin \Phi$ and $p(t) = e^{-t}$ is positive and bounded from above. Since $\varphi(x) = e(e^x - e^{-x})$ is an odd function on $\mathbb{R}$, it is easy to check that

$$\begin{aligned}\hat{\varphi}_k(t) &= \int_{\mathbb{R}} \frac{1}{2\sqrt{\pi t}} \exp\left(-\frac{|x|^2}{4t}\right) e(e^x - e^{-x})\, dx \\ &= 0, \quad t > 0.\end{aligned}$$

In this case, the solution $u = \left(e^{x-k} - e^{-(x-k)}\right) \exp(t + e^{-t})$ is an oscillatory solution of the problem (2.2.10), (2.2.11) satisfying $u(k, t) = 0$, $t \geq 0$. Obviously, the solution u has the zeros which tend to the initial line $t = 0$.

2.3 Notes

It seems that oscillation results for parabolic equations were first obtained by Bykov and Kultaev [40] in 1983. But they treated parabolic equations with functional arguments, and their results cannot be applied to parabolic equations without functional arguments.

In 1985 Kreith and Ladas [143] derived oscillation criteria for delay parabolic equations with positive and negative coefficients. Analogously we cannot apply their results to parabolic equations without delays.

In 1987 forced oscillation of parabolic equations was investigated by Yoshida [302], and parabolic equations studied contain the nonlinear term with functional arguments, whereas oscillation results obtained are applicable to parabolic equations without functional arguments. Thé results in Section 2.1 are extracted from Yoshida [302].

There are few oscillation results for initial value problems for parabolic equations, and there appears to be no known oscillation results for initial value problems for nonlinear parabolic equations. For initial value problems for linear parabolic equations we refer to Kobayashi and Yoshida [132]. Section 2.2 is taken from Kobayashi and Yoshida [132].

Chapter 3

Oscillation of Hyperbolic Equations

3.1 Boundary Value Problems

Oscillation results for boundary value problems for hyperbolic equations were established in 1969 by Kreith [137] in the case of one space variable. The case of several space variables was studied by several authors. We refer to Kahane [121] and Travis [264] for linear hyperbolic equations, and to Cazenave and Haraux [43–45], Haraux and Zuazua [96], Kreith, Kusano and Yoshida [142], Uesaka [269–272], Yoshida [301] and Zuazua [327] for nonlinear hyperbolic equations.

In this section we limit ourselves to C^2-solutions of certain boundary value problems for hyperbolic equations.

We consider the linear hyperbolic equation

$$\frac{\partial^2 u}{\partial t^2} - \sum_{i,j=1}^{n} \frac{\partial}{\partial x_i}\left(a_{ij}(x,t)\frac{\partial u}{\partial x_j}\right) + c(x,t)u = 0 \tag{3.1.1}$$

in a cylindrical domain $\Omega := G \times (0,\infty)$, where G is a bounded domain in $\mathbb{R}^n$ with piecewise smooth boundary ∂G. We use ν to denote the unit exterior normal vector to ∂G.

The boundary condition to be considered is the following:

(B_0) $\quad u = 0 \quad \text{on} \quad \partial G \times (0,\infty)$.

It is shown in Kahane [121] that there is a function $\hat{\Phi}(x) \in C^2(\overline{G};(0,\infty))$ such that

$$-\sum_{i,j=1}^{n} \frac{\partial}{\partial x_i}\left(a_{ij}(x,t)\frac{\partial \hat{\Phi}(x)}{\partial x_j}\right) \geq \varepsilon^2 \hat{\Phi}(x) \quad \text{in } \Omega \tag{3.1.2}$$

for some positive constant ε.

Theorem 3.1.1. *Assume that:*

(i) $a_{ij}(x,t) \in C^1(\overline{G} \times (0,\infty); \mathbb{R})$, $\frac{\partial}{\partial x_k} a_{ij}(x,t)$ $(i,j,k = 1,2,...,n)$ *are uniformly bounded in* $\overline{G} \times (0,\infty)$, *and the matrix* $\left(a_{ij}(x,t)\right)_{i,j=1}^n$ *is symmetric and uniformly positive definite in* Ω;

(ii) $c(x,t) \in C(\overline{\Omega}; \mathbb{R})$, *and* $c(x,t) \geq p(t)$ *in* Ω *for some function* $p(t) \in C((0,\infty); \mathbb{R})$.

Every solution $u \in C^2(\overline{G} \times (0,\infty); \mathbb{R})$ *of the boundary value problem* (3.1.1), (B_0) *is oscillatory in* Ω *if the ordinary differential inequality*

$$y'' + \left(p(t) + \varepsilon^2\right) y \leq 0 \tag{3.1.3}$$

has no eventually positive solution for every positive constant ε.

Proof. Assume on the contrary, that there is a solution $u \in C^2(\overline{G} \times (0,\infty); \mathbb{R})$ of the problem (3.1.1), (B_0) which has no zero in $G \times (t_0, \infty)$ for some $t_0 > 0$. Without loss of generality we may suppose that $u > 0$ in $G \times (t_0,\infty)$. Multiplying (3.1.1) by $\hat{\Phi}(x)$ and then integrating over G, we obtain

$$\int_G \frac{\partial^2 u}{\partial t^2} \hat{\Phi}(x)\, dx - \int_G \sum_{i,j=1}^n \frac{\partial}{\partial x_i}\left(a_{ij}(x,t)\frac{\partial u}{\partial x_j}\right)\hat{\Phi}(x)\, dx + \int_G c(x,t) u\, \hat{\Phi}(x)\, dx = 0. \tag{3.1.4}$$

An integration by parts shows that

$$\int_G \sum_{i,j=1}^n \frac{\partial}{\partial x_i}\left(a_{ij}(x,t)\frac{\partial u}{\partial x_j}\right)\hat{\Phi}(x)\, dx$$

$$= \int_{\partial G}\left[\hat{\Phi}(x)\frac{\partial u}{\partial \zeta} - u\,\frac{\partial \hat{\Phi}(x)}{\partial \zeta}\right] dS + \int_G u \sum_{i,j=1}^n \frac{\partial}{\partial x_j}\left(a_{ij}(x,t)\frac{\partial \hat{\Phi}(x)}{\partial x_i}\right) dx, \tag{3.1.5}$$

where $\frac{\partial}{\partial \zeta}$ denotes the conormal derivative

$$\frac{\partial}{\partial \zeta} = \sum_{i,j=1}^n a_{ij}(x,t)\nu_j \frac{\partial}{\partial x_i},$$

ν_j being the jth components of the unit exterior normal vector ν to ∂G. Because of the hypothesis (i), $\frac{\partial}{\partial \zeta}$ is outwardly directed from ∂G. Since $u > 0$ in $G \times (t_0,\infty)$ and $u = 0$ on $\partial G \times (0,\infty)$, it follows that $\frac{\partial u}{\partial \zeta} \leq 0$ on $\partial G \times (t_0,\infty)$. Therefore we have

$$\int_{\partial G}\left[\hat{\Phi}(x)\frac{\partial u}{\partial \zeta} - u\,\frac{\partial \hat{\Phi}(x)}{\partial \zeta}\right] dS \leq 0, \quad t > t_0. \tag{3.1.6}$$

Combining (3.1.2), (3.1.5) and (3.1.6), we obtain

$$\int_G \sum_{i,j=1}^{n} \frac{\partial}{\partial x_i}\left(a_{ij}(x,t)\frac{\partial u}{\partial x_j}\right)\hat{\Phi}(x)\,dx \le -\varepsilon^2 \int_G u\,\hat{\Phi}(x)\,dx, \quad t > t_0. \tag{3.1.7}$$

The hypothesis (ii) implies that

$$\int_G c(x,t)u\,\hat{\Phi}(x)\,dx \ge p(t)\int_G u\,\hat{\Phi}(x)\,dx, \quad t > t_0. \tag{3.1.8}$$

Combining (3.1.4), (3.1.7) and (3.1.8) yields

$$\hat{U}''(t) + (p(t)+\varepsilon^2)\hat{U}(t) \le 0, \quad t > t_0,$$

where

$$\hat{U}(t) = \int_G u\,\hat{\Phi}(x)\,dx.$$

Hence, $\hat{U}(t)$ is a positive solution of (3.1.3) in (t_0,∞). This contradicts the hypothesis and completes the proof. □

Corollary 3.1.1. *Assume that:*

(i) *the hypothesis* (i) *of Theorem* 3.1.1 *is satisfied;*
(ii) $c(x,t) \in C(\overline{\Omega};\mathbb{R})$ *and* $c(x,t) \ge 0$ *in* Ω.

Then every solution $u \in C^2(\overline{G}\times(0,\infty);\mathbb{R})$ *of the problem* (3.1.1), (B_0) *is oscillatory in* Ω.

Proof. Since $c(x,t) \ge 0$ in Ω, we can choose $p(t) = 0$. Hence, (3.1.3) reduces to

$$y'' + \varepsilon^2 y \le 0. \tag{3.1.9}$$

Suppose that (3.1.9) has an eventually positive solution $y(t)$. Then $y(t) > 0$ in (t_0,∞) for some $t_0 > 0$. We multiply (3.1.9) by $\sin\varepsilon(t-s)$ $(t_0 < s < t < s+(\pi/\varepsilon))$ and then integrate over $[s, s+(\pi/\varepsilon)]$ to obtain

$$\int_s^{s+(\pi/\varepsilon)} y''(t)\sin\varepsilon(t-s)\,dt + \varepsilon^2\int_s^{s+(\pi/\varepsilon)} y(t)\sin\varepsilon(t-s)\,dt \le 0, \quad s > t_0. \tag{3.1.10}$$

Integrating (3.1.10) by parts becomes

$$\varepsilon\left[y(s+(\pi/\varepsilon)) + y(s)\right] \le 0, \quad s > t_0$$

which contradicts the positivity of $y(t)$ $(t > t_0)$. Therefore, (3.1.9) has no eventually positive solution. The conclusion follows from Theorem 3.1.1. □

Remark 3.1.1. If $y'' + p(t)\,y \leq 0$ has no eventually positive solution, then (3.1.3) has no eventually positive solution.

Remark 3.1.2. It is easy to see that Corollary 3.1.1 can be extended to the nonlinear hyperbolic equation

$$\frac{\partial^2 u}{\partial t^2} - \sum_{i,j=1}^{n} \frac{\partial}{\partial x_i}\left(a_{ij}(x,t)\frac{\partial u}{\partial x_j}\right) + c(x,t,u) = 0. \tag{3.1.11}$$

Suppose that $a_{ij}(x,t)$ satisfy the condition (i) of Theorem 3.1.1 and that $c(x,t,\xi) \in C(\overline{\Omega} \times \mathbb{R};\mathbb{R})$ satisfies one of the following conditions:

(i) $\xi c(x,t,\xi) \geq 0$ for $(x,t) \in \Omega,\ \xi \in \mathbb{R}$;
(ii) $c(x,t,\xi) \geq 0$ for $(x,t) \in \Omega,\ \xi > 0$, and $c(x,t,-\xi) = -c(x,t,\xi)$ for $(x,t) \in \Omega,\ \xi > 0$.

Then it follows that every solution $u \in C^2(\overline{G} \times (0,\infty);\mathbb{R})$ of the problem (3.1.11), (B_0) is oscillatory in Ω. We note that (i) is weaker than (ii).

Next we deal with the nonlinear hyperbolic equation with a forcing term

$$\frac{\partial^2 u}{\partial t^2} - \sum_{i,j=1}^{n} \frac{\partial}{\partial x_i}\left(a_{ij}(x,t)\frac{\partial u}{\partial x_j}\right) + c(x,t,u) = f(x,t), \quad (x,t) \in \Omega. \tag{3.1.12}$$

Theorem 3.1.2. *Assume that:*

(i) *the hypothesis* (i) *of Theorem* 3.1.1 *is satisfied;*
(ii) $c(x,t,\xi) \in C(\overline{\Omega} \times \mathbb{R};\mathbb{R})$ *and* $\xi c(x,t,\xi) \geq 0$ *for* $(x,t) \in \Omega,\ \xi \in \mathbb{R}$;
(iii) $f(x,t) \in C(\overline{\Omega};\mathbb{R})$.

Every solution $u \in C^2(\overline{G} \times (0,\infty);\mathbb{R})$ *of the problem* (3.1.12), (B_0) *is oscillatory in* Ω *if the ordinary differential inequalities*

$$y'' + \varepsilon^2 y \leq \pm \hat{F}(t) \tag{3.1.13}$$

have no eventually positive solutions for every positive constant ε, *where*

$$\hat{F}(t) = \int_G f(x,t)\hat{\Phi}(x)\,dx.$$

Proof. Suppose that there exists a solution $u \in C^2(\overline{G} \times (0,\infty);\mathbb{R})$ of (3.1.12), (B_0) which has no zero in $G \times (t_0,\infty)$ for some $t_0 > 0$. First we assume that $u > 0$ in $G \times (t_0,\infty)$. Multiplying (3.1.12) by $\hat{\Phi}(x)$ and then integrating over G, we obtain

$$\int_G \frac{\partial^2 u}{\partial t^2}\hat{\Phi}(x)\,dx - \int_G \sum_{i,j=1}^{n} \frac{\partial}{\partial x_i}\left(a_{ij}(x,t)\frac{\partial u}{\partial x_j}\right)\hat{\Phi}(x)\,dx + \int_G c(x,t,u)\hat{\Phi}(x)\,dx$$
$$= \int_G f(x,t)\hat{\Phi}(x)\,dx. \tag{3.1.14}$$

Since $u > 0$ in $G \times (t_0, \infty)$, we see that $c(x,t,u) \geq 0$ in $G \times (t_0, \infty)$, and therefore

$$\int_G c(x,t,u)\hat{\Phi}(x)\,dx \geq 0, \quad t > t_0. \tag{3.1.15}$$

Proceeding as in the proof of Theorem 3.1.1, we find that (3.1.7) holds. Hence, it follows from (3.1.14) and (3.1.15) that $\hat{U}(t)$ satisfies

$$\hat{U}''(t) + \varepsilon^2 \hat{U}(t) \leq \hat{F}(t), \quad t > t_0. \tag{3.1.16}$$

Consequently, $\hat{U}(t)$ is a positive solution of (3.1.16) in (t_0, ∞). This is a contradiction. If $u < 0$ in $G \times (t_0, \infty)$, then $v := -u$ satisfies

$$\frac{\partial^2 v}{\partial t^2} - \sum_{i,j=1}^{n} \frac{\partial}{\partial x_i}\left(a_{ij}(x,t)\frac{\partial v}{\partial x_j}\right) \leq -f(x,t) \quad \text{in } G \times (t_0, \infty).$$

Arguing as in the case where $u > 0$, we observe that $\hat{V}(t) := \int_G v\,\hat{\Phi}(x)\,dx$ is a positive solution of

$$\hat{V}''(t) + \varepsilon^2 \hat{V}(t) \leq -\hat{F}(t) \quad \text{in } (t_0, \infty).$$

This contradicts the hypothesis. The proof is complete. □

An important special case of (3.1.12) is the following wave equation

$$\frac{\partial^2 u}{\partial t^2} - \Delta u + c(x,t,u) = f(x,t), \quad (x,t) \in \Omega, \tag{3.1.17}$$

where Δ is the Laplacian in $\mathbb{R}^n$.

We consider two kinds of boundary conditions:

(B_1) $\quad u = \psi \quad$ on $\partial G \times (0, \infty)$,

(B_2) $\quad \dfrac{\partial u}{\partial \nu} + \mu u = \tilde{\psi} \quad$ on $\partial G \times (0, \infty)$,

where ψ, $\tilde{\psi} \in C(\partial G \times (0,\infty); \mathbb{R})$ and $\mu \in C(\partial G \times (0,\infty); [0,\infty))$.

Let λ_1 (> 0) be the first eigenvalue of the eigenvalue problem (2.1.4), (2.1.5) and $\Phi(x)$ be the corresponding eigenfunction which is chosen so that $\Phi(x) > 0$ in G.

Theorem 3.1.3. *Assume that the hypotheses* (ii) *and* (iii) *of Theorem* 3.1.2 *are satisfied. Every solution* $u \in C^2(\overline{G} \times (0,\infty); \mathbb{R})$ *of the problem* (3.1.17), (B_1) *is oscillatory in* Ω *if the ordinary differential inequalities*

$$y'' + \lambda_1 y \leq \pm\bigl(-\Psi(t) + F(t)\bigr) \tag{3.1.18}$$

have no eventually positive solutions, where

$$\Psi(t) = \int_{\partial G} \psi \frac{\partial \Phi(x)}{\partial \nu}\,dS,$$

$$F(t) = \int_G f(x,t)\Phi(x)\,dx.$$

Proof. Let $u \in C^2(\overline{G} \times (0,\infty);\mathbb{R})$ be a solution of the problem (3.1.17), (B_1) which has no zero in $G \times (t_0,\infty)$ for some $t_0 > 0$. First we suppose that $u > 0$ in $G \times (t_0,\infty)$. Then we see that $c(x,t,u) \geq 0$ in $G \times (t_0,\infty)$, and hence

$$\frac{\partial^2 u}{\partial t^2} - \Delta u \leq f(x,t) \quad \text{in } G \times (t_0,\infty). \tag{3.1.19}$$

We multiply (3.1.19) by $\Phi(x)$ and then integrating over G to obtain

$$\int_G \frac{\partial^2 u}{\partial t^2}\Phi(x)\,dx - \int_G (\Delta u)\Phi(x)\,dx \leq \int_G f(x,t)\Phi(x)\,dx, \quad t > t_0. \tag{3.1.20}$$

Application of Green's formula yields

$$\int_G (\Delta u)\Phi(x)\,dx = -\Psi(t) - \lambda_1 U(t), \tag{3.1.21}$$

where

$$U(t) = \int_G u\,\Phi(x)\,dx$$

(see (2.1.9)). Combining (3.1.20) with (3.1.21) yields

$$U''(t) + \lambda_1 U(t) \leq -\Psi(t) + F(t), \quad t > t_0. \tag{3.1.22}$$

Hence, $U(t)$ is a positive solution of

$$y'' + \lambda_1 y \leq -\Psi(t) + F(t)$$

in (t_0,∞). This contradicts the hypothesis. If $u < 0$ in $G \times (t_0,\infty)$, we let $v = -u$. As in the case where $u > 0$, we find that $V(t) := \int_G v\,\Phi(x)\,dx$ is a positive solution of

$$y'' + \lambda_1 y \leq -(-\Psi(t) + F(t))$$

in (t_0,∞). This is a contradiction and the proof is complete. □

Theorem 3.1.4. *Assume that:*

(i) *the hypothesis* (iii) *of Theorem* 3.1.2 *is satisfied;*
(ii) $c(x,t,\xi) \in C(\overline{\Omega} \times \mathbb{R};\mathbb{R})$ *and* $c(x,t,\xi) \geq p(t)\varphi(\xi)$ *for* $(x,t,\xi) \in \Omega \times (0,\infty)$, *where* $p(t) \in C((0,\infty);[0,\infty))$ *and* $\varphi(\xi)$ *is a convex function of class* $C((0,\infty);(0,\infty))$;
(iii) $c(x,t,-\xi) = -c(x,t,\xi)$ *for* $(x,t,\xi) \in \Omega \times (0,\infty)$.

Every solution $u \in C^2(\overline{G} \times (0,\infty); \mathbb{R})$ of the problem (3.1.17), (B_2) is oscillatory in Ω if the ordinary differential inequalities

$$y'' + p(t)\varphi(y) \le \pm\big(\tilde{\Psi}(t) + \tilde{F}(t)\big) \tag{3.1.23}$$

have no eventually positive solutions, where

$$\tilde{\Psi}(t) = \frac{1}{|G|}\int_{\partial G} \psi \, dS,$$
$$\tilde{F}(t) = \frac{1}{|G|}\int_{G} f(x,t)\, dx$$

$\Big(|G|$ *denotes the volume of* G, *that is,* $|G| = \displaystyle\int_G dx\Big)$.

Proof. Suppose that $u \in C^2(\overline{G} \times (0,\infty); \mathbb{R})$ is a nonoscillatory solution of the problem (3.1.17), (B_2). Let $u > 0$ in $G \times (t_0, \infty)$ for some $t_0 > 0$. Integrating (3.1.17) over G, we have

$$\int_G \frac{\partial^2 u}{\partial t^2}\, dx - \int_G \Delta u \, dx + \int_G c(x,t,u)\, dx = \int_G f(x,t)\, dx. \tag{3.1.24}$$

The divergence theorem implies that

$$\int_G \Delta u\, dx = \int_{\partial G} \frac{\partial u}{\partial \nu}\, dS = \int_{\partial G} (-\mu u + \tilde{\psi})\, dS \le \int_{\partial G} \tilde{\psi}\, dS, \quad t > t_0. \tag{3.1.25}$$

An application of Jensen's inequality shows that

$$\int_G c(x,t,u)\, dx \ge p(t)\int_G \varphi(u)\, dx \ge p(t)|G|\varphi\left(\frac{1}{|G|}\int_G u\, dx\right), \quad t > t_0. \tag{3.1.26}$$

Dividing (3.1.24) by $|G|$ and using (3.1.25), (3.1.26), we obtain

$$\tilde{U}''(t) + p(t)\varphi(\tilde{U}(t)) \le \tilde{\Psi}(t) + \tilde{F}(t), \quad t > t_0,$$

where

$$\tilde{U}(t) = \frac{1}{|G|}\int_G u\, dx.$$

Hence, $\tilde{U}(t)$ is a positive solution of (3.1.23) with $+\big(\tilde{\Psi}(t) + \tilde{F}(t)\big)$ in (t_0, ∞). This contradicts the hypothesis. If $u < 0$ in $G \times (t_0, \infty)$, the same arguments as in the case where $u > 0$ leads us to a contradiction. The proof is complete. □

Applying Theorems 1.4.3–1.4.5 to Theorems 3.1.2– 3.1.4, we can establish various oscillation results.

Corollary 3.1.2. *Assume that the hypotheses* (ii) *and* (iii) *of Theorem* 3.1.2 *are satisfied. Every solution* $u \in C^2(\overline{G} \times (0,\infty); \mathbb{R})$ *of the problem* (3.1.17), (B_1) *is oscillatory in* Ω *if*

$$\liminf_{t\to\infty} \int_T^t \left(1 - \frac{s}{t}\right) (-\Psi(s) + F(s))\, ds = -\infty,$$

$$\limsup_{t\to\infty} \int_T^t \left(1 - \frac{s}{t}\right) (-\Psi(s) + F(s))\, ds = \infty$$

for all large T.

Proof. Using Theorem 1.4.2, we see that (3.1.18) have no eventually positive solutions. The conclusion follows from Theorem 3.1.3. □

Corollary 3.1.3. *Assume that the hypotheses* (i)–(iii) *of Theorem* 3.1.4 *are satisfied. Every solution* $u \in C^2(\overline{G} \times (0,\infty); \mathbb{R})$ *of the problem* (3.1.17), (B_2) *is oscillatory in* Ω *if*

$$\liminf_{t\to\infty} \int_T^t \left(1 - \frac{s}{t}\right) (\tilde{\Psi}(s) + \tilde{F}(s))\, ds = -\infty,$$

$$\limsup_{t\to\infty} \int_T^t \left(1 - \frac{s}{t}\right) (\tilde{\Psi}(s) + \tilde{F}(s))\, ds = \infty$$

for all large T.

Proof. The conclusion follows by combining Theorem 1.4.2 with Theorem 3.1.4. □

Corollary 3.1.4. *Assume that the hypotheses* (i)–(iii) *of Theorem* 3.1.4 *are satisfied. Every solution* $u \in C^2(\overline{G} \times (0,\infty); \mathbb{R})$ *of the problem* (3.1.17), (B_2) *is oscillatory in* Ω *if the ordinary differential inequality*

$$y'' + p(t)\varphi(y) \le 0$$

has no eventually positive solution and if there is a function $\theta(t) \in C^2((0,\infty); \mathbb{R})$ *with the following properties:*

(i) $\theta(t)$ *is oscillatory at* $t = \infty$;
(ii) $\theta''(t) = \tilde{\Psi}(t) + \tilde{F}(t), \quad t > 0$;
(iii) $\lim_{t\to\infty} \theta(t) = 0$.

Proof. In view of the fact that $\lim_{\xi \to +0} \varphi(\xi) = 0$ and $\varphi(\xi)$ is convex in $(0, \infty)$, it can be shown that $\varphi(\xi)$ is nondecreasing in $(0, \infty)$. It follows from Theorem 1.4.3 that (3.1.23) have no eventually positive solutions. The conclusion follows from Theorem 3.1.4. □

Next we investigate the existence of zeros of solutions to the problem (3.1.17), (B_1) in bounded domains.

Lemma 3.1.1. *Let L be a positive number. If there is a number $s \geq 0$ such that*

$$\int_s^{s+(\pi/L)} F(t) \sin L(t-s)\, dt \leq 0, \tag{3.1.27}$$

then the ordinary differential inequality

$$y'' + L^2 y \leq F(t) \tag{3.1.28}$$

has no positive solution in $(s, s + (\pi/L)]$.

Proof. Let (3.1.28) have a solution $y(t)$ which is positive in $(s, s+(\pi/L)]$. Multiplying (3.1.28) by $\sin L(t-s)$ and then integrating over $[s, s+(\pi/L)]$, we obtain

$$L\big[y(s) + y\big(s + (\pi/L)\big)\big] \leq \int_s^{s+(\pi/L)} F(t) \sin L(t-s)\, dt \tag{3.1.29}$$

(see Naito and Yoshida [213, Lemma 3]). The right hand side of (3.1.29) is nonpositive, but the left hand side of (3.1.29) is positive. This is a contradiction and the proof is complete. □

Theorem 3.1.5. *Assume that the hypotheses* (ii) *and* (iii) *of Theorem* 3.1.2 *are satisfied. If there is a number $s \geq 0$ such that*

$$\int_s^{s+(\pi/\sqrt{\lambda_1})} \big(-\Psi(t) + F(t)\big) \sin \sqrt{\lambda_1}(t-s)\, dt = 0, \tag{3.1.30}$$

then every solution $u \in C^2(\overline{G} \times (0,\infty); \mathbb{R})$ of the problem (3.1.17), (B_1) *has a zero in $G \times \big(s, s + (\pi/\sqrt{\lambda_1})\big]$.*

Proof. Suppose that the problem (3.1.17), (B_1) has a solution $u \in C^2(\overline{G} \times (0,\infty); \mathbb{R})$ which has no zero in $G \times \big(s, s + (\pi/\sqrt{\lambda_1})\big]$. Let $u > 0$ in $G \times \big(s, s + (\pi/\sqrt{\lambda_1})\big]$. Proceeding as in the proof of Theorem 3.1.3, we observe that $U(t)$ is a positive solution of

$$U''(t) + \lambda_1 U(t) \leq -\Psi(t) + F(t), \quad t \in \big(s, s + (\pi/\sqrt{\lambda_1})\big]. \tag{3.1.31}$$

However, Lemma 3.1.1 implies that (3.1.31) has no positive solution in $\left(s, s+(\pi/\sqrt{\lambda_1})\right]$. This is a contradiction. If $u < 0$ in $G \times \left(s, s+(\pi/\sqrt{\lambda_1})\right]$, then $v := -u$ is a positive solution of

$$\begin{aligned} &\frac{\partial^2 v}{\partial t^2} - \Delta v \le -f(x,t) \quad \text{in } G \times \left(s, s+(\pi/\sqrt{\lambda_1})\right], \\ &v = -\psi \quad \text{on } \partial G \times (0,\infty). \end{aligned}$$

Using the same arguments as in the case where $u > 0$, we are led to a contradiction. The proof is complete. □

Example 3.1.1. We consider the problem

$$\frac{\partial^2 u}{\partial t^2} - \frac{\partial^2 u}{\partial x^2} + \frac{2}{3} e^{4x} e^{-\sqrt{2}t} u^3 = e^{-2x}\left(-4\sin t + \sqrt{2}\cos t - \frac{1}{6}\sin 3t\right) e^{t/\sqrt{2}},$$

$$(x,t) \in (0,1) \times (0,\infty), \tag{3.1.32}$$

$$-\frac{\partial u}{\partial x}(0,t) = 2e^{t/\sqrt{2}} \sin t, \quad t > 0, \tag{3.1.33}$$

$$\frac{\partial u}{\partial x}(1,t) = -\frac{2}{e^2} e^{t/\sqrt{2}} \sin t, \quad t > 0. \tag{3.1.34}$$

Here $n = 1$, $G = (0,1)$, $\Omega = (0,1) \times (0,\infty)$, $c(x,t,\xi) = (2/3)e^{4x}e^{-\sqrt{2}t}\xi^3$ and $f(x,t) = e^{-2x}(-4\sin t + \sqrt{2}\cos t - (1/6)\sin 3t)e^{t/\sqrt{2}}$. It is easily verified that

$$\begin{aligned} \tilde{\Psi}(t) &= 2\left(1 - \frac{1}{e^2}\right) e^{t/\sqrt{2}} \sin t, \\ \tilde{F}(t) &= -\left(1 - \frac{1}{e^2}\right) e^{t/\sqrt{2}} \left(2\sin t - \frac{1}{\sqrt{2}}\cos t + \frac{1}{12}\sin 3t\right). \end{aligned}$$

A simple computation shows that

$$\begin{aligned} &\int_T^t \left(1 - \frac{s}{t}\right)\left(\tilde{\Psi}(s) + \tilde{F}(s)\right) ds \\ &= \left(1 - \frac{1}{e^2}\right) e^{t/\sqrt{2}} \left(\frac{3}{\sqrt{2}}\sin(t+\alpha_1) - \frac{1}{114}\sin(3t+\alpha_2)\right) + B(t,T), \end{aligned}$$

where α_1, α_2 are constants and $B(t,T)$ is bounded as $t \to \infty$. We note that Corollary 3.1.3 holds true if the hypotheses (ii), (iii) of Theorem 3.1.4 are replaced by the hypothesis (i) of Remark 3.1.2. Hence, every solution $u \in C^2(\overline{G} \times (0,\infty); \mathbb{R})$ of the problem (3.1.32)–(3.1.34) is oscillatory in $(0,1) \times (0,\infty)$. One such solution is $u = e^{-2x}e^{t/\sqrt{2}} \sin t$.

Example 3.1.2. We consider the problem

$$\frac{\partial^2 u}{\partial t^2} - \frac{\partial^2 u}{\partial x^2} + 2u = (\cos x)e^{-t}(-2\cos t + 3\sin t),$$
$$(x,t) \in (0, \tfrac{\pi}{2}) \times (0,\infty), \quad (3.1.35)$$
$$-\frac{\partial u}{\partial x}(0,t) = 0, \quad t > 0, \quad (3.1.36)$$
$$\frac{\partial u}{\partial x}(\tfrac{\pi}{2}, t) = -e^{-t}\sin t, \quad t > 0. \quad (3.1.37)$$

Here $n = 1, G = (0, \pi/2), \Omega = (0, \pi/2) \times (0, \infty), c(x,t,\xi) = 2\xi$ and $f(x,t) = (\cos x)e^{-t}(-2\cos t + 3\sin t)$. We easily see that

$$\tilde{\Psi}(t) = -\frac{2}{\pi} e^{-t} \sin t,$$
$$\tilde{F}(t) = \frac{2}{\pi} e^{-t}(-2\cos t + 3\sin t).$$

It is known that $y'' + 2y \le 0$ has no eventually positive solution (see, for example, the proof of Corollary 3.1.1). Letting $\theta(t) = (2/\pi)e^{-t}(\cos t + \sin t)$, we observe that

$$\theta(t) = \frac{2\sqrt{2}}{\pi} e^{-t} \sin(t + \tfrac{\pi}{4}),$$
$$\theta''(t) = \frac{4}{\pi} e^{-t}(\sin t - \cos t) = \tilde{\Psi}(t) + \tilde{F}(t),$$
$$\lim_{t\to\infty} \theta(t) = 0,$$

and therefore $\theta(t)$ satisfies the conditions (i)–(iii) of Corollary 3.1.4. Hence, every solution $u \in C^2(\overline{G} \times (0,\infty); \mathbb{R})$ of the problem (3.1.35)–(3.1.37) is oscillatory in $(0, \pi/2) \times (0, \infty)$. Indeed, $u = (\cos x)e^{-t} \sin t$ is such a solution.

Example 3.1.3. We consider the problem

$$\frac{\partial^2 u}{\partial t^2} - \frac{\partial^2 u}{\partial x^2} + \frac{13}{4}\left(\frac{\pi}{L}\right)^2 u = 4\left(\sin\frac{\pi}{L}x\right)\cos\frac{\pi}{2L}t,$$
$$(x,t) \in (0, L) \times (0, \infty), \quad (3.1.38)$$
$$u(0,t) = u(L,t) = 0, \quad t > 0, \quad (3.1.39)$$

where L is some positive number. Here $n = 1$, $G = (0, L)$, $\Omega = (0, L) \times (0, \infty)$, $c(x,t,\xi) = (13/4)(\pi/L)^2\xi$ and $f(x,t) = 4(\sin(\pi/L)x) \times \cos(\pi/(2L))t$. It is easy to check that $\lambda_1 = (\pi/L)^2$, $\Phi(x) = \sin(\pi/L)x$, $\Psi(t) = 0$ and

$$F(t) = 4\cos\frac{\pi}{2L}t \int_0^L \sin^2\frac{\pi}{L}x\,dx = 2L\cos\frac{\pi}{2L}t.$$

Since

$$\int_s^{s+L} 2L\left(\cos\frac{\pi}{2L}t\right)\sin\frac{\pi}{L}(t-s)\,dt = -\frac{8\sqrt{2}}{3\pi}L^2\sin\left(\tfrac{\pi}{2L}s-\tfrac{\pi}{4}\right),$$

Theorem 3.1.5 implies that every solution $u \in C^2(\overline{G}\times(0,\infty);\mathbb{R})$ of the problem (3.1.38), (3.1.39) has a zero in $(0,L)\times(s_n, s_n+L]$, where $s_n = L/2 + 2Ln$ $(n = 0, 1, 2, \ldots)$. For example, $u = (L/\pi)^2(\sin(\pi/L)x)\cos(\pi/(2L))t$ is such a solution.

Remark 3.1.3. Corollary 3.1.2 is not applicable to the case where $f(x,t) = 0$ and $\psi = 0$.

Remark 3.1.4. When $f(x,t) = 0$ and $\psi = 0$, Theorem 3.1.5 was established by Cazenave and Haraux [43, Theorem 2.1]. Lemma 3.1.1 and Theorem 3.1.5 hold true if we replace $\left(s, s+(\pi/L)\right]$ and $G\times\left(s, s+(\pi/\sqrt{\lambda_1})\right]$ by $\left[s, s+(\pi/L)\right)$ and $G\times\left[s, s+(\pi/\sqrt{\lambda_1})\right)$, respectively.

Remark 3.1.5. Kahane [121] showed that the continuous generalized solutions have arbitrarily large zeros on any interior line of the cylinder under the stronger hypotheses on the coefficients of Eq. (3.1.1). Similar results for one-dimensional semilinear wave equations were derived by Uesaka [270–272] in which more general wave equations are studied.

3.2 Initial Value Problems

It is easy to see that $u = (\sin x)\sin t$ is an oscillatory solution of the hyperbolic equation

$$\frac{\partial^2 u}{\partial t^2} - \frac{\partial^2 u}{\partial x^2} = 0, \quad (x,t) \in \mathbb{R}\times(0,\infty)$$

in one-dimensional space.

In this section we derive oscillation results which include the above example as a special case. We consider the initial value problem

$$\frac{\partial^2 u}{\partial t^2} - c^2\frac{\partial^2 u}{\partial x^2} + k^2 u = f(x,t), \quad (x,t) \in \mathbb{R}\times(0,\infty), \tag{3.2.1}$$

$$u(x,0) = \varphi(x), \quad x \in \mathbb{R}, \tag{3.2.2}$$

$$\frac{\partial u}{\partial t}(x,0) = \psi(x), \quad x \in \mathbb{R}, \tag{3.2.3}$$

where c (> 0) and k (≥ 0) are constants. It is assumed that $\varphi(x) \in C^2(\mathbb{R};\mathbb{R})$, $\psi(x) \in C^1(\mathbb{R};\mathbb{R})$ and $f(x,t) \in C^1(\mathbb{R}\times[0,\infty);\mathbb{R})$.

Theorem 3.2.1. *The solution of the initial value problem* (3.2.1)–(3.2.3) *is given by*

$$\begin{aligned} u(x,t) &= \frac{1}{2}(\varphi(x-ct)+\varphi(x+ct)) \\ &\quad -\frac{1}{2c}\int_{x-ct}^{x+ct}\varphi(\xi)\frac{\partial B}{\partial \tau}(\xi,0;x,t)\,d\xi + \frac{1}{2c}\int_{x-ct}^{x+ct}\psi(\xi)B(\xi,0;x,t)\,d\xi \\ &\quad +\frac{1}{2c}\int_0^t\int_{x-c(t-\tau)}^{x+c(t-\tau)} f(\xi,\tau)B(\xi,\tau;x,t)\,d\xi d\tau, \end{aligned} \tag{3.2.4}$$

where

$$B(\xi,\tau;x,t) = J_0\left(\frac{k}{c}\sqrt{c^2(t-\tau)^2-(\xi-x)^2}\right),$$

J_0 *being the Bessel function of the first kind of order* 0. *Moreover, the solution is unique.*

In the case where $\psi(x) = 0$ and $f(x,t) = 0$, Theorem 3.2.1 reduces to Problem 2 of John [120, p.111], and hence the proof of Theorem 3.2.1 will be omitted.

Theorem 3.2.2. *Let u be the solution of the initial value problem* (3.2.1)–(3.2.3). *If there exists a point $(X,T) \in \mathbb{R}\times(0,\infty)$ with the following properties*:

$$\varphi(X-cT)+\varphi(X+cT) = 0,$$

$$\int_{X-cT}^{X+cT}\left(\psi(\xi)B(\xi,0;X,T)-\varphi(\xi)\frac{\partial B}{\partial\tau}(\xi,0;X,T)\right)d\xi = 0,$$

$$\int_{D(X,T)} f(\xi,\tau)B(\xi,\tau;X,T)\,d\xi d\tau = 0,$$

where

$$D(X,T) = \{(\xi,\tau);\ 0<\tau<T,\ |\xi-X|<c(T-\tau)\},$$

then we conclude that $u(X,T) = 0$.

Proof. The conclusion follows from Theorem 3.2.1 with $x = X$ and $t = T$. □

An important special case of (3.2.1) is

$$\frac{\partial^2 u}{\partial t^2} - c^2\frac{\partial^2 u}{\partial x^2} + k^2 u = 0, \quad (x,t)\in\mathbb{R}\times(0,\infty). \tag{3.2.5}$$

Theorem 3.2.3. *Let u be the solution of the initial value problem* (3.2.2), (3.2.3), (3.2.5). *If there are a number $X \in \mathbb{R}$ and a sequence $\{t_n\}_{n=1}^{\infty} \subset (0, \infty)$ such that:*

$$\lim_{n\to\infty} t_n = \infty,$$

$$\varphi(X - ct_n) + \varphi(X + ct_n) = 0,$$

$$\int_{X-ct_n}^{X+ct_n} \left(\psi(\xi) B(\xi, 0; X, t_n) - \varphi(\xi) \frac{\partial B}{\partial \tau}(\xi, 0; X, t_n) \right) d\xi = 0,$$

then $u(X,t)$ is oscillatory at $t = \infty$ as a function of t.

Proof. Theorem 3.2.2 implies that $u(X, t_n) = 0$ $(n = 1, 2, \ldots)$. Hence, $u(X,t)$ is oscillatory at $t = \infty$. □

Definition 3.2.1. A real-valued function u on $\mathbb{R} \times (0, \infty)$ is said to be *oscillatory* in $\mathbb{R} \times (0, \infty)$ if u has a zero in $\mathbb{R} \times (t, \infty)$ for any $t > 0$.

Theorem 3.2.4. *Let u be the solution of the initial value problem* (3.2.2), (3.2.3), (3.2.5). *Assume that for any $t > 0$ there exists an interval $J = [\alpha, \beta] \subset \mathbb{R}$ for which*

$$\beta - \alpha > 2ct,$$

$$\varphi(\alpha) + \varphi(\beta) = 0,$$

$$\int_{\alpha}^{\beta} \left(\psi(\xi) B\left(\xi, 0; \frac{\alpha+\beta}{2}, \frac{\beta-\alpha}{2c}\right) - \varphi(\xi) \frac{\partial B}{\partial \tau}\left(\xi, 0; \frac{\alpha+\beta}{2}, \frac{\beta-\alpha}{2c}\right) \right) d\xi = 0.$$

Then u is oscillatory in $\mathbb{R} \times (0, \infty)$.

Proof. It follows from Theorem 3.2.2 that for any $t > 0$ there is a point $((\alpha+\beta)/2, (\beta-\alpha)/(2c)) \in \mathbb{R} \times (t, \infty)$ such that $u((\alpha+\beta)/2, (\beta-\alpha)/(2c)) = 0$. Hence, u is oscillatory in $\mathbb{R} \times (0, \infty)$. □

Example 3.2.1. We consider the initial value problem

$$\frac{\partial^2 u}{\partial t^2} - \frac{\partial^2 u}{\partial x^2} = 0, \quad (x, t) \in \mathbb{R} \times (0, \infty), \tag{3.2.6}$$

$$u(x, 0) = 0, \quad x \in \mathbb{R}, \tag{3.2.7}$$

$$\frac{\partial u}{\partial t}(x, 0) = \sin x, \quad x \in \mathbb{R}. \tag{3.2.8}$$

Here $c = 1$, $k = 0$, $\varphi(x) = 0$ and $\psi(x) = \sin x$. Letting $J_n = [-n\pi, n\pi]$ $(n = 1, 2, \ldots)$, we see that $|J_n|$ (the length of J_n) tends to ∞, and $\int_{-n\pi}^{n\pi} \sin \xi \, d\xi = 0$. It follows from Theorem 3.2.4 with $B(\xi, \tau; x, t) = 1$ that the solution u of the problem (3.2.6)–(3.2.8) is oscillatory in $\mathbb{R} \times (0, \infty)$. Indeed, $u = (\sin x) \sin t$ is a solution of (3.2.6)–(3.2.8) which is oscillatory in $\mathbb{R} \times (0, \infty)$.

Example 3.2.2. We consider the initial conditions

$$u(x,0) = \cos x, \quad x \in \mathbb{R}, \tag{3.2.9}$$

$$\frac{\partial u}{\partial t}(x,0) = \cos x, \quad x \in \mathbb{R}. \tag{3.2.10}$$

Here $\varphi(x) = \psi(x) = \cos x$. Letting $J_n = [-(3/2)\pi - n\pi, (\pi/2) + n\pi]$ ($n = 1, 2, ...$), we observe that $|J_n| = 2\pi + 2n\pi \to \infty$ ($n \to \infty$),

$$\cos(-(3/2)\pi - n\pi) + \cos((\pi/2) + n\pi) = 0,$$

$$\int_{-(3/2)\pi - n\pi}^{(\pi/2)+n\pi} \cos\xi \, d\xi = 0.$$

Theorem 3.2.4 with $B(\xi, \tau; x, t) = 1$ implies that the solution u of the problem (3.2.6), (3.2.9), (3.2.10) is oscillatory in $\mathbb{R} \times (0, \infty)$. In fact, $u = (\cos x)(\sin t + \cos t)$ is an oscillatory solution of (3.2.6), (3.2.9), (3.2.10).

Remark 3.2.1. When $k = 0$ and $f(x,t) = 0$, we set $B(\xi, \tau; x, t) = 1$. Then (3.2.1) reduces to

$$\frac{\partial^2 u}{\partial t^2} - c^2 \frac{\partial^2 u}{\partial x^2} = 0, \quad (x,t) \in \mathbb{R} \times (0, \infty) \tag{3.2.11}$$

and the solution u of the initial value problem (3.2.2), (3.2.3), (3.2.11) is given by

$$u(x,t) = \frac{1}{2}\big(\varphi(x - ct) + \varphi(x + ct)\big) + \frac{1}{2c}\int_{x-ct}^{x+ct} \psi(\xi)\, d\xi$$

which is known as d'Alembert's formula.

3.3 Characteristic Initial Value Problems

Since the pioneering work of Kreith [136], oscillations of solutions of characteristic initial value problems for hyperbolic equations have been studied by numerous authors. Linear equations were investigated in the papers [108, 144, 233–236]. For nonlinear equations we refer to [105–107, 141, 142, 295, 301, 303].

We consider the characteristic initial value problem

$$\frac{\partial^2 u}{\partial x \partial y} + a\frac{\partial u}{\partial x} + b\frac{\partial u}{\partial y} + c(x, y, u) = f(x, y), \quad (x, y) \in Q, \tag{3.3.1}$$

$$\frac{\partial u}{\partial x} + bu + \lambda_1(x)u = \psi_1(x) \quad \text{on } \Gamma_x, \tag{3.3.2}$$

$$\frac{\partial u}{\partial y} + au + \lambda_2(y)u = \psi_2(y) \quad \text{on } \Gamma_y, \tag{3.3.3}$$

where a and b are constants, and

$$Q = \{(x,y) \in \mathbb{R}^2;\ x > 0,\ y > 0\},$$
$$\Gamma_x = \{(x,0) \in \mathbb{R}^2;\ x > 0\},$$
$$\Gamma_y = \{(0,y) \in \mathbb{R}^2;\ y > 0\}.$$

Let L and k be positive constants. We use the notation:

$$Q(t) = \{(x,y) \in \mathbb{R}^2;\ x > 0,\ y > 0,\ x + L^{-2}k^2 y > t\},$$
$$U(t) = \frac{1}{t}\int_0^t u(\tau, L^2k^{-2}(t-\tau))\,d\tau,$$
$$F(t) = \frac{1}{t}\int_0^t f(\tau, L^2k^{-2}(t-\tau))\,d\tau.$$

Definition 3.3.1. A real-valued function u on Q is said to be *oscillatory* in Q if u has a zero in Q_r for any $r > 0$, where

$$Q_r = Q \cap \{(x,y) \in \mathbb{R}^2;\ (x^2+y^2)^{1/2} > r\}.$$

Since for any $r > 0$ there exist the numbers t_1 and t_2 such that $Q(t_1) \supset Q_r \supset Q(t_2)$, we conclude that Definition 3.3.1 is equivalent to the following definition.

Definition 3.3.2. A real-valued function u on Q is said to be *oscillatory* in Q if u has a zero in $Q(t)$ for any $t > 0$.

Lemma 3.3.1. *Assume that:*

(i) $c(x,y,\xi) \in C(\overline{Q} \times \mathbb{R}; \mathbb{R})$;
(ii) $c(x,y,\xi) \geq p(x + L^{-2}k^2y)\varphi(\xi)$ *for* $(x,y,\xi) \in Q \times (0,\infty)$, *where* $p(t) \in C((0,\infty);[0,\infty))$ *and* $\varphi(\xi)$ *is a convex function of class* $C((0,\infty);[0,\infty))$;
(iii) $\lambda_i(t) \in C((0,\infty);[0,\infty))$ $(i = 1,2)$, $\psi_i(t) \in C((0,\infty);\mathbb{R})$ $(i = 1,2)$ *and* $f(x,y) \in C(\overline{Q};\mathbb{R})$.

If $u \in C^2(\overline{Q};\mathbb{R})$ *is a solution of the problem* (3.3.1)–(3.3.3) *such that* $u > 0$ *in* $Q(t_0)$ *for some* $t_0 > 0$, *then we have*

$$\big(tU(t)\big)'' + \big(aL^2k^{-2} + b\big)\big(tU(t)\big)' + L^2k^{-2}tp(t)\varphi\big(U(t)\big) \leq G(t), \quad t > t_0, \tag{3.3.4}$$

where

$$G(t) = \psi_1(t) + L^2k^{-2}\psi_2(L^2k^{-2}t) + L^2k^{-2}tF(t). \tag{3.3.5}$$

Proof. A direct calculation shows that

$$\begin{aligned}\big(tU(t)\big)' &= u(t,0) + L^2k^{-2}\int_0^t \frac{\partial u}{\partial y}(\tau, L^2k^{-2}(t-\tau))\,d\tau \\ &= u(t,0) + L^2k^{-2}\int_0^t \frac{\partial u}{\partial y}(t-\tau, L^2k^{-2}\tau)\,d\tau. \qquad (3.3.6)\end{aligned}$$

Differentiating (3.3.6), we obtain

$$\begin{aligned}\big(tU(t)\big)'' = \frac{\partial u}{\partial x}(t,0) &+ L^2k^{-2}\frac{\partial u}{\partial y}(0, L^2k^{-2}t) \\ &+ L^2k^{-2}\int_0^t \frac{\partial^2 u}{\partial x\partial y}(t-\tau, L^2k^{-2}\tau)\,d\tau. \qquad (3.3.7)\end{aligned}$$

Since $tU(t)$ can be rewritten as

$$tU(t) = \int_0^t u(t-\tau, L^2k^{-2}\tau)\,d\tau,$$

we have

$$\big(tU(t)\big)' = u(0, L^2k^{-2}t) + \int_0^t \frac{\partial u}{\partial x}(t-\tau, L^2k^{-2}\tau)\,d\tau$$

and therefore

$$aL^2k^{-2}\big(tU(t)\big)' = aL^2k^{-2}u(0, L^2k^{-2}t) + L^2k^{-2}\int_0^t a\frac{\partial u}{\partial x}(t-\tau, L^2k^{-2}\tau)\,d\tau. \qquad (3.3.8)$$

It follows from (3.3.6) that

$$b\big(tU(t)\big)' = bu(t,0) + L^2k^{-2}\int_0^t b\,\frac{\partial u}{\partial y}(t-\tau, L^2k^{-2}\tau)\,d\tau. \qquad (3.3.9)$$

Combining (3.3.7)–(3.3.9) yields

$$\begin{aligned}&\big(tU(t)\big)'' + \big(aL^2k^{-2}+b\big)\big(tU(t)\big)' \\ &= \frac{\partial u}{\partial x}(t,0) + bu(t,0) + L^2k^{-2}\left(\frac{\partial u}{\partial y}(0, L^2k^{-2}t) + au(0, L^2k^{-2}t)\right) \\ &\quad + L^2k^{-2}\int_0^t \left(\frac{\partial^2 u}{\partial x\partial y} + a\,\frac{\partial u}{\partial x} + b\,\frac{\partial u}{\partial y}\right)(t-\tau, L^2k^{-2}\tau)\,d\tau. \qquad (3.3.10)\end{aligned}$$

Since $u > 0$ in $Q(t_0)$, the hypothesis (ii) implies

$$\int_0^t c\big(t-\tau, L^2k^{-2}\tau, u(t-\tau, L^2k^{-2}\tau)\big)d\tau \ge p(t)\int_0^t \varphi\big(u(t-\tau, L^2k^{-2}\tau)\big)d\tau \qquad (3.3.11)$$

for $t > t_0$. Applying Jensen's inequality, we obtain

$$\int_0^t \varphi(u(t-\tau, L^2k^{-2}\tau))d\tau \geq t\varphi\left(\frac{1}{t}\int_0^t u(t-\tau, L^2k^{-2}\tau)\,d\tau\right)$$
$$= t\varphi(U(t)), \quad t > t_0. \tag{3.3.12}$$

Combining (3.3.11) with (3.3.12), we see that

$$\int_0^t c(t-\tau, L^2k^{-2}\tau, u(t-\tau, L^2k^{-2}\tau))d\tau \geq tp(t)\varphi(U(t)), \quad t > t_0. \tag{3.3.13}$$

Since u is a positive solution of (3.3.1), we observe, using (3.3.13), that

$$\int_0^t \left(\frac{\partial^2 u}{\partial x \partial y} + a\,\frac{\partial u}{\partial x} + b\,\frac{\partial u}{\partial y}\right)(t-\tau, L^2k^{-2}\tau)\,d\tau$$
$$= \int_0^t \Big(-c(t-\tau, L^2k^{-2}\tau, u(t-\tau, L^2k^{-2}\tau)) + f(t-\tau, L^2k^{-2}\tau)\Big)d\tau$$
$$\leq -tp(t)\varphi(U(t)) + tF(t), \quad t > t_0. \tag{3.3.14}$$

We combine (3.3.10) with (3.3.14) to obtain the inequality (3.3.4). □

Theorem 3.3.1. *Assume that:*

(i) $c(x, y, \xi) \in C(\overline{Q} \times \mathbb{R}; \mathbb{R})$ *and* $c(x, y, -\xi) = -c(x, y, \xi)$ *for* $(x, y, \xi) \in Q \times (0, \infty)$;

(ii) *the hypothesis* (ii) *of Lemma* 3.3.1;

(iii) *the ordinary differential inequalities*

$$(ty)'' + (aL^2k^{-2} + b)(ty)' + L^2k^{-2}tp(t)\varphi(y) \leq \pm G(t) \tag{3.3.15}$$

have no eventually positive solutions.

Then every solution $u \in C^2(\overline{Q}; \mathbb{R})$ *of the problem* (3.3.1)–(3.3.3) *is oscillatory in* Q.

Proof. Suppose to the contrary that there exists a solution $u \in C^2(\overline{Q}; \mathbb{R})$ of the problem (3.3.1)–(3.3.3) which has no zero in $Q(t_0)$ for some $t_0 > 0$. First we assume that $u > 0$ in $Q(t_0)$. Then Lemma 3.3.1 implies that $U(t)$ is a positive solution of (3.3.15) with $+G(t)$ in (t_0, ∞). This contradicts the hypothesis (iii). If $u < 0$ in $Q(t_0)$, then $v := -u$ satisfies

$$\frac{\partial^2 v}{\partial x \partial y} + a\,\frac{\partial v}{\partial x} + b\,\frac{\partial v}{\partial y} + c(x, y, v) = -f(x, y), \quad (x, y) \in Q,$$
$$\frac{\partial v}{\partial x} + bv + \lambda_1(x)v = -\psi_1(x) \quad \text{on } \Gamma_x,$$
$$\frac{\partial v}{\partial y} + av + \lambda_2(y)v = -\psi_2(y) \quad \text{on } \Gamma_y.$$

By the same arguments as in the proof of Lemma 3.3.1, we find that

$$V(t) := \frac{1}{t}\int_0^t v(\tau, L^2k^{-2}(t-\tau))\,d\tau$$

is a positive solution of (3.3.15) with $-G(t)$ in (t_0, ∞). This is a contradiction. The proof is complete. □

An important special case of the problem (3.3.1)–(3.3.3) is the following:

$$\frac{\partial^2 u}{\partial x \partial y} + c(x, y, u) = f(x, y), \quad (x, y) \in Q, \tag{3.3.16}$$

$$\frac{\partial u}{\partial x} + \lambda_1(x)u = \psi_1(x) \quad \text{on } \Gamma_x, \tag{3.3.17}$$

$$\frac{\partial u}{\partial y} + \lambda_2(y)u = \psi_2(y) \quad \text{on } \Gamma_y. \tag{3.3.18}$$

Theorem 3.3.2. *Assume that the hypotheses* (i) *and* (ii) *of Theorem* 3.3.1 *hold. Every solution* $u \in C^2(\overline{Q}; \mathbb{R})$ *of the problem* (3.3.16)–(3.3.18) *is oscillatory in* Q *if*

$$\liminf_{t\to\infty} \int_T^t \left(1 - \frac{s}{t}\right)\left(\psi_1(s) + \psi_2(s) + \int_0^s f(\tau, s-\tau)\,d\tau\right) ds = -\infty,$$

$$\limsup_{t\to\infty} \int_T^t \left(1 - \frac{s}{t}\right)\left(\psi_1(s) + \psi_2(s) + \int_0^s f(\tau, s-\tau)\,d\tau\right) ds = \infty$$

for all large T.

Proof. Theorem 3.3.2 implies that

$$y'' \le \pm(\psi_1(t) + \psi_2(t) + tF(t))$$

have no eventually positive solutions. Hence, (3.3.15) with $a = b = 0$ and $L = k$ have no eventually positive solutions. The conclusion follows from Theorem 3.3.1. □

Theorem 3.3.3. *Assume that the hypotheses* (i) *and* (ii) *of Theorem* 3.3.1 *hold. Every solution* $u \in C^2(\overline{Q}; \mathbb{R})$ *of the problem* (3.3.16)–(3.3.18) *is oscillatory in* Q *if the ordinary differential inequality*

$$(ty)'' + tp(t)\varphi(y) \le 0$$

has no eventually positive solution and if there exists a function $\theta(t) \in C^2((0, \infty); \mathbb{R})$ *with the following properties:*

(i) $\theta(t)$ *is oscillatory at* $t = \infty$;

(ii) $(t\theta(t))'' = \psi_1(t) + \psi_2(t) + \int_0^t f(\tau, t-\tau)\,d\tau$;

(iii) $\lim_{t\to\infty} t\theta(t) = 0$.

Proof. The conclusion follows from Theorem 1.4.3 of Section 1.4 and Theorem 3.3.1 with $L = k$. □

We consider the more special case where $f(x, y) = 0$, $\psi_i(t) = 0$ $(i = 1, 2)$, that is,

$$\frac{\partial^2 u}{\partial x \partial y} + c_1(x, y)u^\gamma = 0, \quad (x, y) \in Q, \tag{3.3.19}$$

$$\frac{\partial u}{\partial x} + \lambda_1(x)u = 0 \quad \text{on } \Gamma_x, \tag{3.3.20}$$

$$\frac{\partial u}{\partial y} + \lambda_2(y)u = 0 \quad \text{on } \Gamma_y. \tag{3.3.21}$$

Theorem 3.3.4. *Let $\gamma \geq 1$ be the quotient of odd integers. Assume that:*

(i) $c_1(x, y) \in C(\overline{Q}; \mathbb{R})$ *and* $c_1(x, y) \geq p(x + y)$ *for* $(x, y) \in Q$, *where* $p(t) \in C((0, \infty); [0, \infty))$;

(ii) $\int_{t_0}^{\infty} t^{2-\gamma} p(t)\, dt = \infty \quad (\gamma > 1)$,

$$\int_{t_0}^{\infty} t^2 \left[\ell_m(t)^\mu \prod_{i=0}^{m} \ell_i(t) \right]^{-1} p(t)\, dt = \infty \quad (\gamma = 1)$$

for some numbers $t_0 > 0$, $\mu > 0$ *and some integer* m, *where* $\ell_0(t) = t$ *and* $\ell_i(t) = \log\big(\ell_{i-1}(t) + 1\big)$.

Then every solution $u \in C^2(\overline{Q}; \mathbb{R})$ *of the problem* (3.3.19)–(3.3.21) *is oscillatory in* Q.

Proof. In the case where $a = b = 0$, $L = k$, $f(x, y) = 0$, $\psi_i(t) = 0$ $(i = 1, 2)$, the inequality (3.3.15) reduces to

$$(ty)'' + tp(t)\varphi(y) \leq 0$$

which is equivalent to

$$(t^2 y')' + t^2 p(t)\varphi(y) \leq 0. \tag{3.3.22}$$

Using Proposition 1.2.2, we conclude that (3.3.22) has no eventually positive solution. Hence, the conclusion follows from Theorem 3.3.1. □

Next we investigate the existence of bounded domains in which every solution of Eq. (3.3.1) with the initial conditions

$$u(x, 0) = \alpha(x), \quad x > 0, \tag{3.3.23}$$

$$u(0, y) = \beta(y), \quad y > 0 \tag{3.3.24}$$

has a zero (or changes sign). The following notation is used:

$$Q(t_1, t_2) = \{(x, y) \in \mathbb{R}^2;\ x > 0,\ y > 0,\ t_1 < x + L^{-2}k^2 y < t_2\},$$
$$Q(t_1, t_2] = \{(x, y) \in \mathbb{R}^2;\ x > 0,\ y > 0,\ t_1 < x + L^{-2}k^2 y \le t_2\}.$$

Lemma 3.3.2. *Assume that:*

(i) $c(x, y, \xi) \in C(\overline{Q} \times \mathbb{R}; \mathbb{R})$,
$c(x, y, \xi) \ge K^2\xi$ *for* $(x, y, \xi) \in Q \times (0, \infty)$,
$c(x, y, \xi) \le K^2\xi$ *for* $(x, y, \xi) \in Q \times (-\infty, 0)$, *where* K *is a positive constant;*

(ii) $\alpha(x) \in C^1([0, \infty); \mathbb{R})$, $\beta(y) \in C^1([0, \infty); \mathbb{R})$ *and* $\alpha(0) = \beta(0)$;

(iii) $f(x, y) \in C(\overline{Q}; \mathbb{R})$.

If $u \in C^2(\overline{Q}; \mathbb{R})$ *is a positive solution of the problem* (3.3.1), (3.3.23), (3.3.24) *in* $Q(t_1, t_2)$, *then* $U(t)$ *satisfies*

$$\big(tU(t)\big)'' + \big(aL^2k^{-2} + b\big)\big(tU(t)\big)' + L^2k^{-2}K^2 tU(t) \le H(t), \quad t_1 < t < t_2, \tag{3.3.25}$$

where

$$H(t) = b\alpha(t) + aL^2k^{-2}\beta(L^2k^{-2}t) + \alpha'(t) + L^2k^{-2}\beta'(L^2k^{-2}t) + L^2k^{-2}tF(t). \tag{3.3.26}$$

Proof. Using the same arguments as were used in Lemma 3.3.1, we see that (3.3.10) holds. From the hypothesis (i) we easily obtain the desired inequality (3.3.25). □

We define

$$J(s) = \int_s^{s+(\pi/\tilde{d})} H(t)e^{d(t-s)} \sin \tilde{d}(t - s)\, dt,$$

where

$$d = \frac{1}{2}\big(aL^2k^{-2} + b\big),$$
$$\tilde{d} = \frac{1}{2}\big(4L^2k^{-2}K^2 - (aL^2k^{-2} + b)^2\big)^{1/2} > 0.$$

Lemma 3.3.3. *Assume that there is a number* $s \ge 0$ *such that* $J(s) \le 0$. *Let* $Y(t) \in C^2\big(s, s + (\pi/\tilde{d})\big) \cap C^1\big[s, s + (\pi/\tilde{d})\big]$ *be a solution of*

$$Y''(t) + 2dY'(t) + (d^2 + \tilde{d}^2)Y(t) \le H(t), \quad t \in \big(s, s + (\pi/\tilde{d})\big). \tag{3.3.27}$$

Then $Y(t)$ *cannot be positive in* $\big(s, s + (\pi/\tilde{d})\big]$.

The proof of Lemma 3.3.3 follows from the inequality due to Naito and Yoshida [213, Lemma 3].

Theorem 3.3.5. *Assume that the hypotheses* (i)–(iii) *of Lemma* 3.3.2 *are satisfied, and that* $4L^2k^{-2}K^2 > \left(aL^2k^{-2}+b\right)^2$. *If there is a number* $s \geq 0$ *such that* $J(s) = 0$, *then every solution* $u \in C^2(\overline{Q};\mathbb{R})$ *of the problem* (3.3.1), (3.3.23), (3.3.24) *has a zero in* $Q\big(s, s+(\pi/\tilde{d})\big]$.

Proof. Let $u \in C^2(\overline{Q};\mathbb{R})$ be a solution of the problem (3.3.1), (3.3.23), (3.3.24) which has no zero in $Q\big(s, s+(\pi/\tilde{d})\big]$. First we assume that $u > 0$ in $Q\big(s, s+(\pi/\tilde{d})\big]$. Lemma 3.3.2 implies that $tU(t)$ satisfies (3.3.27). From Lemma 3.3.3 we see that $tU(t)$ cannot be positive in $\big(s, s+(\pi/\tilde{d})\big]$. This contradicts the positivity of $tU(t)$ $\big(t \in (s, s+(\pi/\tilde{d})]\big)$. If $u < 0$ in $Q\big(s, s+(\pi/\tilde{d})\big]$, then $v := -u > 0$ in $Q(s, s+(\pi/\tilde{d})]$ and v satisfies

$$\frac{\partial^2 v}{\partial x \partial y} + a\frac{\partial v}{\partial x} + b\frac{\partial v}{\partial y} + K^2 v \leq -f(x,y) \quad \text{in } Q\big(s, s+(\pi/\tilde{d})\big].$$

Proceeding as in the case where $u > 0$, we are led to a contradiction. The proof is complete. □

Theorem 3.3.6. *Assume that the hypotheses* (i)–(iii) *of Lemma* 3.3.2 *are satisfied. Assume, moreover, that:*

$$K^2 = k^2 + \frac{1}{4}L^{-2}k^2\left(aL^2k^{-2}+b\right)^2,$$

$$\alpha(x) > 0 \quad \textit{for } x > 0,$$

$$\beta(y) > 0 \quad \textit{for } y > 0,$$

$$b\alpha(t) + aL^2k^{-2}\beta(L^2k^{-2}t) + \alpha'(t) + L^2k^{-2}\beta'(L^2k^{-2}t) \leq 0 \quad \textit{for } t > 0.$$

If there is a number $s \geq 0$ *for which*

$$\int_s^{s+(\pi/L)} \left(\int_0^t f(\tau, L^2k^{-2}(t-\tau))\,d\tau\right) e^{d(t-s)} \sin L(t-s)\,dt \leq 0,$$

then every solution $u \in C^2(\overline{Q};\mathbb{R})$ *of the problem* (3.3.1), (3.3.23), (3.3.24) *changes sign in* $Q\big(s, s+(\pi/L)\big)$.

Proof. Suppose that the problem (3.3.1), (3.3.23), (3.3.24) has a solution $u \in C^2(\overline{Q};\mathbb{R})$ which does not change sign in $Q\big(s, s+(\pi/L)\big)$. Since $\alpha(x) > 0$ $(x > 0)$ and $\beta(y) > 0$ $(y > 0)$, we see that $u \geq 0$ in $Q\big(s, s+(\pi/L)\big]$ and that $tU(t) > 0$ in $\big(s, s+(\pi/L)\big]$. From the inequality (3.3.25) and the hypothesis we observe that

$$\begin{aligned}&\big(tU(t)\big)'' + \big(aL^2k^{-2}+b\big)\big(tU(t)\big)' + L^2k^{-2}K^2tU(t)\\ &\quad \leq L^2k^{-2}\int_0^t f(\tau, L^2k^{-2}(t-\tau))\,d\tau, \quad t \in \big(s, s+(\pi/L)\big). \quad (3.3.28)\end{aligned}$$

Since $K^2 = k^2 + 4^{-1}L^{-2}k^2\left(aL^2k^{-2}+b\right)^2$, we find that $\tilde{d} = L > 0$. It follows from Lemma 3.3.3 that (3.3.28) has no positive solution in $\left(s, s+(\pi/L)\right]$. This contradicts the positivity of $tU(t)$ and completes the proof. □

Corollary 3.3.1. *Let $g(x,y) \in C(\overline{Q};(0,\infty))$ satisfy*

$$g(x,y) \geq k^2 + \frac{1}{4}(a+b)^2$$

and let

$$\begin{aligned}&\alpha(x) > 0 \quad \text{for } x > 0,\\ &\beta(y) > 0 \quad \text{for } y > 0,\\ &b\alpha(t) + a\beta(t) + \alpha'(t) + \beta'(t) \leq 0 \quad \text{for } t > 0.\end{aligned}$$

Then every solution $u \in C^2(\overline{Q};\mathbb{R})$ of

$$\frac{\partial^2 u}{\partial x \partial y} + a\,\frac{\partial u}{\partial x} + b\,\frac{\partial u}{\partial y} + g(x,y)u = 0, \quad (x,y) \in Q$$

satisfying (3.3.23) *and* (3.3.24) *changes sign in $Q\left(s, s+(\pi/k)\right)$ for any $s \geq 0$.*

Proof. The conclusion follows by applying Theorem 3.3.6 with $L = k$ and $f(x,y) = 0$. □

Example 3.3.1. We consider the problem

$$\frac{\partial^2 u}{\partial x \partial y} + u = -\frac{2}{x+y}\,e^{x+y} + \log(x+y)\sin(x+y), \quad (x,y) \in Q, \tag{3.3.29}$$

$$\frac{\partial u}{\partial x}(x,0) = e^x, \quad x > 0, \tag{3.3.30}$$

$$\frac{\partial u}{\partial y}(0,y) = e^y, \quad y > 0. \tag{3.3.31}$$

Here $c(x,y,\xi) = \xi$, $f(x,y) = -\left(2/(x+y)\right)e^{x+y} + \log(x+y)\sin(x+y)$, $\lambda_1(x) = 0$, $\psi_1(x) = e^x$, $\lambda_2(y) = 0$, $\psi_2(y) = e^y$. Since

$$\begin{aligned}&\int_T^t \left(1-\frac{s}{t}\right)\left(e^s + e^s + \int_0^s \left(-\frac{2}{s}\,e^s + (\log s)\sin s\right) d\xi\right) ds\\ &= \int_T^t \left(1-\frac{s}{t}\right)(s\,\log s)\sin s\,ds\\ &= -(\log t)\sin t + O(1) \quad (t \to \infty),\end{aligned}$$

we see that

$$\liminf_{t\to\infty} \int_T^t \left(1-\frac{s}{t}\right)\left(e^s+e^s+\int_0^s \left(-\frac{2}{s}\,e^s+(\log s)\sin s\right)d\xi\right)ds=-\infty,$$

$$\limsup_{t\to\infty} \int_T^t \left(1-\frac{s}{t}\right)\left(e^s+e^s+\int_0^s \left(-\frac{2}{s}\,e^s+(\log s)\sin s\right)d\xi\right)ds=\infty$$

for all large T. It follows from Theorem 3.3.2 that every solution $u \in C^2(\overline{Q};\mathbb{R})$ of the problem (3.3.29)–(3.3.31) is oscillatory in Q. However, we note that the solution u of the problem

$$\begin{aligned} &\frac{\partial^2 u}{\partial x \partial y}+u=0, \quad (x,y)\in Q,\\ &u(x,0)=e^x, \quad x>0,\\ &u(0,y)=e^y, \quad y>0 \end{aligned}$$

satisfies $u \geq 1$ (see Pagan [235, p.270]).

Example 3.3.2. We consider the problem

$$\frac{\partial^2 u}{\partial x \partial y}+2u=-2e^{-(x+y)}\sin(x+y), \quad (x,y)\in Q, \tag{3.3.32}$$

$$\frac{\partial u}{\partial x}(x,0)=e^{-x}\sin x, \quad x>0, \tag{3.3.33}$$

$$\frac{\partial u}{\partial y}(0,y)=e^{-y}\sin y, \quad y>0. \tag{3.3.34}$$

Here $c(x,y,\xi)=2\xi$, $f(x,y)=-2e^{-(x+y)}\sin(x+y)$, $\lambda_1(x)=0$, $\psi_1(x)=e^{-x}\sin x$, $\lambda_2(y)=0$, $\psi_2(y)=e^{-y}\sin y$. Since

$$\begin{aligned} &\int_T^t \left(1-\frac{s}{t}\right)\left(e^{-s}\sin s+e^{-s}\sin s+\int_0^s (-2e^{-s}\sin s)\,d\xi\right)ds\\ &=\int_T^t \left(1-\frac{s}{t}\right)(2e^{-s}\sin s-2se^{-s}\sin s)\,ds<\infty \quad (t\to\infty), \end{aligned}$$

Theorem 3.3.2 does not apply to the problem (3.3.32)–(3.3.34), but Theorem 3.3.3 does. It is easily checked that the ordinary differential inequality $(ty)''+2ty \leq 0$ has no eventually positive solution, and that $\theta(t):=t^{-1}e^{-t}\sin t-e^{-t}\cos t$ satisfies the following:

(i) $\theta(t)$ is oscillatory at $t=\infty$;
(ii) $\left(t\theta(t)\right)''=2e^{-t}\sin t-2te^{-t}\sin t$;
(iii) $\lim_{t\to\infty} t\theta(t)=0$.

It follows from Theorem 3.3.3 that every solution $u \in C^2(\overline{Q}; \mathbb{R})$ of the problem (3.3.32)–(3.3.34) is oscillatory in Q. One such solution is

$$u = -\frac{1}{\sqrt{2}} e^{-(x+y)} \sin\left(x + y + \tfrac{\pi}{4}\right).$$

Example 3.3.3. We consider the problem

$$\frac{\partial^2 u}{\partial x \partial y} + k \frac{\partial u}{\partial x} + k \frac{\partial u}{\partial y} + g(x,y)u = 0, \quad (x,y) \in Q,$$
$$u(x,0) = e^{-kx}, \quad x > 0,$$
$$u(0,y) = e^{-ky}, \quad y > 0,$$

where $g(x,y) \geq 2k^2$. Here $a = b = k$ and $\alpha(t) = \beta(t) = e^{-kt}$. Since $\alpha(t) = \beta(t) > 0$ $(t > 0)$ and $k\alpha(t) + k\beta(t) + \alpha'(t) + \beta'(t) = 0$ $(t > 0)$, Corollary 3.3.1 implies that every solution $u \in C^2(\overline{Q}; \mathbb{R})$ of the above problem changes sign in $Q(s, s + (\pi/k))$ for any $s \geq 0$.

Remark 3.3.1. As is well known, the characteristic initial value problem

$$\frac{\partial^2 u}{\partial x \partial y} + k^2 u = 0, \quad (x,y) \in Q,$$
$$u(x,0) = 1, \quad x > 0,$$
$$u(0,y) = 1, \quad y > 0$$

has an oscillatory solution $J_0(2k\sqrt{xy})$, where J_0 is the Bessel function of the first kind of order 0 (see, for example, Kreith [136]). We conclude that $J_0(2k\sqrt{xy})$ is also an oscillatory solution of the problem (3.3.19)–(3.3.21) with $\gamma = 1$, $c_1(x,y) = k^2$, $\lambda_i(t) = 0$ $(i = 1, 2)$.

Remark 3.3.2. In the case where $s = n\pi/k$ $(n = 1, 2, ...)$, Corollary 3.3.1 was established by Pagan [233, p.311].

3.4 Ultrahyperbolic Equations

Extensions of oscillation results of hyperbolic equations to ultrahyperbolic equations were investigated by several authors. We refer to Altin [18], Domshlak and Tamoev [67], Narita and Yoshida [217] for comparison theorems of linear equations, and to Narita [216] and Narita and Yoshida [218] for oscillation theorems of nonlinear equations.

We consider the nonlinear ultrahyperbolic equation

$$\Delta_x u - \sum_{k,l=1}^{m} \frac{\partial}{\partial y_k}\left(a_{kl}(x,y)\frac{\partial u}{\partial y_l}\right) + c(x,y,u) = 0 \tag{3.4.1}$$

for $(x, y) \in E \times G \subset \mathbb{R}^n \times \mathbb{R}^m$, where E is an exterior domain of $\mathbb{R}^n$ and G is a bounded domain of $\mathbb{R}^m$ with piecewise smooth boundary ∂G. Let

$$E \supset \{x \in \mathbb{R}^n;\ |x| \geq r_0\}$$

for some $r_0 > 0$.

Definition 3.4.1. A real-valued function u on $E \times G$ is said to be *oscillatory* in $E \times G$ if u has a zero in $E_r \times G$ for any $r > 0$, where

$$E_r = E \cap \{x \in \mathbb{R}^n;\ |x| > r\}.$$

We note that $E_r = \{x \in \mathbb{R}^n;\ |x| > r\}$ for $r \geq r_0$. Points in $\mathbb{R}^n$ and $\mathbb{R}^m$ are denoted by $x = (x_1, x_2, ..., x_n)$ and $y = (y_1, y_2, ..., y_m)$, respectively. The Laplacian in $\mathbb{R}^n$ is denoted by Δ_x. It is assumed that $a_{kl}(x, y) \in C^1(\overline{E \times G}; \mathbb{R})$ and the matrix $\left(a_{kl}(x, y)\right)_{k,l=1}^{m}$ is symmetric and positive definite in $E \times G$. We use $\frac{\partial}{\partial \zeta}$ to denote the conormal derivative

$$\frac{\partial}{\partial \zeta} = \sum_{k,l=1}^{m} a_{kl}(x, y)\nu_k \frac{\partial}{\partial y_l}$$

on $E \times \partial G$, where $\nu = (\nu_1, \nu_2, ..., \nu_m)$ is the unit exterior normal vector to ∂G.

Lemma 3.4.1. *Assume that:*

(i) $c(x, y, \xi) \in C(\overline{E \times G} \times \mathbb{R}; \mathbb{R})$;

(ii) $c(x, y, \xi) \geq c(x)\varphi(\xi)$ *for* $(x, y, \xi) \in E \times G \times (0, \infty)$, *where* $c(x)$ *is a function of class* $C(E; [0, \infty))$ *and* $\varphi(\xi) \in C((0, \infty); (0, \infty))$ *is a convex function;*

(iii) $u \in C^2(E \times \overline{G}; \mathbb{R})$ *is a positive solution of* (3.4.1) *in* $E_{r_1} \times G$ *and satisfies the boundary condition*

$$\frac{\partial u}{\partial \zeta} + \mu u = 0 \quad \text{on } E \times \partial G \tag{3.4.2}$$

for some $r_1 > r_0$, *where* $\mu \in C(E \times \partial G; [0, \infty])$ ($\mu = \infty$ *denotes the boundary condition* $u = 0$ *on* $E \times \partial G$).

Then the function $w(x)$ *defined by*

$$w(x) = \frac{1}{|G|} \int_G u(x, y)\, dy$$

satisfies the elliptic differential inequality

$$\Delta_x w + c(x)\varphi(w) \leq 0 \quad \text{in } E_{r_1}. \tag{3.4.3}$$

Proof. It follows from (3.4.1) that

$$\begin{aligned}\Delta_x w &= \frac{1}{|G|}\int_G \Delta_x u\,dy \\ &= \frac{1}{|G|}\int_G \left(\sum_{k,l=1}^{m} \frac{\partial}{\partial y_k}\left(a_{kl}(x,y)\frac{\partial u}{\partial y_l}\right) - c(x,y,u)\right) dy. \quad (3.4.4)\end{aligned}$$

Consider the case where $\mu = \infty$. Since $u > 0$ in $E_{r_1} \times G$ and $u = 0$ on $E_{r_1} \times \partial G$, it can be shown that $\frac{\partial u}{\partial \zeta} \le 0$ on $E_{r_1} \times \partial G$. In case $0 \le \mu < \infty$, it is easy to see that $\frac{\partial u}{\partial \zeta} = -\mu u \le 0$ on $E_{r_1} \times \partial G$. An application of the divergence theorem shows that

$$\int_G \sum_{k,l=1}^{m} \frac{\partial}{\partial y_k}\left(a_{kl}(x,y)\frac{\partial u}{\partial y_l}\right) dy = \int_{\partial G} \frac{\partial u}{\partial \zeta}\,dS \le 0 \quad \text{in } E_{r_1}. \quad (3.4.5)$$

Taking account of the hypothesis (ii) and using Jensen's inequality, we obtain

$$\frac{1}{|G|}\int_G c(x,y,u)\,dy \ge c(x)\frac{1}{|G|}\int_G \varphi(u)\,dy \ge c(x)\varphi(w) \quad \text{in } E_{r_1}. \quad (3.4.6)$$

Combining (3.4.4)–(3.4.6) yields

$$\Delta_x w \le -\frac{1}{|G|}\int_G c(x,y,u)\,dy \le -c(x)\varphi(w) \quad \text{in } E_{r_1}$$

which is the desired inequality (3.4.3). □

Theorem 3.4.1. *Assume that:*

(i) *the hypotheses* (i) *and* (ii) *of Lemma* 3.4.1 *are satisfied;*
(ii) $c(x,y,-\xi) = -c(x,y,\xi)$ *for* $(x,y,\xi) \in E \times G \times (0,\infty)$.

Every solution $u \in C^2(E \times \overline{G}; \mathbb{R})$ *of* (3.4.1) *satisfying the boundary condition* (3.4.2) *is oscillatory in* $E \times G$ *if the elliptic differential inequality* (3.4.3) *has no positive solution in* E_r *for any* $r > 0$.

Proof. Suppose to the contrary that there exists a solution $u \in C^2(E \times \overline{G}; \mathbb{R})$ of the boundary value problem (3.4.1), (3.4.2) which has no zero in E_{r_1} for some $r_1 > r_0$. We may assume that $u > 0$ in E_{r_1}. Then Lemma 3.4.1 implies that the function $w(x)$ is a positive solution of (3.4.3). This contradicts the hypothesis and completes the proof. □

Corollary 3.4.1. *Let γ (> 1) be the quotient of odd integers. Every solution $u \in C^2(E \times \overline{G}; \mathbb{R})$ of the boundary value problem*

$$\Delta_x u - \sum_{k,l=1}^{m} \frac{\partial}{\partial y_k}\left(a_{kl}(x,y)\frac{\partial u}{\partial y_l}\right) + p(|x|)u^\gamma = 0 \quad in\ E \times G,$$

$$\frac{\partial u}{\partial \zeta} + \mu u = 0 \quad on\ E \times \partial G \quad (0 \le \mu \le \infty)$$

is oscillatory in $E \times G$ if

$$\int_{r_1}^{\infty} (r \log r) p(r)\, dr = \infty \quad (n = 2),$$

$$\int_{r_1}^{\infty} r^{n-1-\gamma(n-2)} p(r)\, dr = \infty \quad (n \ge 3)$$

for some $r_1 \ge r_0$, where $p(r) \in C([r_0, \infty); [0, \infty))$.

Proof. Proceeding as in Corollary 1.2.1, we conclude that

$$\Delta_x w + p(|x|)w^\gamma \le 0$$

has no positive solution in E_r for any $r > 0$. Hence, the conclusion follows from Theorem 3.4.1. □

Remark 3.4.1. In the case where E is a conical domain or a cylindrical domain, the analogous results were obtained in [218]. Narita [216] derived oscillation results for nonlinear ultrahyperbolic equations when E has the form

$$\{x \in \mathbb{R}^n;\ 0 < x_i < \infty\ (i = 1, 2, ..., n)\}.$$

Remark 3.4.2. Comparison theorems were employed to obtain oscillation results for linear (singular) ultrahyperbolic equations (see [18, 217]).

3.5 Bianchi Equations

In 1895 Bianchi [31] studied the partial differential equation

$$\frac{\partial^n u}{\partial x_1 \partial x_2 \cdots \partial x_n} = Mu$$

with the aid of the methods of Riemann. The existence of solutions of the so-called Bianchi equation

$$\frac{\partial^n u}{\partial x_1 \partial x_2 \cdots \partial x_n} = f$$

was investigated in [29, 53, 89].

We are concerned with the oscillatory behavior of solutions to characteristic initial value problems for the nonlinear Bianchi equation

$$\frac{\partial^3 u}{\partial x_1 \partial x_2 \partial x_3} + c(x_1, x_2, x_3, u) = 0, \quad (x_1, x_2, x_3) \in H_\rho, \tag{3.5.1}$$

where

$$H_\rho = \left\{(x_1, x_2, x_3) \in \mathbb{R}^3;\ x_i > 0\ (i = 1, 2, 3),\ \sum_{i=1}^{3} x_i > \rho\right\}\ (\rho \geq 0).$$

Points in $\mathbb{R}^3$ will be denoted by $x = (x_1, x_2, x_3)$, and $|x|$ will be used for the Euclidean length of $x \in \mathbb{R}^3$.

Definition 3.5.1. A real-valued function u on H_ρ is said to be *oscillatory* in H_ρ if u has a zero in $H_\rho(r)$ for any $r > 0$, where

$$H_\rho(r) = H_\rho \cap \{x \in \mathbb{R}^3;\ |x| > r\}.$$

We note that $H_\rho(r) = H_\rho$ for sufficiently large $\rho > r$.

We use the notation:

$$\begin{aligned}
&H_t = \left\{x \in H_\rho;\ \sum_{i=1}^{3} x_i > t\right\}\ (t \geq \rho),\\
&\Gamma_i = \{x \in \mathbb{R}^3;\ x_i > 0, x_j = 0 \text{ for } j \neq i\} \quad (i = 1, 2, 3),\\
&\Gamma_{ij} = \{x \in \mathbb{R}^3;\ x_i > 0,\ x_j > 0,\ x_k = 0 \text{ for } k \neq i, j\} \quad (1 \leq i < j \leq 3).
\end{aligned}$$

We define the function $U(t)$ by

$$U(t) = \frac{1}{(1/2)t^2} \int_0^t dx_3 \int_0^{t-x_3} u(t - x_2 - x_3, x_2, x_3)\, dx_2, \quad t > \rho.$$

Lemma 3.5.1. *If $u \in C^3(H_\rho; \mathbb{R}) \cap C^2(\overline{H_\rho}; \mathbb{R})$, then u satisfies the identity*

$$\begin{aligned}
&\left((1/2)t^2 U(t)\right)'''\\
&= \frac{\partial u}{\partial x_1}(t, 0, 0) + \frac{\partial u}{\partial x_2}(0, t, 0) + \frac{\partial u}{\partial x_3}(0, 0, t)\\
&\quad + \int_0^t \frac{\partial^2 u}{\partial x_1 \partial x_2}(t - x_2, x_2, 0)\, dx_2 + \int_0^t \frac{\partial^2 u}{\partial x_2 \partial x_3}(0, t - x_3, x_3)\, dx_3\\
&\quad + \int_0^t \frac{\partial^2 u}{\partial x_1 \partial x_3}(x_1, 0, t - x_1)\, dx_1\\
&\quad + \int_0^t dx_3 \int_0^{t-x_3} \frac{\partial^3 u}{\partial x_1 \partial x_2 \partial x_3}(t - x_2 - x_3, x_2, x_3)\, dx_2, \quad t > \rho. \qquad (3.5.2)
\end{aligned}$$

Proof. We define the function $\hat{U}(t)$ by

$$\hat{U}(t) = (1/2)t^2 U(t) = \int_0^t dx_3 \int_0^{t-x_3} u(t - x_2 - x_3, x_2, x_3)\, dx_2. \quad (3.5.3)$$

By making a change of variables, we see that

$$\int_0^t v(t - x_i, x_i)\, dx_i = \int_0^t v(x_j, t - x_j)\, dx_j \quad (i, j = 1, 2, 3) \quad (3.5.4)$$

and

$$\begin{aligned}
&\int_0^t dx_3 \int_0^{t-x_3} w(t - x_2 - x_3, x_2, x_3)\, dx_2 \\
&= \int_0^t dx_1 \int_0^{t-x_1} w(x_1, t - x_1 - x_3, x_3)\, dx_3 \\
&= \int_0^t dx_2 \int_0^{t-x_2} w(x_1, x_2, t - x_1 - x_2)\, dx_1 \quad (3.5.5)
\end{aligned}$$

for any continuous functions $v(x_i, x_j)$ and $w(x_1, x_2, x_3)$. Differentiating (3.5.3), we observe, using Leibniz' rule, (3.5.4) and (3.5.5), that

$$\begin{aligned}
\hat{U}'(t) &= \int_0^t dx_3 \int_0^{t-x_3} \frac{\partial u}{\partial x_1}(t - x_2 - x_3, x_2, x_3)\, dx_2 + \int_0^t u(0, t - x_3, x_3)\, dx_3 \\
&= \int_0^t dx_1 \int_0^{t-x_1} \frac{\partial u}{\partial x_1}(x_1, t - x_1 - x_3, x_3)\, dx_3 + \int_0^t u(0, t - x_3, x_3)\, dx_3.
\end{aligned}$$

Analogously we obtain

$$\begin{aligned}
\hat{U}''(t) &= \int_0^t dx_1 \int_0^{t-x_1} \frac{\partial^2 u}{\partial x_1 \partial x_2}(x_1, t - x_1 - x_3, x_3)\, dx_3 \\
&\quad + \int_0^t \frac{\partial u}{\partial x_1}(x_1, 0, t - x_1)\, dx_1 + \int_0^t \frac{\partial u}{\partial x_2}(0, t - x_3, x_3)\, dx_3 + u(0, 0, t) \\
&= \int_0^t dx_2 \int_0^{t-x_2} \frac{\partial^2 u}{\partial x_1 \partial x_2}(x_1, x_2, t - x_1 - x_2)\, dx_1 \\
&\quad + \int_0^t \frac{\partial u}{\partial x_1}(x_1, 0, t - x_1)\, dx_1 + \int_0^t \frac{\partial u}{\partial x_2}(0, x_2, t - x_2)\, dx_2 + u(0, 0, t).
\end{aligned}$$

Now a third differentiation yields

$$
\begin{aligned}
\hat{U}'''(t) = &\int_0^t dx_2 \int_0^{t-x_2} \frac{\partial^3 u}{\partial x_1 \partial x_2 \partial x_3}(x_1, x_2, t - x_1 - x_2)\, dx_1 \\
&+ \int_0^t \frac{\partial^2 u}{\partial x_1 \partial x_2}(t - x_2, x_2, 0)\, dx_2 \\
&+ \int_0^t \frac{\partial^2 u}{\partial x_1 \partial x_3}(x_1, 0, t - x_1)\, dx_1 + \frac{\partial u}{\partial x_1}(t, 0, 0) \\
&+ \int_0^t \frac{\partial^2 u}{\partial x_2 \partial x_3}(0, x_2, t - x_2)\, dx_2 + \frac{\partial u}{\partial x_2}(0, t, 0) \\
&+ \frac{\partial u}{\partial x_3}(0, 0, t).
\end{aligned} \tag{3.5.6}
$$

Using (3.5.5), we conclude that (3.5.6) is equivalent to (3.5.2). □

Theorem 3.5.1. *Assume that:*

(i) $c(x_1, x_2, x_3, \xi) \in C(\overline{H_\rho} \times \mathbb{R}; \mathbb{R})$ *and* $c(x_1, x_2, x_3, -\xi) = -c(x_1, x_2, x_3, \xi)$ *for* $(x_1, x_2, x_3, \xi) \in H_\rho \times (0, \infty)$;

(ii) $c(x_1, x_2, x_3, \xi) \geq p\Big(\sum_{i=1}^{3} x_i\Big)\varphi(\xi)$ *for* $(x_1, x_2, x_3, \xi) \in H_\rho \times (0, \infty)$, *where* $p(t) \in C((\rho, \infty); [0, \infty))$ *and* $\varphi(\xi) \in C((0, \infty); [0, \infty))$ *is a convex function;*

(iii) *every eventually positive solution* $w(t)$ *of*

$$
\big(t^2 w(t)\big)''' + t^2 p(t) \varphi\big(w(t)\big) \leq 0 \tag{3.5.7}
$$

satisfies $\lim_{t\to\infty} t^\alpha w(t) = 0$ *for some* $\alpha \in \mathbb{R}$.

Then every solution $u \in C^3(H_\rho; \mathbb{R}) \cap C^2(\overline{H_\rho}; \mathbb{R})$ *of* (3.5.1) *satisfying*

$$
\frac{\partial u}{\partial x_i} + \lambda_i(x_i) u = 0 \quad \text{on } \Gamma_i \cap \{x_i > \rho\}\ (i = 1, 2, 3), \tag{3.5.8}
$$

$$
\frac{\partial^2 u}{\partial x_i \partial x_j} + \mu_{ij}(x_i, x_j) u = 0 \quad \text{on } \Gamma_{ij} \cap \{x_i + x_j > \rho\}\ (1 \leq i < j \leq 3) \tag{3.5.9}
$$

is oscillatory in H_ρ, *or satisfies*

$$
\lim_{t\to\infty} t^\alpha U(t) = 0, \tag{3.5.10}
$$

where λ_i *and* μ_{ij} *are nonnegative continuous functions.*

Proof. Assume on the contrary, that there exists a nonoscillatory solution $u \in C^3(H_\rho; \mathbb{R}) \cap C^2(\overline{H_\rho}; \mathbb{R})$ of the problem (3.5.1), (3.5.8), (3.5.9) which does not satisfy (3.5.10). Let $u > 0$ in $H_\rho(t_0)$ for some some $t_0 > \rho$. Then we see that $H_\rho(t_0) = H_{t_0}$ for large $t_0 > \rho$. Hence we have

$$\frac{\partial u}{\partial x_i} \le 0 \quad \text{on } \Gamma_i \cap \{x_i > t_0\} \quad (i = 1, 2, 3),$$
$$\frac{\partial^2 u}{\partial x_i \partial x_j} \le 0 \quad \text{on } \Gamma_{ij} \cap \{x_i + x_j > t_0\} \quad (1 \le i < j \le 3).$$

Therefore, from (3.5.2) we find that

$$\hat{U}'''(t) \le \int_0^t dx_3 \int_0^{t-x_3} \frac{\partial^3 u}{\partial x_1 \partial x_2 \partial x_3}(t - x_2 - x_3, x_2, x_3)\, dx_2, \quad t > t_0. \tag{3.5.11}$$

Since u is a solution of (3.5.1), the hypothesis (ii) implies that

$$\begin{aligned}
&\int_0^t dx_3 \int_0^{t-x_3} \frac{\partial^3 u}{\partial x_1 \partial x_2 \partial x_3}(t - x_2 - x_3, x_2, x_3)\, dx_2 \\
&= -\int_0^t dx_3 \int_0^{t-x_3} c(t - x_2 - x_3, x_2, x_3, u(t - x_2 - x_3, x_2, x_3))\, dx_2 \\
&\le -p(t) \int_0^t dx_3 \int_0^{t-x_3} \varphi(u(t - x_2 - x_3, x_2, x_3))\, dx_2, \quad t > t_0.
\end{aligned} \tag{3.5.12}$$

An application of Jensen's inequality shows that

$$\int_0^t dx_3 \int_0^{t-x_3} \varphi(u(t - x_2 - x_3, x_2, x_3))\, dx_2 \ge (1/2)t^2 \varphi(U(t)), \quad t > t_0. \tag{3.5.13}$$

Combining (3.5.12) with (3.5.13) yields

$$\int_0^t dx_3 \int_0^{t-x_3} \frac{\partial^3 u}{\partial x_1 \partial x_2 \partial x_3}(t - x_2 - x_3, x_2, x_3)\, dx_2 \le -p(t)(1/2)t^2 \varphi(U(t)) \tag{3.5.14}$$

for $t > t_0$. From (3.5.11) and (3.5.14) we see that

$$\left(t^2 U(t)\right)''' + t^2 p(t) \varphi(U(t)) \le 0, \quad t > t_0$$

and therefore $U(t)$ is a positive solution of (3.5.7) in (t_0, ∞) which does not satisfy (3.5.10). This contradicts the hypothesis (iii) and completes the proof. □

Corollary 3.5.1. *Let $\gamma \ge 1$ be the quotient of odd integers and u be a solution of the equation*

$$\frac{\partial^3 u}{\partial x_1 \partial x_2 \partial x_3} + c(x_1, x_2, x_3) u^\gamma = 0 \quad \text{in } H_\rho \tag{3.5.15}$$

which satisfies (3.5.8), (3.5.9) with $\alpha = 2$. Assume that:

(i) $c(x_1, x_2, x_3) \geq p\Big(\sum_{i=1}^{3} x_i\Big)$ *in* H_ρ, *where* $c(x_1, x_2, x_3) \in C(\overline{H_\rho}; \mathbb{R})$ *and* $p(t) \in C((\rho, \infty); (0, \infty))$;

(ii) $\int_{t_0}^{\infty} \psi_\gamma(t) p(t)\, dt = \infty$ *for some* $t_0 > \rho$, *where*

$$\psi_\gamma(t) = \begin{cases} t^{4-2\gamma} & (\gamma > 1) \\ t^{2-\varepsilon} & (\gamma = 1) \end{cases}$$

for some $\varepsilon > 0$.

Then u *is oscillatory in* H_ρ, *or satisfies* $\lim_{t\to\infty} t^2 U(t) = 0$.

Proof. In this case, (3.5.7) can be written in the form

$$\big(t^2 w(t)\big)''' + t^2 p(t)\big(w(t)\big)^\gamma \leq 0. \tag{3.5.16}$$

By the substitution $h(t) = t^2 w(t)$, (3.5.16) reduces to

$$h'''(t) + t^{2-2\gamma} p(t)\big(h(t)\big)^\gamma \leq 0. \tag{3.5.17}$$

The hypothesis (ii) implies

$$\int_{r_0}^{\infty} t^{3-1}\big(t^{2-2\gamma} p(t)\big)\, dt = \infty \quad (\gamma > 1),$$

$$\int_{r_0}^{\infty} t^{3-1-\varepsilon} p(t)\, dt = \infty \quad \text{for some } \varepsilon > 0 \quad (\gamma = 1).$$

Then every eventually positive solution $h(t)$ of the equation

$$h'''(t) + t^{2-2\gamma} p(t)\big(h(t)\big)^\gamma = 0$$

satisfies $\lim_{t\to\infty} h(t) = 0$. (For the case $\gamma > 1$ see [124, 179], and also [198] for the case $\gamma = 1$.) It follows from the results of [123, 229] that every eventually positive solution $h(t)$ of (3.5.17) satisfies $\lim_{t\to\infty} h(t) = 0$, that is, every eventually positive solution $w(t)$ of (3.5.16) satisfies $\lim_{t\to\infty} t^2 w(t) = 0$. Hence, the conclusion follows from Theorem 3.5.1 with $\alpha = 2$. □

The following example shows that the conclusion of Corollary 3.5.1 is in general false if initial conditions (3.5.8), (3.5.9) are not required.

Example 3.5.1. We consider the equation

$$\frac{\partial^3 u}{\partial x_1 \partial x_2 \partial x_3} + 2\Big(3 + \sum_{i=1}^{3} x_i\Big)\Big(\sum_{i=1}^{3} x_i\Big)^{-11/5} e^{(4/5)x_3} u^{9/5} = 0 \quad \text{in } H_{\rho_0}, \tag{3.5.18}$$

where ρ_0 is a positive number. Since $e^{(4/5)x_3} \geq 1$ in H_{ρ_0} and

$$\int_{\rho_0}^{\infty} t^{4-2(9/5)} 2(3+t) t^{-11/5}\, dt = \int_{\rho_0}^{\infty} 2(3t^{-9/5} + t^{-4/5})\, dt = \infty,$$

the hypotheses (i), (ii) of Corollary 3.5.1 are satisfied. Equation (3.5.18) has a nonoscillatory solution $u = \left(\sum_{i=1}^{3} x_i\right)^{-1} e^{-x_3}$ which does not satisfy the condition $\lim_{t\to\infty} t^2 U(t) = 0$. In fact, we have

$$\lim_{t\to\infty} t^2 U(t) = \lim_{t\to\infty} 2(1 - t^{-1} + t^{-1}e^{-t}) = 2.$$

We obtain

$$\frac{\partial^2 u}{\partial x_1 \partial x_2} + \mu_{12}(x_1, x_2)u = \left(\frac{2}{(x_1 + x_2)^2} + \mu_{12}(x_1, x_2)\right)\frac{1}{x_1 + x_2} > 0$$

on $\Gamma_{12} \cap \{x_1 + x_2 > \rho_0\}$ for any $\mu_{12}(x_1, x_2) \geq 0$, and therefore the solution u does not satisfy the initial condition (3.5.9).

We show an example which illustrates that there exists an oscillatory solution which does not satisfy the initial conditions (3.5.8), (3.5.9).

Example 3.5.2. We consider the equation

$$\frac{\partial^3 u}{\partial x_1 \partial x_2 \partial x_3} + 8u = 0 \quad \text{in } H_0. \tag{3.5.19}$$

Since

$$\int_0^\infty 8t^{2-\varepsilon} dt = \infty \quad (0 < \varepsilon \leq 3),$$

the hypotheses (i), (ii) of Corollary 3.5.1 are satisfied. Equation (3.5.19) has an oscillatory solution

$$u = \exp\left(\sum_{i=1}^{3} x_i\right) \sin\left(\sqrt{3} \sum_{i=1}^{3} x_i\right)$$

which does not satisfy the condition $\lim_{t\to\infty} t^2 U(t) = 0$. In fact, since

$$\lim_{t\to\infty} t^2 U(t) = \lim_{t\to\infty} t^2 e^t \sin(\sqrt{3}\, t),$$

there does not exist $\lim_{t\to\infty} t^2 U(t)$. It is easily verified that the solution u does not satisfy the initial conditions (3.5.8), (3.5.9).

Remark 3.5.1. Oscillations of the nth order Bianchi equation

$$\frac{\partial^n u}{\partial x_1 \partial x_2 \cdots \partial x_n} + c(x_1, x_2, \cdots, x_n, u) = 0$$

were studied by Travis and Yoshida [266].

3.6 Higher Order Hyperbolic Equations

Characteristic initial value problems for the hyperbolic equation

$$\frac{\partial^2 u}{\partial x \partial y} + c(x, y, u) = 0, \quad (x, y) \in Q$$

were discussed in Section 3.3, where

$$Q = \{(x, y) \in \mathbb{R}^2;\ x > 0,\ y > 0\}.$$

In this section we investigate the oscillatory behavior of solutions of the higher order hyperbolic equation

$$\frac{\partial^{k+l} u}{\partial x^k \partial y^l} + c(x, y, u) = 0, \quad (x, y) \in Q, \tag{3.6.1}$$

where k and l are positive integers.

We define the function space $\mathcal{D}(Q)$ by

$$\mathcal{D}(Q) = C^{k+l}(Q; \mathbb{R}) \cap C^{k+l-1}(\overline{Q}; \mathbb{R})$$

and we use the notation:

$$U(t) = \frac{1}{t} \int_0^t u(t - \tau, \tau)\, d\tau,$$
$$\Gamma_x = \{(x, 0) \in \mathbb{R}^2;\ x > 0\},$$
$$\Gamma_y = \{(0, y) \in \mathbb{R}^2;\ y > 0\}.$$

Lemma 3.6.1. *If $u \in \mathcal{D}(Q)$ and $v \in C^k(Q; \mathbb{R}) \cap C^{k-1}(\overline{Q}; \mathbb{R})$, then we obtain the following two identities:*

$$\int_0^t \frac{\partial^k v}{\partial x^k}(t - \tau, \tau)\, d\tau = \frac{d^k}{dt^k} \int_0^t v(t - \tau, \tau)\, d\tau - \sum_{i=0}^{k-1} \frac{\partial^{k-1} v}{\partial x^i \partial y^{k-1-i}}(0, t), \tag{3.6.2}$$

$$\int_0^t \frac{\partial^l u}{\partial y^l}(t - \tau, \tau)\, d\tau = \frac{d^l}{dt^l} \int_0^t u(t - \tau, \tau)\, d\tau - \sum_{j=0}^{l-1} \frac{\partial^{l-1} u}{\partial x^{l-1-j} \partial y^j}(t, 0). \tag{3.6.3}$$

Proof. A direct calculation yields

$$\int_0^t \frac{\partial v}{\partial x}(t - \tau, \tau)\, d\tau = \frac{d}{dt} \int_0^t v(t - \tau, \tau)\, d\tau - v(0, t). \tag{3.6.4}$$

Differentiating both sides of (3.6.4) with respect to t, we have

$$\int_0^t \frac{\partial^2 v}{\partial x^2}(t - \tau, \tau)\, d\tau = \frac{d^2}{dt^2} \int_0^t v(t - \tau, \tau)\, d\tau - \frac{d}{dt} v(0, t) - \frac{\partial v}{\partial x}(0, t). \tag{3.6.5}$$

Repetition of the argument that led from (3.6.4) to (3.6.5) yields finally the identity

$$\int_0^t \frac{\partial^k v}{\partial x^k}(t-\tau,\tau)\,d\tau = \frac{d^k}{dt^k}\int_0^t v(t-\tau,\tau)\,d\tau - \sum_{i=0}^{k-1} \frac{d^{k-1-i}}{dt^{k-1-i}}\left(\frac{\partial^i v}{\partial x^i}(0,t)\right)$$

which is the desired identity (3.6.2).

It is known that the following identity holds:

$$\int_0^t \frac{\partial u}{\partial y}(t-\tau,\tau)\,d\tau = \frac{d}{dt}\int_0^t u(t-\tau,\tau)\,d\tau - u(t,0) \tag{3.6.6}$$

(see, for example, [295, p.96]). Replacing u by $\frac{\partial u}{\partial y}$, we obtain

$$\int_0^t \frac{\partial^2 u}{\partial y^2}(t-\tau,\tau)\,d\tau = \frac{d}{dt}\int_0^t \frac{\partial u}{\partial y}(t-\tau,\tau)\,d\tau - \frac{\partial u}{\partial y}(t,0). \tag{3.6.7}$$

Combining (3.6.6) with (3.6.7) yields

$$\int_0^t \frac{\partial^2 u}{\partial y^2}(t-\tau,\tau)\,d\tau = \frac{d^2}{dt^2}\int_0^t u(t-\tau,\tau)\,d\tau - \frac{d}{dt}u(t,0) - \frac{\partial u}{\partial y}(t,0).$$

Proceeding in this fashion, we arrive at (3.6.3). □

Lemma 3.6.2. *The following identity holds for any* $u \in \mathcal{D}(Q)$:

$$\int_0^t \frac{\partial^{k+l} u}{\partial x^k \partial y^l}(t-\tau,\tau)\,d\tau = \frac{d^{k+l}}{dt^{k+l}}\int_0^t u(t-\tau,\tau)\,d\tau - \sum_{i=0}^{k-1} \frac{\partial^{k+l-1} u}{\partial x^i \partial y^{k+l-1-i}}(0,t)$$
$$- \sum_{j=0}^{l-1} \frac{\partial^{k+l-1} u}{\partial x^{k+l-1-j}\partial y^j}(t,0), \quad t > 0. \tag{3.6.8}$$

Proof. Letting $v = \frac{\partial^l u}{\partial y^l}$ in (3.6.2), we obtain

$$\int_0^t \frac{\partial^{k+l} u}{\partial x^k \partial y^l}(t-\tau,\tau)\,d\tau$$
$$= \frac{d^k}{dt^k}\int_0^t \frac{\partial^l u}{\partial y^l}(t-\tau,\tau)\,d\tau - \sum_{i=0}^{k-1} \frac{\partial^{k-1}}{\partial x^i \partial y^{k-1-i}}\left(\frac{\partial^l u}{\partial y^l}\right)(0,t). \tag{3.6.9}$$

Differentiating (3.6.3) k times yields

$$\frac{d^k}{dt^k}\int_0^t \frac{\partial^l u}{\partial y^l}(t-\tau,\tau)\,d\tau$$
$$= \frac{d^{k+l}}{dt^{k+l}}\int_0^t u(t-\tau,\tau)\,d\tau - \sum_{j=0}^{l-1} \frac{\partial^{k+l-1} u}{\partial x^{k+l-1-j}\partial y^j}(t,0). \tag{3.6.10}$$

Combining (3.6.9) with (3.6.10) yields the desired identity (3.6.8). □

Theorem 3.6.1. *Let $k+l$ be an even integer, and assume that:*

(i) $c(x,y,\xi) \in C(\overline{Q} \times \mathbb{R}; \mathbb{R})$ *and* $c(x,y,-\xi) = -c(x,y,\xi)$ *for* $(x,y,\xi) \in Q \times (0,\infty)$;

(ii) $c(x,y,\xi) \geq p(x+y)\varphi(\xi)$ *for* $(x,y,\xi) \in Q \times (0,\infty)$, *where* $p(t)$ *is a function of class* $C((0,\infty);[0,\infty))$ *and* $\varphi(\xi)$ *is a convex function of class* $C((0,\infty);[0,\infty))$;

(iii) *the ordinary differential inequality*

$$\frac{d^{k+l}}{dt^{k+l}}(tz(t)) + tp(t)\varphi(z(t)) \leq 0 \tag{3.6.11}$$

has no eventually positive solution.

Then every solution $u \in \mathcal{D}(Q)$ *of* (3.6.1) *satisfying the conditions*

$$\sum_{j=0}^{l-1} \frac{\partial^{k+l-1}u}{\partial x^{k+l-1-j}\partial y^j} + \sigma_1(x)u = 0 \quad on\ \Gamma_x, \tag{3.6.12}$$

$$\sum_{i=0}^{k-1} \frac{\partial^{k+l-1}u}{\partial x^i \partial y^{k+l-1-i}} + \sigma_2(y)u = 0 \quad on\ \Gamma_y \tag{3.6.13}$$

is oscillatory in Q, *where* $\sigma_i(s) \in C([0,\infty);[0,\infty))$ *and* $\sigma_1(0) = \sigma_2(0)$.

Proof. Suppose to the contrary that there exists a solution $u \in \mathcal{D}(Q)$ of (3.6.1) satisfying (3.6.12) and (3.6.13) which is nonoscillatory in Q. Then there is a positive number t_0 such that u has no zero in $Q(t_0)$, where

$$Q(t_0) = \{(x,y) \in \mathbb{R}^2;\ x > 0,\ y > 0,\ x+y > t_0\}.$$

Without loss of generality we may suppose that $u > 0$ in $Q(t_0)$. Then we see that

$$\sum_{j=0}^{l-1} \frac{\partial^{k+l-1}u}{\partial x^{k+l-1-j}\partial y^j} = -\sigma_1(x)u \leq 0 \quad \text{on } \Gamma_x \cap \{x > t_0\}, \tag{3.6.14}$$

$$\sum_{i=0}^{k-1} \frac{\partial^{k+l-1}u}{\partial x^i \partial y^{k+l-1-i}} = -\sigma_2(y)u \leq 0 \quad \text{on } \Gamma_y \cap \{y > t_0\}. \tag{3.6.15}$$

Using Lemma 3.6.2 and taking account of (3.6.14), (3.6.15), we observe that

$$\frac{d^{k+l}}{dt^{k+l}} \int_0^t u(t-\tau,\tau)\,d\tau \leq \int_0^t \frac{\partial^{k+l}u}{\partial x^k \partial y^l}(t-\tau,\tau)\,d\tau, \quad t > t_0. \tag{3.6.16}$$

An application of Jensen's inequality yields

$$\int_0^t \varphi(u(t-\tau,\tau))\,d\tau \geq t\varphi(U(t)), \quad t > t_0. \tag{3.6.17}$$

Since u is a solution of (3.6.1), we see from the hypothesis (ii) and (3.6.17) that

$$\int_0^t \frac{\partial^{k+l} u}{\partial x^k \partial y^l}(t-\tau,\tau)\,d\tau \le -p(t)\int_0^t \varphi(u(t-\tau,\tau))\,d\tau \le -tp(t)\varphi(U(t)), \quad t > t_0. \tag{3.6.18}$$

Combining (3.6.16) with (3.6.18), we conclude that $U(t)$ is a positive solution of (3.6.11) in (t_0, ∞). This contradicts the hypothesis (iii) and completes the proof. □

Corollary 3.6.1. *Let $k+l$ be an even integer and $\gamma \ge 1$ be the quotient of odd integers. Assume that $u \in \mathcal{D}(Q)$ is a solution of the equation*

$$\frac{\partial^{k+l} u}{\partial x^k \partial y^l} + c(x,y)u^\gamma = 0, \quad (x,y) \in Q \tag{3.6.19}$$

which satisfies the conditions (3.6.12), (3.6.13). *Then u is oscillatory in Q if the following conditions hold:*

(i) *$c(x,y) \ge p(x+y)$ for $(x,y) \in Q$, where $p(t) \in C([0,\infty);(0,\infty))$;*

(ii) *$\int_{t_0}^\infty \psi_\gamma(t)p(t)\,dt = \infty$, where*

$$\psi_\gamma(t) = \begin{cases} t^{k+l-\gamma} & (\gamma > 1) \\ t^{k+l-1-\varepsilon} & (\gamma = 1) \end{cases}$$

for some $t_0 > 0$ and some $\varepsilon > 0$.

Proof. Since $\varphi(\xi) = \xi^\gamma$, the differential inequality (3.6.11) reduces to

$$\frac{d^{k+l} w(t)}{dt^{k+l}} + t^{1-\gamma}p(t)w(t)^\gamma \le 0, \tag{3.6.20}$$

where $w(t) = tz(t)$. Combining the results of [124, 179, 198] and the results of [123, 229], we observe that the hypothesis (ii) implies that (3.6.20) has no eventually positive solution. The conclusion follows from Theorem 3.6.1. □

Theorem 3.6.2. *Let $k+l$ be an odd integer, and assume that the hypotheses* (i) *and* (ii) *of Theorem* 3.6.1 *hold. If every eventually positive solution $z(t)$ of* (3.6.11) *satisfies $\lim_{t\to\infty} tz(t) = 0$, then every solution $u \in \mathcal{D}(Q)$ of the problem* (3.6.1), (3.6.12), (3.6.13) *is oscillatory in Q, or satisfies $\lim_{t\to\infty} tU(t) = 0$.*

Proof. Suppose to the contrary that there is a solution of $u \in \mathcal{D}(Q)$ of the problem (3.6.1), (3.6.12), (3.6.13) which has no zero in $Q(t_0)$ for some $t_0 > 0$ and does not satisfy $\lim_{t\to\infty} tU(t) = 0$. Arguing as in the proof of Theorem 3.6.1, we find that $U(t)$ is a positive solution of (3.6.11) in (t_0, ∞) which does not satisfy $\lim_{t\to\infty} tU(t) = 0$. This contradicts the hypothesis and the proof is complete. □

Corollary 3.6.2. *Let $k+l$ be an odd integer and $\gamma \geq 1$ be the quotient of odd integers. Assume that the hypotheses* (i) *and* (ii) *of Corollary* 3.6.1 *hold. Then every solution $u \in \mathcal{D}(Q)$ of the problem* (3.6.19), (3.6.12), (3.6.13) *is oscillatory in Q, or satisfies* $\lim\limits_{t\to\infty} tU(t) = 0$.

Proof. The conclusion follows by the same arguments as were used in the proof of Corollary 3.6.1. □

We give an example which shows that the conclusion of Corollary 3.6.1 is in general false if the condition (3.6.12) is not required.

Example 3.6.1. We consider the problem

$$\frac{\partial^4 u}{\partial x^2 \partial y^2} + \left(\frac{15}{16}(x+y+1)^{-14/3} + \frac{9}{2}(x+y+1)^{-11/3} + 9(x+y+1)^{-8/3} \right) e^{8y} u^{7/3} = 0 \quad \text{in } Q, \tag{3.6.21}$$

$$\frac{\partial^3 u}{\partial y^3} + \frac{\partial^3 u}{\partial x \partial y^2} + \left(216 - \frac{3}{4}(y+1)^{-3} - \frac{15}{2}(y+1)^{-2} - 72(y+1)^{-1} \right) u = 0 \quad \text{on } \Gamma_y. \tag{3.6.22}$$

Here $k = l = 2$, $\gamma = 7/3$ and

$$\sigma_2(y) = 216 - \frac{3}{4}(y+1)^{-3} - \frac{15}{2}(y+1)^{-2} - 72(y+1)^{-1} \geq 0 \quad \text{in } (0, \infty).$$

Since $e^{8y} \geq 1$ in Q and

$$\int_0^\infty t^{4-7/3} \left(\frac{15}{16}(t+1)^{-14/3} + \frac{9}{2}(t+1)^{-11/3} + 9(t+1)^{-8/3} \right) dt = \infty,$$

we find that (i) and (ii) of Corollary 3.6.1 are satisfied. There is a nonoscillatory solution $u = (x+y+1)^{1/2} e^{-6y}$ of the problem (3.6.21), (3.6.22) which does not satisfy the condition (3.6.12). In fact, we obtain

$$\frac{\partial^3 u}{\partial x^3} + \frac{\partial^3 u}{\partial x^2 \partial y} + \sigma_1(x) u = \frac{3}{4}(x+1)^{-5/2} + \frac{3}{2}(x+1)^{-3/2} + \sigma_1(x)(x+1)^{1/2} > 0 \quad \text{on } \Gamma_x$$

for any $\sigma_1(x)$ such that $\sigma_1(x) \geq 0$.

Remark 3.6.1. When $k = l = 1$, Theorem 3.6.1 reduces to a result of Yoshida [295, Theorem 3].

Remark 3.6.2. Kiguradze classes for characteristic initial value problems associated with (3.6.1) were discussed by Kreith and Swanson [145]. We note that oscillations of the higher order linear hyperbolic equation

$$\frac{\partial^{k+l}u}{\partial x^k \partial y^l} = p_0(x,y)u + p_1(x,y)\frac{\partial^k u}{\partial x^k} + p_2(x,y)\frac{\partial^l u}{\partial y^l}$$

were studied by Kiguradze and Stavroulakis [127].

3.7 Notes

Oscillation properties of weak solutions of wave equations were studied in the papers [43–45, 95, 96, 327]. Corollary 3.1.1 was established by Kahane [121], and Theorem 3.1.1 is based on Travis [264]. Theorem 3.1.2 can be found in Kreith, Kusano and Yoshida [142]. Lemma 3.1.1 and Theorem 3.1.5 are extracted from Yoshida [301].

Theorems 3.2.1–3.2.4 are based on Yoshida [315]. Oscillation property of initial value problems for semilinear wave equations was investigated by Uesaka [273].

Theorem 3.3.3 is obtained by Yoshida [295]. Theorems 3.3.4 and 3.3.5 are extracted from Yoshida [303]. Equation (3.3.1) was also studied by Petrova [238].

The results in Section 3.4 are due to Narita and Yoshida [218].

Lemma 3.5.1, Theorem 3.5.1 and Corollary 3.5.1 were extracted from Travis and Yoshida [265].

Section 3.6 is taken from Kusano and Yoshida [158].

Chapter 4

Oscillation of Beam Equations

4.1 Extensible Beam Equations

Oscillations of extensible beam equations have been investigated by several authors. We refer to Feireisl and Herrmann [78], Timoshenko, Young and Weaver [261], Yoshida [298, 299, 305]. The existence of solutions of beam equations was studied in the papers [27, 79, 135, 197, 215]. We consider the extensible beam equation

$$\frac{\partial^2 u}{\partial t^2} + \alpha \frac{\partial^4 u}{\partial x^4} - \left(\beta + \gamma \int_0^L \left(\frac{\partial u(\xi, t)}{\partial \xi} \right)^2 d\xi \right) \frac{\partial^2 u}{\partial x^2} + c(x, t, u) = f(x, t) \tag{4.1.1}$$

in a cylindrical domain $\Omega = I \times (0, \infty)$, where α, β, γ, L are constants such that $\alpha > 0$, $\gamma \geq 0$, $L > 0$ and $I = (0, L)$. We deal with the case of hinged ends for which

$$u(0, t) = \frac{\partial^2 u}{\partial x^2}(0, t) = u(L, t) = \frac{\partial^2 u}{\partial x^2}(L, t) = 0, \quad t > 0. \tag{4.1.2}$$

Theorem 4.1.1. *Assume that:*

(i) $c(x, t, \xi) \in C(\overline{\Omega} \times \mathbb{R}; \mathbb{R})$ *and* $\xi c(x, t, \xi) \geq 0$ *for* $(x, t) \in \Omega,\ \xi \in \mathbb{R}$;

(ii) $f(x, t) \in C(\overline{\Omega}; \mathbb{R})$;

(iii) *there exists a function* $\theta(x) \in C^4(I; (0, \infty))$ *such that*

$$\alpha \theta^{(4)}(x) - \beta \theta''(x) \geq k\, \theta(x) \quad \text{in } I \text{ for some constant } k \geq 0,$$

$$\theta''(x) \leq 0 \quad \text{in } I,$$

$$\theta(0) = \theta(L) = \theta''(0) = \theta''(L) = 0.$$

Every solution $u \in C^4(\overline{\Omega}; \mathbb{R})$ *of the boundary value problem* (4.1.1), (4.1.2) *is oscillatory in* Ω *if the ordinary differential inequalities*

$$y'' + k\, y \leq \pm \int_0^L f(x, t) \theta(x)\, dx \tag{4.1.3}$$

have no eventually positive solutions.

Proof. Suppose to the contrary that there is a solution $u \in C^4(\overline{\Omega}; \mathbb{R})$ of the problem (4.1.1), (4.1.2) which has no zero in $I \times (t_0, \infty)$ for some $t_0 > 0$. First we assume that $u > 0$ in $I \times (t_0, \infty)$. Since $c(x, t, u) \geq 0$ in $I \times (t_0, \infty)$, we have

$$\frac{\partial^2 u}{\partial t^2} + \alpha \frac{\partial^4 u}{\partial x^4} - \left(\beta + \gamma \int_0^L \left(\frac{\partial u(\xi, t)}{\partial \xi} \right)^2 d\xi \right) \frac{\partial^2 u}{\partial x^2} \leq f(x, t),$$
$$(x, t) \in I \times (t_0, \infty). \quad (4.1.4)$$

We multiply (4.1.4) by $\theta(x)$ and then integrate over I to obtain

$$\int_0^L \frac{\partial^2 u}{\partial t^2} \theta(x)\, dx + \alpha \int_0^L \frac{\partial^4 u}{\partial x^4} \theta(x)\, dx - \beta \int_0^L \frac{\partial^2 u}{\partial x^2} \theta(x)\, dx$$
$$- \gamma \int_0^L \left(\frac{\partial u(\xi, t)}{\partial \xi} \right)^2 d\xi \int_0^L \frac{\partial^2 u}{\partial x^2} \theta(x)\, dx$$
$$\leq \int_0^L f(x, t)\theta(x)\, dx, \quad t > t_0. \quad (4.1.5)$$

Integration by parts yields

$$\int_0^L \frac{\partial^2 u}{\partial x^2} \theta(x)\, dx = \int_0^L u\, \theta''(x)\, dx, \quad t > t_0, \quad (4.1.6)$$

$$\int_0^L \frac{\partial^4 u}{\partial x^4} \theta(x)\, dx = \int_0^L u\, \theta^{(4)}(x)\, dx, \quad t > t_0. \quad (4.1.7)$$

Combining (4.1.5)–(4.1.7), we obtain

$$\frac{d^2}{dt^2} \int_0^L u\, \theta(x)\, dx + \int_0^L u \left(\alpha \theta^{(4)}(x) - \beta \theta''(x) \right) dx$$
$$- \gamma \int_0^L \left(\frac{\partial u(\xi, t)}{\partial \xi} \right)^2 d\xi \int_0^L u\, \theta''(x)\, dx$$
$$\leq \int_0^L f(x, t)\theta(x)\, dx, \quad t > t_0$$

and therefore

$$U''(t) + k\, U(t) \leq \int_0^L f(x, t)\theta(x)\, dx, \quad t > t_0, \quad (4.1.8)$$

where

$$U(t) = \int_0^L u\, \theta(x)\, dx.$$

Consequently, we observe that $U(t)$ is a positive solution of

$$y'' + k\,y \le \int_0^L f(x,t)\theta(x)\,dx, \quad t > t_0.$$

This contradicts the hypothesis. If $u < 0$ in $I \times (t_0, \infty)$, $v := -u$ satisfies

$$\frac{\partial^2 v}{\partial t^2} + \alpha \frac{\partial^4 v}{\partial x^4} - \left(\beta + \gamma \int_0^L \left(\frac{\partial v(\xi,t)}{\partial \xi}\right)^2 d\xi\right) \frac{\partial^2 v}{\partial x^2} \le -f(x,t),$$
$$(x,t) \in I \times (t_0, \infty).$$

Proceeding as in the case where $u > 0$, we are led to a contradiction. The proof is complete. □

If $\alpha(\pi/L)^4 + \beta(\pi/L)^2 \ge 0$, it is easy to see that $\theta(x) = \sin(\pi/L)x$ satisfies the hypothesis (iii) of Theorem 4.1.1 with $k = \alpha(\pi/L)^4 + \beta(\pi/L)^2$. Combining Theorem 4.1.1 with Theorems 1.4.2 and 1.4.3, we can establish the following corollaries.

Corollary 4.1.1. *Assume that the hypotheses* (i) *and* (ii) *of Theorem* 4.1.1 *are satisfied, and that* $\alpha(\pi/L)^4 + \beta(\pi/L)^2 \ge 0$. *Every solution* $u \in C^4(\overline{\Omega}; \mathbb{R})$ *of the problem* (4.1.1), (4.1.2) *is oscillatory in* Ω *if*

$$\liminf_{t\to\infty} \int_T^t \left(1 - \frac{s}{t}\right) F(s)\,ds = -\infty,$$

$$\limsup_{t\to\infty} \int_T^t \left(1 - \frac{s}{t}\right) F(s)\,ds = \infty$$

for all large T, *where*

$$F(t) = \int_0^L f(x,t) \sin\frac{\pi}{L}x\,dx.$$

Corollary 4.1.2. *Assume that the hypotheses* (i) *and* (ii) *of Theorem* 4.1.1 *are satisfied, and that* $\alpha(\pi/L)^4 + \beta(\pi/L)^2 > 0$. *Every solution* $u \in C^4(\overline{\Omega}; \mathbb{R})$ *of the problem* (4.1.1), (4.1.2) *is oscillatory in* Ω *if for some* $T > 0$ *there exists a function* $\rho(t) \in C^2([T,\infty); \mathbb{R})$ *with the following properties*:

(i) $\rho(t)$ *is oscillatory at* $t = \infty$;
(ii) $\rho''(t) = F(t), \quad t \ge T$;
(iii) $\lim_{t\to\infty} \rho(t) = 0$.

Theorem 4.1.2. *Assume that the hypotheses* (i) *and* (ii) *of Theorem* 4.1.1 *hold, and that* $\alpha(\pi/L)^4 + \beta(\pi/L)^2 > 0$. *If there is a number* $s \ge 0$ *such that*

$$\int_s^{s+(\pi/\omega)} F(t) \sin\omega(t-s)\,dt = 0,$$

then every solution $u \in C^4(\overline{\Omega};\mathbb{R})$ of the problem (4.1.1), (4.1.2) *has a zero in $I \times (s, s+(\pi/\omega)]$, where*

$$\omega = \left(\alpha\left(\frac{\pi}{L}\right)^4 + \beta\left(\frac{\pi}{L}\right)^2\right)^{1/2}.$$

Proof. Suppose that there exists a solution $u \in C^4(\overline{\Omega};\mathbb{R})$ of the problem (4.1.1), (4.1.2) which has no zero in $I \times (s, s+(\pi/\omega)]$. If $u > 0$ in $I \times (s, s+(\pi/\omega)]$, we have $c(x,t,u) \geq 0$ in $I \times (s, s+(\pi/\omega)]$. Arguing as in the proof of Theorem 4.1.1, we observe that $U(t) := \int_0^L u \sin(\pi/L)x\, dx$ is a positive solution of

$$y'' + \omega^2 y \leq F(t), \quad t \in (s, s+(\pi/\omega)].$$

This contradicts Lemma 3.1.1. In case $u < 0$ in $I \times (s, s+(\pi/\omega)]$, we find that $V(t) := \int_0^L (-u) \sin(\pi/L)x\, dx$ is a positive solution of

$$y'' + \omega^2 y \leq -F(t), \quad t \in (s, s+(\pi/\omega)].$$

We are also led to a contradiction and the proof is complete. □

The following corollary is an immediate consequence of Theorem 4.1.2.

Corollary 4.1.3. *Assume that $f(x,t) = 0$, that is,*

$$\frac{\partial^2 u}{\partial t^2} + \alpha\frac{\partial^4 u}{\partial x^4} - \left(\beta + \gamma\int_0^L \left(\frac{\partial u(\xi,t)}{\partial \xi}\right)^2 d\xi\right)\frac{\partial^2 u}{\partial x^2} + c(x,t,u) = 0, \quad (x,t) \in \Omega. \tag{4.1.9}$$

Moreover, assume that the hypothesis (i) *of Theorem* 4.1.1 *is satisfied, and that $\alpha(\pi/L)^4 + \beta(\pi/L)^2 > 0$. Then every solution $u \in C^4(\overline{\Omega};\mathbb{R})$ of the problem* (4.1.9), (4.1.2) *has a zero in $I \times (s, s+(\pi/\omega)]$ for any $s \geq 0$.*

Example 4.1.1. We consider the equation

$$\frac{\partial^2 u}{\partial t^2} + \alpha\frac{\partial^4 u}{\partial x^4} - \left(\beta + \gamma\int_0^L \left(\frac{\partial u(\xi,t)}{\partial \xi}\right)^2 d\xi\right)\frac{\partial^2 u}{\partial x^2} = 0,$$
$$(x,t) \in (0,L) \times (0,\infty), \tag{4.1.10}$$

where $\gamma \geq 0$ and $\alpha(\pi/L)^4 + \beta(\pi/L)^2 > 0$. We choose $\theta(x) = \sin(\pi/L)x$. Since $f(x,t) = 0$, (4.1.3) reduces to

$$y'' + \left(\alpha\left(\frac{\pi}{L}\right)^4 + \beta\left(\frac{\pi}{L}\right)^2\right)y \leq 0$$

which has no eventually positive solution. Hence, Theorem 4.1.1 implies that every solution $u \in C^4(\overline{\Omega}; \mathbb{R})$ of the problem (4.1.10), (4.1.2) is oscillatory in Ω. In fact, there exists an oscillatory solution $u = \left(\sin{(\pi/L)x}\right)T(t)$, where $T(t)$ is an oscillatory solution of the Duffing's equation

$$T'' + \left(\alpha\left(\frac{\pi}{L}\right)^4 + \beta\left(\frac{\pi}{L}\right)^2\right)T + \gamma\left(\frac{\pi}{L}\right)^4\left(\frac{L}{2}\right)T^3 = 0.$$

Example 4.1.2. We consider the equation

$$\frac{\partial^2 u}{\partial t^2} + \alpha\frac{\partial^4 u}{\partial x^4} - \left(\beta + \gamma\int_0^L \left(\frac{\partial u(\xi,t)}{\partial \xi}\right)^2 d\xi\right)\frac{\partial^2 u}{\partial x^2} + c(x,t,u)$$
$$= \left(\sin\frac{\pi}{L}x\right)e^{-t}\cos t, \quad (x,t) \in \Omega, \qquad (4.1.11)$$

where $\gamma \geq 0$ and $\alpha(\pi/L)^4 + \beta(\pi/L)^2 > 0$. Since

$$\left|\int_T^t \left(1 - \frac{s}{t}\right)\left(\int_0^L \left(\sin\frac{\pi}{L}x\right)^2 e^{-s}\cos s\, dx\right) ds\right|$$
$$\leq \frac{L}{2}\int_T^t \left|\left(1-\frac{s}{t}\right)e^{-s}\cos s\right| ds$$
$$\leq \frac{L}{2}\int_0^\infty e^{-s}ds = \frac{L}{2} < \infty,$$

Corollary 4.1.1 is not applicable to (4.1.11). Letting $\rho(t) = -(L/4)e^{-t}\sin t$, we find that $\rho(t)$ satisfies (i)–(iii) of Corollary 4.1.2. Hence, every solution $u \in C^4(\overline{\Omega}; \mathbb{R})$ of the problem (4.1.11), (4.1.2) is oscillatory in Ω.

Remark 4.1.1. In the case of sliding ends for which

$$\frac{\partial u}{\partial x}(0,t) = \frac{\partial^3 u}{\partial x^3}(0,t) = \frac{\partial u}{\partial x}(L,t) = \frac{\partial^3 u}{\partial x^3}(L,t) = 0, \quad t > 0,$$

we can establish various oscillation results (see [298, 305]).

Remark 4.1.2. Oscillation results can be established for the following various end conditions:

$$u(0,t) = \frac{\partial^2 u}{\partial x^2}(0,t) = \frac{\partial u}{\partial x}(L,t) = \frac{\partial^3 u}{\partial x^3}(L,t) = 0 \text{ (hinged-sliding ends)},$$
$$u(0,t) = \frac{\partial u}{\partial x}(0,t) = u(L,t) = \frac{\partial u}{\partial x}(L,t) = 0 \qquad \text{(clamped-clamped ends)},$$
$$u(0,t) = \frac{\partial u}{\partial x}(0,t) = u(L,t) = \frac{\partial^2 u}{\partial x^2}(L,t) = 0 \qquad \text{(clamped-hinged ends)},$$
$$u(0,t) = \frac{\partial u}{\partial x}(0,t) = \frac{\partial^2 u}{\partial x^2}(L,t) = \frac{\partial^3 u}{\partial x^3}(L,t) = 0 \text{ (clamped-free ends)}.$$

4.2 Timoshenko Beam Equations

Timoshenko beam equations were studied by several authors, see, for example, Kusano and Yoshida [159], Timoshenko, Young and Weaver [261], Wang and Gagnon [282], Yoshida [305] and the references cited therein.

We investigate the Timoshenko beam equation

$$\frac{\partial^4 u}{\partial t^4} + pqr\frac{\partial^2 u}{\partial t^2} - (q+r)\frac{\partial^4 u}{\partial x^2 \partial t^2} + qr\frac{\partial^4 u}{\partial x^4} + c(x,t,u) = f(x,t), \quad (x,t) \in \Omega, \tag{4.2.1}$$

where $\Omega = I \times (0, \infty)$, $I = (0, L)$ and p, q, r are positive constants.

Theorem 4.2.1. *Assume that the hypotheses* (i) *and* (ii) *of Theorem* 4.1.1 *are satisfied. Every solution* $u \in C^4(\overline{\Omega}; \mathbb{R})$ *of the boundary value problem* (4.2.1), (4.1.2) *is oscillatory in* Ω *if the fourth order ordinary differential inequalities*

$$y^{(4)} + \left(pqr + (q+r)\left(\frac{\pi}{L}\right)^2\right)y'' + qr\left(\frac{\pi}{L}\right)^4 y \le F(t), \tag{4.2.2}$$

$$y^{(4)} + \left(pqr + (q+r)\left(\frac{\pi}{L}\right)^2\right)y'' + qr\left(\frac{\pi}{L}\right)^4 y \le -F(t) \tag{4.2.3}$$

have no eventually positive solutions.

Proof. Suppose that there is a solution $u \in C^4(\overline{\Omega}; \mathbb{R})$ of the problem (4.2.1), (4.1.2) which has no zero in $I \times (t_0, \infty)$ for some $t_0 > 0$. In case $u > 0$ in $I \times (t_0, \infty)$, we obtain

$$\frac{\partial^4 u}{\partial t^4} + pqr\frac{\partial^2 u}{\partial t^2} - (q+r)\frac{\partial^4 u}{\partial x^2 \partial t^2} + qr\frac{\partial^4 u}{\partial x^4} \le f(x,t), \quad (x,t) \in I \times (t_0, \infty) \tag{4.2.4}$$

in view of the fact that $c(x,t,u) \ge 0$ in $I \times (t_0, \infty)$. Multiplying (4.2.4) by $\psi(x) = \sin(\pi/L)x$ and then integrating over I, we have

$$U^{(4)}(t) + pqr\,U''(t) - (q+r)\frac{d^2}{dt^2}\int_0^L \frac{\partial^2 u}{\partial x^2}\psi(x)\,dx + qr\int_0^L \frac{\partial^4 u}{\partial x^4}\psi(x)\,dx \le F(t) \tag{4.2.5}$$

for $t > t_0$, where

$$U(t) = \int_0^L u \sin\frac{\pi}{L}x\,dx.$$

Integrating by parts and using (4.1.2), we obtain

$$\int_0^L \frac{\partial^2 u}{\partial x^2}\,\psi(x)\,dx = -\left(\frac{\pi}{L}\right)^2 U(t),$$

$$\int_0^L \frac{\partial^4 u}{\partial x^4}\,\psi(x)\,dx = \left(\frac{\pi}{L}\right)^4 U(t).$$

Hence, (4.2.5) reduces to

$$U^{(4)}(t) + \left(pqr + (q+r)\left(\frac{\pi}{L}\right)^2\right)U''(t) + qr\left(\frac{\pi}{L}\right)^4 U(t) \le F(t), \quad t > t_0. \tag{4.2.6}$$

Consequently, we observe that $U(t)$ is an eventually positive solution of (4.2.2). This contradicts the hypothesis. If $u < 0$ in $I \times (t_0, \infty)$, we are also led to a contradiction by using the same arguments as in the case where $u > 0$. The proof is complete. □

Corollary 4.2.1. *Assume that the hypotheses* (i) *and* (ii) *of Theorem* 4.1.1 *are satisfied. Every solution* $u \in C^4(\overline{\Omega}; \mathbb{R})$ *of the problem* (4.2.1), (4.1.2) *is oscillatory in* Ω *if*

$$\liminf_{t\to\infty} \int_T^t \left(1 - \frac{s}{t}\right)^3 F(s)\,ds = -\infty,$$

$$\limsup_{t\to\infty} \int_T^t \left(1 - \frac{s}{t}\right)^3 F(s)\,ds = \infty$$

for all large T.

Proof. Let $y(t)$ be a solution of (4.2.2) such that $y(t) > 0$ on $[t_0, \infty)$ for some $t_0 > 0$. Integration of (4.2.2) over $[t_0, t]$ twice yields

$$y''(t) + \left(pqr + (q+r)\left(\frac{\pi}{L}\right)^2\right)y(t) \le c_0 + c_1 t + \int_{t_0}^t (t-s)F(s)\,ds \tag{4.2.7}$$

for some constants c_0 and c_1. Integrating (4.2.7) over $[t_0, t]$ twice, we obtain

$$y(t) \le d_0 + d_1 t + d_2 t^2 + d_3 t^3 + \int_{t_0}^t (t-s)\left(\int_{t_0}^s (s-\xi)F(\xi)\,d\xi\right)ds \tag{4.2.8}$$

for some constants d_i $(i = 0, 1, 2, 3)$. Since

$$\int_{t_0}^t (t-s)\left(\int_{t_0}^s (s-\xi)F(\xi)\,d\xi\right)ds = \frac{1}{6}\int_{t_0}^t (t-\xi)^3 F(\xi)\,d\xi,$$

we see from (4.2.8) that

$$y(t) \le t^3\left(B(t, t_0) + \frac{1}{6}\int_{t_0}^t \left(1 - \frac{s}{t}\right)^3 F(s)\,ds\right), \tag{4.2.9}$$

where $B(t, t_0)$ is bounded as t tends to infinity. By the hypothesis we find that the right hand side of (4.2.9) is not bounded from below, whereas the left hand side $y(t)$ is positive. This is a contradiction, and hence (4.2.2) has no eventually positive solution. Analogously, we conclude that (4.2.3) has no eventually positive solution. The conclusion follows from Theorem 4.2.1. □

We define the positive constants $k_\pm$ by

$$k_\pm^2 = \frac{1}{2}\left(pqr + (q+r)\left(\frac{\pi}{L}\right)^2 \pm \sqrt{\left(pqr + (q+r)\left(\frac{\pi}{L}\right)^2\right)^2 - 4qr\left(\frac{\pi}{L}\right)^4} \right).$$

We note that

$$\begin{aligned}&\left(pqr + (q+r)\left(\frac{\pi}{L}\right)^2\right)^2 - 4qr\left(\frac{\pi}{L}\right)^4 \\ &= (pqr)^2 + 2pqr(q+r)\left(\frac{\pi}{L}\right)^2 + (q-r)^2\left(\frac{\pi}{L}\right)^4 > 0.\end{aligned}$$

Theorem 4.2.2. *Assume that the hypotheses* (i) *and* (ii) *of Theorem* 4.1.1 *are satisfied. If there is a number $s \geq 0$ such that*

$$\int_s^{s+(\pi/k_-)} \left(\int_\tau^{\tau+(\pi/k_+)} F(t) \sin k_+(t-\tau)\, dt \right) \sin k_-(\tau - s)\, d\tau = 0,$$

then every solution $u \in C^4(\overline{\Omega};\mathbb{R})$ of the problem (4.2.1), (4.1.2) *has a zero in $I \times (s, s + (\pi/k_+) + (\pi/k_-))$.*

Proof. Suppose that there exists a solution $u \in C^4(\overline{\Omega};\mathbb{R})$ of the problem (4.2.1), (4.1.2) which has no zero in $I \times (s, s + (\pi/k_+) + (\pi/k_-))$. Assume that $u > 0$ in $I \times (s, s + (\pi/k_+) + (\pi/k_-))$. Proceeding as in the proof of Theorem 4.2.1, we see that $U(t) = \int_0^L u \sin(\pi/L)x\, dx$ is a positive solution of (4.2.6) in $(s, s + (\pi/k_+) + (\pi/k_-))$, which can be written in the form

$$\left(\frac{d^2}{dt^2} + k_+^2\right)\left(\frac{d^2}{dt^2} + k_-^2\right) U(t) \leq F(t), \quad t \in (s, s + (\pi/k_+) + (\pi/k_-)). \tag{4.2.10}$$

Using the arguments used in the proof of Lemma 3.1.1, we observe that

$$\begin{aligned}&U(s + (\pi/k_+) + (\pi/k_-)) + U(s + (\pi/k_+)) + U(s + (\pi/k_-)) + U(s) \\ &\leq \frac{1}{k_-}\frac{1}{k_+}\int_s^{s+(\pi/k_-)} \left(\int_\tau^{\tau+(\pi/k_+)} F(t) \sin k_+(t-\tau)\, dt \right) \times \\ &\qquad \times \sin k_-(\tau - s)\, d\tau.\end{aligned} \tag{4.2.11}$$

Since $U(s+(\pi/k_+)+(\pi/k_-)) \geq 0$, $U(s+(\pi/k_\pm)) > 0$ and $U(s) \geq 0$, the left hand side of (4.2.11) is positive. However, the right hand side of (4.2.11) is zero by the hypothesis. This is a contradiction. In the case where $u < 0$ in $I \times (s, s + (\pi/k_+) + (\pi/k_-))$, we are also led to a contradiction by using the same arguments as in the case where $u > 0$. The proof is complete. □

Example 4.2.1. We consider the equation

$$\frac{\partial^4 u}{\partial t^4}+\frac{4}{3}\frac{\partial^2 u}{\partial t^2}-\frac{2}{3}\left(\frac{L}{\pi}\right)^2\frac{\partial^4 u}{\partial x^2\partial t^2}+\frac{1}{9}\left(\frac{L}{\pi}\right)^4\frac{\partial^4 u}{\partial x^4}=-\frac{8}{9}\left(\sin\frac{\pi}{L}x\right)t^4\sin t \tag{4.2.12}$$

for $(x,t)\in\Omega$. Here $\alpha=12(\pi/L)^4$, $\beta=\gamma=(1/3)(L/\pi)^2$ and $c(x,t,u)=0$. A simple computation shows that

$$\begin{aligned}
&\int_T^t\left(1-\frac{s}{t}\right)^3\left(\int_0^L-\frac{8}{9}s^4(\sin s)\left(\sin\frac{\pi}{L}x\right)^2 dx\right)ds\\
&=-\frac{4}{9}L\int_T^t\left(1-\frac{s}{t}\right)^3 s^4\sin s\,ds\\
&=-\frac{8}{3}Lt\sin t+B(t,T),
\end{aligned}$$

where $B(t,T)$ is bounded as $t\to\infty$. Therefore we obtain

$$\liminf_{t\to\infty}\int_T^t\left(1-\frac{s}{t}\right)^3\left(-\frac{8}{9}\int_0^L s^4(\sin s)\left(\sin\frac{\pi}{L}x\right)^2 dx\right)ds=-\infty,$$

$$\limsup_{t\to\infty}\int_T^t\left(1-\frac{s}{t}\right)^3\left(-\frac{8}{9}\int_0^L s^4(\sin s)\left(\sin\frac{\pi}{L}x\right)^2 dx\right)ds=\infty$$

for all large T. It follows from Corollary 4.2.1 that every solution $u\in C^4(\overline{\Omega};\mathbb{R})$ of the problem (4.2.12), (4.1.2) is oscillatory in Ω. In fact

$$u=\left(\sin\frac{\pi}{L}x\right)(t^4\sin t-54t^2\sin t+108t\cos t+513\sin t)$$

is such a solution.

Example 4.2.2. We consider the equation

$$\frac{\partial^4 u}{\partial t^4}+\frac{\partial^2 u}{\partial t^2}-4\left(\frac{L}{\pi}\right)^2\frac{\partial^4 u}{\partial x^2\partial t^2}+4\left(\frac{L}{\pi}\right)^4\frac{\partial^4 u}{\partial x^4}=40\left(\sin\frac{\pi}{L}x\right)\sin 3t \tag{4.2.13}$$

for $(x,t)\in\Omega$. Here $p=4^{-1}(\pi/L)^4$ and $q=r=2(L/\pi)^2$. It is readily seen that $k_+=2,\ k_-=1$ and $F(t)=20L\sin 3t$. An easy computation yields

$$\int_s^{s+\pi}\left(\int_\tau^{\tau+(\pi/2)}F(t)\sin 2(t-\tau)\,dt\right)\sin(\tau-s)\,d\tau=0$$

for any $s\geq 0$. Theorem 4.2.2 implies that every solution $u\in C^4(\overline{\Omega};\mathbb{R})$ of the problem (4.2.13), (4.1.2) has a zero in $I\times(s,s+(3/2)\pi)$ for any $s\geq 0$. One such solution is $u=(\sin(\pi/L)x)\sin 3t$.

Remark 4.2.1. The system of coupled beam equations

$$\mu \frac{\partial^4 u}{\partial x^4} = -\lambda \frac{\partial^2}{\partial t^2}(u - Mv) + g(x,t),$$
$$K \frac{\partial^2 v}{\partial x^2} - K_1 \frac{\partial^4 v}{\partial x^4} = -\lambda M \frac{\partial^2}{\partial t^2}(u - Mv) + \nu \frac{\partial^2 v}{\partial t^2} + h(x,t)$$

was studied in [261, 305], where λ, μ, ν, K, K_1 and M are positive constants. We note that the system of coupled beam equations

$$2\left(\frac{L}{\pi}\right)^4 \frac{\partial^4 u}{\partial x^4} = -\frac{\partial^2}{\partial t^2}(u - v) + \left(\sin \frac{\pi}{L} x\right)(\sin t + \cos t),$$
$$2\left(\frac{L}{\pi}\right)^2 \frac{\partial^2 v}{\partial x^2} - 2\left(\frac{L}{\pi}\right)^4 \frac{\partial^4 v}{\partial x^4}$$
$$= -\frac{\partial^2}{\partial t^2}(u - v) + 2\frac{\partial^2 v}{\partial t^2} - \left(\sin \frac{\pi}{L} x\right)(\sin t + \cos t)$$

has an oscillatory solution $u = (\sin(\pi/L)x)\sin t$, $v = (\sin(\pi/L)x)\cos t$ (*cf.* [305, Example 3]).

4.3 Notes

Theorem 4.1.1, Corollaries 4.1.1, 4.1.2 are taken form Yoshida [298]. Theorem 4.1.2 and Corollary 4.1.3 are adopted from Yoshida [305]. The oscillation results in Section 4.1 can be generalized to the more general equation

$$\frac{\partial^2 u}{\partial t^2} + \Delta^2 u - Q(\|\nabla u(\cdot,t)\|^2_{L^2})\Delta u + c(x,t,u) = f(x,t)$$

(*cf.* [298, Remark 1]).

Oscillation results in Section 4.2 were extracted from Kusano and Yoshida [159], Yoshida [298, 305]. In Section 4.2 we treated the case of hinged ends, however we can derive oscillation results for the cases of hinged-sliding ends or sliding ends (*cf.* [159]).

Chapter 5

Functional Elliptic Equations

5.1 Equations with Deviating Arguments

In 1984 Tramov [263] treated the elliptic equation with deviating arguments

$$\Delta u(x) + p(x)u(x - \tau) = 0 \tag{5.1.1}$$

and studied the existence of zeros of solutions in a ball, where Δ is the Laplacian in $\mathbb{R}^n$, $p(x) \in C(\mathbb{R}^n; [0, \infty))$ and τ is a constant vector of $\mathbb{R}^n$.

We consider the elliptic equation with deviating arguments

$$\Delta u(x - \sigma) + p(x)u(x - \tau) = f(x) \tag{5.1.2}$$

and present conditions which imply that every solution of (5.1.2) has a zero in bounded domains of $\mathbb{R}^n$. We consider two kinds of bounded domains, that is, balls and rectangular domains.

The following notation will be used:

$$B_r = \{x \in \mathbb{R}^n;\ |x| \le r\},$$
$$R(s_1, ..., s_n) = \{x \in \mathbb{R}^n;\ |x_i| \le s_i\ (i = 1, 2, ..., n)\},$$

where $|x|$ denotes the Euclidean length of $x \in \mathbb{R}^n$.

Theorem 5.1.1. *Assume that:*

(i) $p(x) \in C(\mathbb{R}^n; [0, \infty))$, $f(x) \in C(\mathbb{R}^n; \mathbb{R})$;
(ii) *σ and τ are constant vectors of $\mathbb{R}^n$.*

If there exist a positive number r and an integer $m \ge 2$ with the following properties that:

(iii) $\displaystyle\int_{B_r} f(x)(r^2 - |x|^2)^m\, dx = 0$;

(iv) *there is a number* $\alpha \geq 0$ *such that*

$$|\sigma - \tau| \leq \alpha < \left(1 - \sqrt{\frac{n}{n+2m-2}}\right) r;$$

(v) $$p(x) \geq \frac{2mnr^{2m-2}}{\left(r^2 - \left(\sqrt{\frac{n}{n+2m-2}}\, r + \alpha\right)^2\right)^m}, \quad x \in B_r,$$

then every solution $u \in C^2(B_{r+\min\{|\sigma|,|\tau|\}+|\sigma-\tau|}; \mathbb{R})$ *of* (5.1.2) *has a zero in* $B_{r+\min\{|\sigma|,|\tau|\}+|\sigma-\tau|}$.

Proof. Assume, for the sake of contradiction, that there is a C^2-solution $u(x)$ of (5.1.2) which has no zero in $B_{r+\min\{|\sigma|,|\tau|\}+|\sigma-\tau|}$. First we assume that $u(x) > 0$ in $B_{r+\min\{|\sigma|,|\tau|\}+|\sigma-\tau|}$. Define the function $\psi_m(x)$ $(m \geq 2)$ by

$$\psi_m(x) = \begin{cases} (r^2 - |x|^2)^m, & x \in B_r \\ 0, & x \notin B_r. \end{cases}$$

Multiplying (5.1.2) by $\psi_m(x)$ and then integrating over B_r yield

$$\int_{B_r} \Delta u(x-\sigma)\psi_m(x)\, dx + \int_{B_r} p(x)u(x-\tau)\psi_m(x)\, dx = \int_{B_r} f(x)\psi_m(x)\, dx.$$

From Green's formula [242, p.82] and the hypothesis (iii) it follows that

$$\int_{B_r} u(x-\sigma)\Delta\psi_m(x)\, dx + \int_{B_r} p(x)u(x-\tau)\psi_m(x)\, dx = 0. \tag{5.1.3}$$

By making changes of variables, (5.1.3) can be written in the form

$$\int_{|y+\sigma|\leq r} u(y)\Delta\psi_m(y+\sigma)\, dy + \int_{|y+\tau|\leq r} u(y)p(y+\tau)\psi_m(y+\tau)\, dy = 0. \tag{5.1.4}$$

It is clear that $|\sigma| \leq \min\{|\sigma|, |\tau|\} + |\sigma - \tau|$ and $|\tau| \leq \min\{|\sigma|, |\tau|\} + |\sigma - \tau|$. Since $\{y \in \mathbb{R}^n;\ |y + \sigma| \leq r\} \subset B_{r+|\sigma|} \subset B_{r+\min\{|\sigma|,|\tau|\}+|\sigma-\tau|}$, we find that $u(y) > 0$ on $\{y \in \mathbb{R}^n;\ |y+\sigma| \leq r\}$. Similarly, we see that $u(y) > 0$ on $\{y \in \mathbb{R}^n;\ |y + \tau| \leq r\}$. Since

$$\Delta\psi_m(x) = \begin{cases} 2m\left(r^2 - |x|^2\right)^{m-2}\left((n+2m-2)|x|^2 - nr^2\right) & \text{for} \quad |x| \leq r \\ 0 & \text{for} \quad |x| > r, \end{cases}$$

it can be shown that

$$\Delta\psi_m(x) > 0 \qquad \text{for} \quad \sqrt{\frac{n}{n+2m-2}}\, r < |x| < r,$$

$$\Delta\psi_m(x) \geq -2mnr^{2m-2} \qquad \text{for} \quad |x| \leq \sqrt{\frac{n}{n+2m-2}}\, r.$$

Therefore we have

$$\int_{|y+\sigma|\le r} u(y)\Delta\psi_m(y+\sigma)\,dy > \int_{|y+\sigma|\le\sqrt{\frac{n}{n+2m-2}}\,r} u(y)\Delta\psi_m(y+\sigma)\,dy$$

$$\ge \int_{|y+\sigma|\le\sqrt{\frac{n}{n+2m-2}}\,r} u(y)\left(-2mnr^{2m-2}\right)dy. \quad (5.1.5)$$

If $|y+\sigma| \le \sqrt{\frac{n}{n+2m-2}}\,r$, then

$$|y+\tau| \le |y+\sigma| + |\sigma-\tau| \le \sqrt{\frac{n}{n+2m-2}}\,r + \alpha < r. \quad (5.1.6)$$

Hence we obtain

$$\left\{y\in\mathbb{R}^n;\ |y+\sigma| \le \sqrt{\frac{n}{n+2m-2}}\,r\right\} \subset \{y\in\mathbb{R}^n;\ |y+\tau|\le r\}.$$

In view of the fact that $u(y)p(y+\tau)\psi_m(y+\tau) \ge 0$ on $\{y\in\mathbb{R}^n;\ |y+\tau|\le r\}$, we find that

$$\int_{|y+\tau|\le r} u(y)p(y+\tau)\psi_m(y+\tau)\,dy$$

$$\ge \int_{|y+\sigma|\le\sqrt{\frac{n}{n+2m-2}}\,r} u(y)p(y+\tau)\psi_m(y+\tau)\,dy. \quad (5.1.7)$$

We observe, using the inequality (5.1.6), that

$$\psi_m(y+\tau) = \left(r^2 - |y+\tau|^2\right)^m \ge \left(r^2 - \left(\sqrt{\frac{n}{n+2m-2}}\,r+\alpha\right)^2\right)^m > 0$$

on $\left\{y\in\mathbb{R}^n;\ |y+\sigma| \le \sqrt{\frac{n}{n+2m-2}}\,r\right\}$, and therefore

$$\int_{|y+\sigma|\le\sqrt{\frac{n}{n+2m-2}}\,r} u(y)p(y+\tau)\psi_m(y+\tau)\,dy$$

$$\ge \int_{|y+\sigma|\le\sqrt{\frac{n}{n+2m-2}}\,r} u(y)p(y+\tau)\left(r^2 - \left(\sqrt{\frac{n}{n+2m-2}}\,r+\alpha\right)^2\right)^m dy. \quad (5.1.8)$$

Combining (5.1.4), (5.1.5), (5.1.7) and (5.1.8) yields

$$\int_{|y+\sigma|\le\sqrt{\frac{n}{n+2m-2}}\,r} u(y)\,\times$$

$$\times\left(p(y+\tau)\left(r^2 - \left(\sqrt{\frac{n}{n+2m-2}}\,r+\alpha\right)^2\right)^m - 2mnr^{2m-2}\right)dy < 0. \quad (5.1.9)$$

Taking into account the hypothesis (v), we conclude that the integrand of (5.1.9) is nonnegative. This yields a contradiction. If $u(x) < 0$ in $B_{r+\min\{|\sigma|,|\tau|\}+|\sigma-\tau|}$, $v(x) := -u(x)$ satisfies

$$\Delta v(x-\sigma) + p(x)v(x-\tau) = -f(x).$$

Arguing as in the case where $u(x) > 0$, we are led to a contradiction. The proof is complete. □

Corollary 5.1.1. *Assume that $f(x) = 0$, that is,*

$$\Delta u(x-\sigma) + p(x)u(x-\tau) = 0. \tag{5.1.10}$$

If $p(x) \geq \varepsilon$ in $\mathbb{R}^n$ for some $\varepsilon > 0$, then every solution $u \in C^2(B_r;\mathbb{R})$ of (5.1.10) *has a zero in B_r for a sufficiently large r.*

Proof. Since we have

$$\lim_{r\to\infty}\left(1-\sqrt{\frac{n}{n+2m-2}}\right)r = \infty,$$

$$\lim_{r\to\infty}\frac{2mnr^{2m-2}}{\left(r^2-\left(\sqrt{\frac{n}{n+2m-2}}\,r+\alpha\right)^2\right)^m} = 0$$

for some $\alpha \geq |\sigma-\tau|$, we see that the hypotheses of Theorem 5.1.1 are satisfied for a sufficiently large r. The conclusion follows from Theorem 5.1.1. □

Corollary 5.1.2. (Tramov [263, Theorem 1]) *Assume that $\sigma = 0$ and $f(x) = 0$, that is,*

$$\Delta u(x) + p(x)u(x-\tau) = 0. \tag{5.1.11}$$

Moreover, assume that $p(x) \geq \varepsilon$ in $\mathbb{R}^n$ for some $\varepsilon > 0$. If there exist positive numbers τ_0 and r for which

$$|\tau| \leq \tau_0 < \left(1-\sqrt{\frac{n}{n+2}}\right)r,$$

$$\varepsilon\left(r^2-\left(\sqrt{\frac{n}{n+2}}\,r+\tau_0\right)^2\right)^2 \geq 4nr^2,$$

then every solution $u \in C^2(B_{r+\tau_0};\mathbb{R})$ of (5.1.11) *has a zero in $B_{r+\tau_0}$.*

Proof. Since

$$p(x) \geq \varepsilon \geq \frac{4nr^2}{\left(r^2-\left(\sqrt{\frac{n}{n+2}}\,r+\tau_0\right)^2\right)^2}, \quad x \in B_r,$$

the hypotheses of Theorem 5.1.1 are satisfied with $m = 2$, and hence the conclusion follows from Theorem 5.1.1. □

By using the same arguments as were used in the proof of Theorem 5.1.1 we can establish the following (see Yoshida [310]).

Theorem 5.1.2. *Assume that the hypotheses* (i), (ii) *of Theorem* 5.1.1 *are satisfied. If there exist an integer* $m \geq 2$ *and positive numbers* $s_1, ..., s_n$ *with the properties that:*

(iii) $\displaystyle\int_{R(s_1,...,s_n)} f(x) \prod_{i=1}^{n} \left(s_i^2 - x_i^2\right)^m dx = 0;$

(iv) *there are numbers* $\alpha_i \geq 0$ *such that*

$$|\sigma_i - \tau_i| \leq \alpha_i < \left(1 - \frac{1}{\sqrt{2m-1}}\right) s_i \quad (i = 1, 2, ..., n);$$

(v) $$p(x) \geq \frac{2m\left(\prod_{j=1}^{n} s_j^{2m}\right) \sum_{i=1}^{n} s_i^{-2}}{\prod_{i=1}^{n}\left(s_i^2 - \left(\frac{s_i}{\sqrt{2m-1}} + \alpha_i\right)^2\right)^m}, \quad x \in R(s_1, ..., s_n),$$

then every solution $u \in C^2(R(s_1 + k_1, ..., s_n + k_n); \mathbb{R})$ *of* (5.1.2) *has a zero in* $R(s_1 + k_1, ..., s_n + k_n)$, *where*

$$k_i = \min\{|\sigma_i|, |\tau_i|\} + |\sigma_i - \tau_i| \quad (i = 1, 2, ..., n).$$

Corollary 5.1.3. *Assume that* $p(x) \geq \varepsilon$ *in* $\mathbb{R}^n$ *for some* $\varepsilon > 0$, *then every solution* $u \in C^2(R(s, ..., s); \mathbb{R})$ *of* (5.1.10) *has a zero in* $R(s, ..., s)$ *for a sufficiently large* s.

Proof. It is easy to check that

$$\lim_{s\to\infty} \left(1 - \frac{1}{\sqrt{2m-1}}\right) s = \infty,$$

$$\lim_{s\to\infty} \frac{2ms^{2mn} ns^{-2}}{\prod_{i=1}^{n}\left(s^2 - \left(\frac{s}{\sqrt{2m-1}} + \alpha_i\right)^2\right)^m} = 0$$

for some $\alpha_i \geq |\sigma_i - \tau_i|$. The conclusion follows from Theorem 5.1.2. □

Remark 5.1.1. If $f(x) = \prod_{i=1}^{n} f_i(x_i)$ and each $f_i(x_i)$ is an odd function in $\mathbb{R}$, that is, $f_i(-x_i) = -f_i(x_i)$ in $\mathbb{R}$, then the condition (iii) of Theorem 5.1.2 is satisfied.

Remark 5.1.2. In [263] Tramov used the function $\psi_2(x)$. In this case, $\Delta\psi_2(x)$ is discontinuous on ∂B_r. We note that $\Delta\psi_m(x)$ $(m \geq 3)$ are continuous functions in $\mathbb{R}^n$.

Remark 5.1.3. By making a change of variables $x - \sigma = y$, we observe that (5.1.2) can be written as

$$\Delta u(y) + p(y + \sigma)u(y - (\tau - \sigma)) = f(y + \sigma). \qquad (5.1.12)$$

Analogously, (5.1.2) reduces to

$$\Delta u(y - (\sigma - \tau)) + p(y + \tau)u(y) = f(y + \tau), \quad (y = x - \tau). \qquad (5.1.13)$$

If we can establish any criteria for (5.1.12) or (5.1.13), we can also derive the results about zeros of solutions of (5.1.2).

Example 5.1.1. We consider the elliptic equation with deviating arguments

$$\Delta u\left(x_1 - \tfrac{\pi}{2}, x_2 - \tfrac{\pi}{2}\right) + 175u\left(x_1 - \tfrac{\pi}{4}, x_2 - \tfrac{\pi}{4}\right) = 143(\sin 4x_1)\sin 4x_2 \qquad (5.1.14)$$

for $(x_1, x_2) \in \mathbb{R}^2$. Here $n = 2, \sigma = (\pi/2, \pi/2), \tau = (\pi/4, \pi/4)$ and $f(x_1, x_2) = 143(\sin 4x_1)\sin 4x_2$. We easily see that

$$\int_{R(\pi,\pi)} f(x_1, x_2)\left(\pi^2 - x_1^2\right)^3\left(\pi^2 - x_2^2\right)^3 dx_1 dx_2 = 0$$

(see Remark 5.1.1). It is clear that

$$|\sigma_i - \tau_i| = \frac{\pi}{4} < \left(1 - \frac{1}{\sqrt{5}}\right)\pi \quad (i = 1, 2).$$

An easy computation yields

$$\frac{6\pi^{12}2\pi^{-2}}{\left(\pi^2 - \left(\frac{\pi}{\sqrt{5}} + \frac{\pi}{4}\right)^2\right)^3\left(\pi^2 - \left(\frac{\pi}{\sqrt{5}} + \frac{\pi}{4}\right)^2\right)^3}$$
$$\leq \frac{12\pi^{10}}{\left(\pi^2 - \left(\frac{\pi}{2} + \frac{\pi}{4}\right)^2\right)^6} = 12\left(\frac{16}{7}\right)^6 \pi^{-2} \leq 175.$$

It follows from Theorem 5.1.2 with $m = 3$ that every solution u of (5.1.14) has a zero in $R(\pi + (\pi/2), \pi + (\pi/2))$. For example, $u = (\sin 4x_1)\sin 4x_2$ is such a solution.

5.2 Notes

Corollary 5.1.2 is due to Tramov [263], and the other results in Section 5.1 are extracted from Yoshida [310]. There are some open questions:

(I) Can Theorem 5.1.1 be generalized to the nonlinear elliptic equation

$$\Delta u(x-\sigma)+p(x)\varphi(u(x-\tau))=f(x)?$$

(II) Are similar results obtainable for the higher order elliptic equation

$$\Delta^m u(x-\sigma)+p(x)u(x-\tau)=f(x)?$$

(III) Is it possible to establish conditions which imply that every solution of certain elliptic equation with deviating arguments has a zero in any neighborhood of infinity?

Chapter 6

Functional Parabolic Equations

6.1 Equations with Functional Arguments

In 1983 oscillation of parabolic equations with functional arguments was first treated by Bykov and Kultaev [40].

Mishev [204] studied delay parabolic equations with constant coefficients, and derived necessary and sufficient conditions for the oscillation of solutions. Oscillation of partial difference equations of parabolic type was investigated by Cheng and Zhang [51]. Logistic equations with delay and diffusion were treated by Xie and Cheng [286]. Kusano and Yoshida [160] obtained oscillation results for functional parabolic equations with oscillating coefficients.

We consider the parabolic equation with functional arguments

$$\frac{\partial u}{\partial t}(x,t) - a(t)\Delta u(x,\tau(t)) - b(x,t)u(x,\tau(t))$$
$$-c(x,t,z[u](x,t),(z_i[u](x,t))_{i=1}^{\tilde{m}}) = 0, \quad (x,t) \in \Omega := G \times (0,\infty), \tag{6.1.1}$$

where G is a bounded domain in $\mathbb{R}^n$ with piecewise smooth boundary ∂G and Δ is the Laplacian in $\mathbb{R}^n$.

Assume that the following conditions are satisfied:

(H6.1-1) $a(t) \in C([0,\infty);[0,\infty))$, $\tau(t) \in C([0,\infty);\mathbb{R})$, $\lim\limits_{t\to\infty} \tau(t) = \infty$ and $b(x,t) \in C(\overline{G} \times [0,\infty);[0,\infty))$;

(H6.1-2) $c(x,t,\xi,(\xi_i)_{i=1}^{\tilde{m}}) \in C(\overline{\Omega} \times \mathbb{R} \times \mathbb{R}^{\tilde{m}};\mathbb{R})$,
$c(x,t,\xi,(\xi_i)_{i=1}^{\tilde{m}}) \geq p(x,t)\varphi(\xi)$ for $(x,t,\xi,(\xi_i)_{i=1}^{\tilde{m}}) \in \Omega \times [0,\infty)^{\tilde{m}+1}$,
$c(x,t,-\xi,(-\xi_i)_{i=1}^{\tilde{m}}) \leq -p(x,t)\varphi(\xi)$ for $(x,t,\xi,(\xi_i)_{i=1}^{\tilde{m}}) \in \Omega \times [0,\infty)^{\tilde{m}+1}$,
where $[0,\infty)^j = [0,\infty) \times [0,\infty)^{j-1}$ $(j = 1,2,...,\tilde{m}+1)$, $p(x,t) \in C(\overline{G} \times [0,\infty);[0,\infty))$, $\varphi(\xi) \in C([0,\infty);[0,\infty))$, $\varphi(\xi) > 0$ for $\xi > 0$ and $\varphi(\xi)$ is nondecreasing for $\xi > 0$;

(H6.1-3)

$$z[u](x,t) = \int_G K(x,t,\tilde{x})u(\tilde{x},\sigma(t))\,d\tilde{x},$$

where $\sigma(t) \in C([0,\infty);\mathbb{R})$, $\lim_{t\to\infty} \sigma(t) = \infty$ and $K(x,t,\tilde{x}) \in C(\overline{G} \times [0,\infty) \times \overline{G}; [0,\infty))$;

(H6.1-4)

$$z_i[u](x,t) = \begin{cases} u(x,\sigma_i(t)) & (i = 1,2,...,m), \\ \max_{s\in B_i(t)} u(x,s) & (i = m+1, m+2,..., m_1), \\ \displaystyle\sum_{j=1}^{N_i} \int_G K_{ij}(x,t,y)\omega_{ij}\big(u(y,\sigma_{ij}(t))\big)\,dy & \\ & (i = m_1+1, m_1+2,..., \tilde{m}), \end{cases}$$

where $\sigma_i(t) \in C([0,\infty);\mathbb{R})$ $(i = 1,2,...,m)$, $\lim_{t\to\infty} \sigma_i(t) = \infty$, $B_i(t)$ $(i = m+1, m+2,..., m_1)$ are closed bounded sets of $[0,\infty)$ such that $\lim_{t\to\infty} \min_{s\in B_i(t)} s = \infty$, $\sigma_{ij}(t) \in C([0,\infty);\mathbb{R})$ $(i = m_1+1, m_1+2,..., \tilde{m};\ j = 1,2,..., N_i)$, $\lim_{t\to\infty} \sigma_{ij}(t) = \infty$, $K_{ij}(x,t,y) \in C(\overline{\Omega} \times \overline{G}; [0,\infty))$, and $\omega_{ij}(s) \in C(\mathbb{R};\mathbb{R})$ are odd functions with the property that $\omega_{ij}(s) \geq 0$ for $s > 0$.

Definition 6.1.1. By a *solution* of Eq. (6.1.1) we mean a function $u(x,t) \in C^2(\overline{G}\times[t_{-1},\infty);\mathbb{R}) \cap C^1(\overline{G}\times[0,\infty);\mathbb{R}) \cap C(\overline{G}\times[\tilde{t}_{-1},\infty);\mathbb{R})$ which satisfies (6.1.1), where

$$t_{-1} = \inf_{t\geq 0} \tau(t),$$

$$\tilde{t}_{-1} = \min\left\{0,\ t_{-1},\ \inf_{t\geq 0}\sigma(t),\ \min_{1\leq i\leq m}\left\{\inf_{t\geq 0}\sigma_i(t)\right\},\ \min_{\substack{m_1+1\leq i\leq \tilde{m}\\ 1\leq j\leq N_i}}\left\{\inf_{t\geq 0}\sigma_{ij}(t)\right\}\right\}.$$

Definition 6.1.2. A real-valued function u on Ω is said to be *oscillatory* in Ω if u has a zero in $G \times (t,\infty)$ for any $t > 0$.

We recall that the first eigenvalue λ_1 of the eigenvalue problem (EVP)

$$-\Delta w = \lambda w \quad \text{in } G,$$

$$w = 0 \quad \text{on } \partial G$$

is positive and the corresponding eigenfunction $\Phi(x)$ may be chosen so that $\Phi(x) > 0$ in G (*cf.* Section 2.1 of Chapter 1.8).

Theorem 6.1.1. *Assume that the hypotheses* (H6.1-1)–(H6.1-4) *are satisfied, and assume, moreover, that:*

(i) $b(x,t) - \lambda_1 a(t) \geq 0$ *in* $G \times [T_1, \infty)$ *for some* $T_1 > 0$;

(ii) *for some* $T_2 > 0$ *there exists a function* $p(t) \in C([T_2, \infty); [0, \infty))$ *such that*

$$\int_G p(x,t)\Phi(x)\,dx \geq p(t), \quad t \geq T_2;$$

(iii) *there is a constant* $c_0 > 0$ *such that*

$$K(x,t,\tilde{x}) \geq c_0 \Phi(\tilde{x}) \quad \text{in } G \times [T_3, \infty) \times G$$

for some $T_3 > 0$.

If every eventually positive solution $y(t)$ *of the functional differential inequality*

$$y'(t) - p(t)\varphi(c_0 y(\sigma(t))) \geq 0 \tag{6.1.2}$$

satisfies $\lim_{t\to\infty} y(t) = \infty$, *then every solution* u *of* (6.1.1) *satisfies either*

(1) u *is oscillatory in* Ω

or

(2) $\lim_{t\to\infty} |U(t)| = \infty$, *where*

$$U(t) = \int_G u(x,t)\Phi(x)\,dx.$$

Proof. Suppose to the contrary that there exists a solution $u(x,t)$ of (6.1.1) which satisfies neither (1) nor (2). First we assume that $u(x,t) > 0$ in $G \times [t_0, \infty)$ for some $t_0 > 0$. The hypotheses (H6.1-1), (H6.1-3), (H6.1-4) imply that there is a constant $T > \max\{t_0, T_1, T_2, T_3\}$ for which $u(x,\tau(t)) > 0$, $u(x,\sigma(t)) > 0$, $z_i[u](x,t) \geq 0$ in $G \times [T, \infty)$ $(i = 1, 2, ..., \tilde{m})$. It is easy to see that

$$\begin{aligned} z[u](x,t) &= \int_G K(x,t,\tilde{x}) u(\tilde{x}, \sigma(t))\,d\tilde{x} \\ &\geq c_0 \int_G u(\tilde{x}, \sigma(t))\Phi(\tilde{x})\,d\tilde{x} \\ &= c_0 U(\sigma(t)) \geq 0, \quad t \geq T. \end{aligned} \tag{6.1.3}$$

Multiplying (6.1.1) by $\Phi(x)$ and then integrating over G, we obtain

$$\begin{aligned} U'(t) - a(t) \int_G \Delta u(x,\tau(t))\Phi(x)\,dx - \int_G b(x,t) u(x,\tau(t))\Phi(x)\,dx \\ - \int_G c(x,t,z[u](x,t),(z_i[u](x,t))_{i=1}^{\tilde{m}})\Phi(x)\,dx = 0, \quad t \geq 0. \end{aligned} \tag{6.1.4}$$

An application of Green's formula shows that

$$\int_G \Delta u(x,\tau(t))\Phi(x)\,dx = \int_{\partial G}\left[\frac{\partial u}{\partial \nu}(x,\tau(t))\Phi(x) - u(x,\tau(t))\frac{\partial \Phi(x)}{\partial \nu}\right] dS + \int_G u(x,\tau(t))\Delta\Phi(x)\,dx, \tag{6.1.5}$$

where ν denotes the unit exterior normal vector to ∂G. Since $u(x,\tau(t)) \geq 0$ on $\partial G \times [T,\infty)$ and $\frac{\partial\Phi(x)}{\partial\nu} \leq 0$ on ∂G, we observe that

$$u(x,\tau(t))\frac{\partial\Phi(x)}{\partial\nu} \leq 0 \quad \text{on } \partial G \times [T,\infty)$$

and therefore (6.1.5) implies that

$$\int_G \Delta u(x,\tau(t))\Phi(x)\,dx \geq -\lambda_1 \int_G u(x,\tau(t))\Phi(x)\,dx, \quad t \geq T. \tag{6.1.6}$$

Using the hypothesis (ii) and (6.1.3), we see from (H6.1-2) that

$$\begin{aligned} &\int_G c(x,t,z[u](x,t),(z_i[u](x,t))_{i=1}^{\tilde{m}})\Phi(x)\,dx \\ &\geq \int_G p(x,t)\varphi\big(z[u](x,t)\big)\Phi(x)\,dx \\ &\geq \int_G p(x,t)\varphi\big(c_0U(\sigma(t))\big)\Phi(x)\,dx \\ &\geq p(t)\varphi\big(c_0U(\sigma(t))\big), \quad t \geq T. \end{aligned} \tag{6.1.7}$$

Combining (6.1.4), (6.1.6) and (6.1.7), we obtain

$$U'(t)+\int_G (\lambda_1 a(t)-b(x,t))u(x,\tau(t))\Phi(x)\,dx-p(t)\varphi\big(c_0U(\sigma(t))\big) \geq 0, \quad t \geq T$$

and hence

$$U'(t) - p(t)\varphi\big(c_0U(\sigma(t))\big) \geq 0, \quad t \geq T$$

in view of the hypothesis (i). Since $U(t)$ is an eventually positive solution of (6.1.2), the hypothesis implies that $\lim_{t\to\infty} U(t) = \infty$. This contradicts the hypothesis. In the case where $u(x,t) < 0$ in $G \times [t_0,\infty)$ for some $t_0 > 0$, we are also led to a contradiction. The proof is complete. □

Lemma 6.1.1. *If*

$$\int_{t_0}^{\infty} p(t)\,dt = \infty \quad \textit{for some } t_0 > 0, \tag{6.1.8}$$

then every eventually positive solution $y(t)$ of (6.1.2) satisfies $\lim_{t\to\infty} y(t) = \infty$.

Proof. Let $y(t)$ be an eventually positive solution of (6.1.2). Then there exists a constant $T > 0$ such that $y(t) > 0$ and $y(\sigma(t)) > 0$ for $t \geq T$. Since

$$y'(t) \geq p(t)\varphi\big(c_0 y(\sigma(t))\big) \geq 0, \quad t \geq T,$$

we see that $y(t)$ is nondecreasing for $t \geq T$. In view of the fact that $y(\sigma(t)) \geq \beta$ $(t \geq \tilde{T})$ for some $\beta > 0$ and some $\tilde{T} \geq T$, we find that

$$\varphi\big(c_0 y(\sigma(t))\big) \geq \varphi(c_0\beta) > 0, \quad t \geq \tilde{T}.$$

Since

$$y'(t) \geq p(t)\varphi\big(c_0 y(\sigma(t))\big) \geq p(t)\varphi(c_0\beta), \quad t \geq \tilde{T},$$

an integration of the above inequality over $[\tilde{T}, t]$ yields

$$y(t) - y(\tilde{T}) \geq \varphi(c_0\beta) \int_{\tilde{T}}^{t} p(s)\, ds.$$

Letting $t \to \infty$, we conclude that $\lim_{t\to\infty} y(t) = \infty$. This completes the proof. □

Theorem 6.1.2. *Assume that the hypotheses* (H6.1-1)–(H6.1-4) *and* (i)–(iii) *of Theorem* 6.1.1 *hold. If* (6.1.8) *is satisfied, then the conclusion of Theorem* 6.1.1 *is valid.*

Proof. The conclusion follows from Theorem 6.1.1 and Lemma 6.1.1. □

Theorem 6.1.3. *Assume that the hypotheses* (H6.1-1)–(H6.1-4) *and* (i)–(iii) *of Theorem* 6.1.1 *hold. If* (6.1.2) *has no eventually positive solution, then every solution* u *of* (6.1.1) *is oscillatory in* Ω.

Proof. Let $u(x,t)$ be an eventually positive solution of (6.1.1). Then $u(x,t) > 0$ in $G \times [t_0, \infty)$ for some $t_0 > 0$. Proceeding as in the proof of Theorem 6.1.1, we observe that $U(t)$ is an eventually positive solution of (6.1.2). This contradicts the hypothesis. The case where $u(x,t) < 0$ in $G \times [t_0, \infty)$ can be handled similarly, and we are also led to a contradiction. The proof is complete. □

Lemma 6.1.2. *If*

$$\int_{A[\sigma]} p(t)\, dt = \infty, \tag{6.1.9}$$

$$\int_{\xi_0}^{\infty} \frac{1}{\varphi(\xi)}\, d\xi < \infty \quad \textit{for any } \xi_0 > 0, \tag{6.1.10}$$

then (6.1.2) *has no eventually positive solution, where* $A[\sigma]$ *denotes the advanced part of* $\sigma(t)$*, that is,*

$$A[\sigma] = \{t \in [0, \infty);\ t \leq \sigma(t)\}.$$

Proof. Suppose that $y(t)$ is an eventually positive solution of (6.1.2). Let $y(t) > 0$ and $y(\sigma(t)) > 0$ on $[T, \infty)$ for some $T > 0$. It follows from (6.1.2) that

$$\frac{y'(t)}{\varphi(c_0 y(t))} \geq p(t) \frac{\varphi(c_0 y(\sigma(t)))}{\varphi(c_0 y(t))}, \quad t \geq T. \tag{6.1.11}$$

Integrating (6.1.11) over $[T, t]$, we obtain

$$\int_T^t \frac{y'(s)}{\varphi(c_0 y(s))} ds \geq \int_T^t p(s) \frac{\varphi(c_0 y(\sigma(s)))}{\varphi(c_0 y(s))} ds, \quad t \geq T. \tag{6.1.12}$$

Since

$$y'(t) \geq p(t)\varphi(c_0 y(\sigma(t))) \geq 0, \quad t \geq T,$$

we see that $y(t)$ is nondecreasing for $t \geq T$. Hence we have

$$\int_T^t p(s) \frac{\varphi(c_0 y(\sigma(s)))}{\varphi(c_0 y(s))} ds \geq \int_{A[\sigma]\cap[T,t]} p(s)\, ds. \tag{6.1.13}$$

It is easily seen that

$$\begin{aligned} \int_T^t \frac{y'(s)}{\varphi(c_0 y(s))} ds &= \frac{1}{c_0} \int_{c_0 y(T)}^{c_0 y(t)} \frac{1}{\varphi(\xi)} d\xi \\ &\leq \frac{1}{c_0} \int_{c_0 y(T)}^{\infty} \frac{1}{\varphi(\xi)} d\xi < \infty. \end{aligned} \tag{6.1.14}$$

Combining (6.1.12)–(6.1.14) yields

$$\int_{A[\sigma]\cap[T,t]} p(s)\, ds \leq \frac{1}{c_0} \int_{c_0 y(T)}^{\infty} \frac{1}{\varphi(\xi)} d\xi < \infty$$

which contradicts the hypothesis (6.1.9). This completes the proof. □

Theorem 6.1.4. *Assume that the hypotheses* (H6.1-1)–(H6.1-4) *and* (i)–(iii) *of Theorem* 6.1.1 *hold. If* (6.1.9) *and* (6.1.10) *are satisfied, then every solution* u *of* (6.1.1) *is oscillatory in* Ω.

Proof. The conclusion follows by combining Theorem 6.1.3 with Lemma 6.1.2. □

Example 6.1.1. We consider the equation

$$\begin{aligned} &\frac{\partial u}{\partial t}(x,t) - \frac{\partial^2 u}{\partial x^2}(x, t - \log 2) - u(x, t - \log 2) \\ &- \frac{1}{4e} \int_0^{\pi} (\sin \tilde{x}) u(\tilde{x}, t+1)\, d\tilde{x} = 0, \quad (x,t) \in (0,\pi) \times (0,\infty). \end{aligned} \tag{6.1.15}$$

Here $n = 1$, $G = (0, \pi)$, $\Omega = (0, \pi) \times (0, \infty)$, $\tau(t) = t - \log 2$, $\sigma(t) = t + 1$, $a(t) = 1$, $b(x,t) = 1$, $p(x,t) = 1/(4e)$ and $\varphi(\xi) = \xi$. It is readily seen that $\lambda_1 = 1$ and $\Phi(x) = \sin x$. Since $b(x,t) - \lambda_1 a(t) = 0$,

$$\int_0^\pi p(x,t)\Phi(x)\,dx = \int_0^\pi \frac{1}{4e}\sin x\,dx = \frac{1}{2e}$$

and $K(x,t,\tilde{x}) = \sin\tilde{x} = \Phi(\tilde{x})$, we see that the hypotheses (i)–(iii) of Theorem 6.1.1 are satisfied with $p(t) = 1/(2e)$ and $c_0 = 1$. It is obvious that

$$\int_1^\infty p(t)\,dt = \int_1^\infty \frac{1}{2e}\,dt = \infty.$$

Hence, the hypotheses of Theorem 6.1.2 are satisfied, and therefore the conclusion of Theorem 6.1.2 is valid. In fact, $u(x,t) = e^t$ is a solution of (6.1.15) such that

$$\lim_{t\to\infty} |U(t)| = \lim_{t\to\infty} 2e^t = \infty.$$

Example 6.1.2. We consider the equation

$$\frac{\partial u}{\partial t}(x,t) - \frac{\partial^2 u}{\partial x^2}(x, t+\pi) - u(x,t+\pi)$$
$$-\frac{2}{\pi}(\sin x)\int_0^\pi (\sin\tilde{x})u(\tilde{x}, t+\tfrac{\pi}{2})\,d\tilde{x} = 0, \quad (x,t) \in (0,\pi)\times(0,\infty). \quad (6.1.16)$$

Here $n = 1$, $G = (0,\pi)$, $\Omega = (0,\pi)\times(0,\infty)$, $\tau(t) = t+\pi$, $\sigma(t) = t + (\pi/2)$, $a(t) = 1$, $b(x,t) = 1$, $p(x,t) = (2/\pi)\sin x$, $\varphi(\xi) = \xi$. Since $\lambda_1 = 1$ and $\Phi(x) = \sin x$, we observe that $b(x,t) - \lambda_1 a(t) = 0$,

$$\int_0^\pi p(x,t)\Phi(x)\,dx = \int_0^\pi \frac{2}{\pi}\sin^2 x\,dx = 1$$

and $K(x,t,\tilde{x}) = \sin\tilde{x} = \Phi(\tilde{x})$. It is easy to check that the hypotheses (i)–(iii) of Theorem 6.1.1 are satisfied with $p(t) = 1$ and $c_0 = 1$. Since

$$\int_1^\infty p(t)\,dt = \int_1^\infty dt = \infty,$$

the hypotheses of Theorem 6.1.2 are satisfied, and hence the conclusion of Theorem 6.1.2 is valid. In fact, $u(x,t) = (\sin x)\sin t$ is an oscillatory solution of (6.1.16).

Remark 6.1.1. In the case where $\tilde{m} = 2$, $m = 1$ and $\sigma(t) = t$, Theorems 6.1.1–6.1.4 were established by Bykov and Kultaev [40].

Remark 6.1.2. In Theorems 6.1.1 and 6.1.2, $\varphi(\xi)$ is only assumed to be a positive and nondecreasing function for $\xi > 0$. If $\varphi(\xi) = \xi^\gamma$ in Theorem 6.1.4, then (6.1.10) implies that $\gamma > 1$, that is, $\varphi(\xi)$ is a convex function.

6.2 Boundary Value Problems

In this section we deal with the boundary value problems for the parabolic equation with functional arguments

$$\frac{\partial u}{\partial t}(x,t) - a(t)\Delta u(x,t) + c(x,t,u(x,\sigma(t)),(z_i[u](x,t))_{i=1}^{\tilde{m}}) = 0, \quad (x,t) \in \Omega := G \times (0,\infty), \tag{6.2.1}$$

where G is a bounded domain in $\mathbb{R}^n$ with piecewise smooth boundary ∂G.

It is assumed that:

(H6.2-1) $a(t) \in C([0,\infty);[0,\infty))$, $\sigma(t) \in C([0,\infty);\mathbb{R})$ and $\lim_{t\to\infty} \sigma(t) = \infty$,

and that the hypotheses (H6.1-2), (H6.1-4) of Section 6.1 are satisfied.

Definition 6.2.1. By a *solution* of Eq. (6.2.1) we mean a function $u(x,t) \in C^2(\overline{G} \times [0,\infty);\mathbb{R}) \cap C(\overline{G} \times [\tilde{t}_{-1},\infty);\mathbb{R})$ which satisfies (6.2.1), where

$$\tilde{t}_{-1} = \min\left\{0,\ \inf_{t\geq 0} \sigma(t),\ \min_{1\leq i\leq m}\left\{\inf_{t\geq 0} \sigma_i(t)\right\},\ \min_{\substack{m_1+1\leq i\leq \tilde{m}\\ 1\leq j\leq N_i}}\left\{\inf_{t\geq 0} \sigma_{ij}(t)\right\}\right\}.$$

We consider two kinds of boundary conditions:

$$(\mathrm{B}_0) \qquad u = 0 \quad \text{on } \partial G \times (0,\infty),$$

$$(\tilde{\mathrm{B}}_0) \qquad \frac{\partial u}{\partial \nu} + \mu u = 0 \quad \text{on } \partial G \times (0,\infty),$$

where $\mu \in C(\partial G \times (0,\infty);[0,\infty))$ and ν denotes the unit exterior normal vector to ∂G.

The following notation will be used:

$$A(t) = \int_0^t a(s)\,ds,$$

$$p(t) = \min_{x\in\overline{G}} p(x,t).$$

Theorem 6.2.1. *Assume that the hypotheses* (H6.1-2), (H6.1-4) *and* (H6.2-1) *are satisfied, and assume, moreover, that:*

(i) $\varphi(\xi)$ *is convex in* $(0,\infty)$;

(ii) $\varphi(\xi_1\xi_2) \geq \varphi_1(\xi_1)\varphi_2(\xi_2)$ *for* $\xi_1 > 0$, $\xi_2 > 0$, *where* $\varphi_1(\xi_1) \geq 0$, $\varphi_2(\xi_2) > 0$ *and* $\varphi_2(\xi_2)$ *is nondecreasing for* $\xi_2 > 0$.

If every eventually positive solution $y(t)$ of the functional differential inequality

$$y'(t) + p(t)e^{\lambda_1 A(t)}\varphi_1\left(e^{-\lambda_1 A(\sigma(t))}\right)\varphi_2\big(y(\sigma(t))\big) \le 0 \tag{6.2.2}$$

satisfies $\lim_{t\to\infty} y(t) = 0$, *then every solution* u *of the boundary value problem* (6.2.1), (B_0) *is oscillatory in* Ω *or satisfies*

$$\lim_{t\to\infty} e^{\lambda_1 A(t)}U(t) = 0,$$

where

$$U(t) = \left(\int_G \Phi(x)\,dx\right)^{-1}\int_G u(x,t)\Phi(x)\,dx. \tag{6.2.3}$$

Proof. Suppose the contrary and let $u(x,t)$ be a nonoscillatory solution of the problem (6.2.1), (B_0) which does not satisfy $\lim_{t\to\infty} e^{\lambda_1 A(t)}U(t) = 0$. First we assume that $u(x,t) > 0$ in $G\times[t_0,\infty)$ for some $t_0 > 0$. There exists a constant $T > t_0$ such that $u(x,\sigma(t)) > 0$ and $z_i[u](x,t) \ge 0$ in $G\times[T,\infty)$ $(i = 1,2,...,\tilde{m})$. We multiply (6.2.1) by $\Phi(x)$ and then integrate over G to obtain

$$\frac{d}{dt}\int_G u(x,t)\Phi(x)\,dx - a(t)\int_G \Delta u(x,t)\Phi(x)\,dx$$
$$+ \int_G c(x,t,u(x,\sigma(t)),(z_i[u](x,t))_{i=1}^{\tilde{m}})\Phi(x)\,dx = 0. \tag{6.2.4}$$

It follows from Green's formula that

$$\int_G \Delta u(x,t)\Phi(x)\,dx = \int_G u(x,t)\Delta\Phi(x)\,dx = -\lambda_1\int_G u(x,t)\Phi(x)\,dx. \tag{6.2.5}$$

Using (H6.1-2) and Jensen's inequality [228, p.160], we have

$$\begin{aligned}
&\int_G c(x,t,u(x,\sigma(t)),(z_i[u](x,t))_{i=1}^{\tilde{m}})\Phi(x)\,dx\\
&\ge \int_G p(x,t)\varphi\big(u(x,\sigma(t))\big)\Phi(x)\,dx\\
&\ge p(t)\int_G \varphi\big(u(x,\sigma(t))\big)\Phi(x)\,dx\\
&\ge p(t)\left(\int_G \Phi(x)\,dx\right)\varphi\left(\left(\int_G \Phi(x)\,dx\right)^{-1}\int_G u(x,\sigma(t))\Phi(x)\,dx\right)
\end{aligned} \tag{6.2.6}$$

for $t \ge T$. Combining (6.2.4)–(6.2.6) yields

$$U'(t) + \lambda_1 a(t)U(t) + p(t)\varphi\big(U(\sigma(t))\big) \le 0, \quad t \ge T,$$

that is,

$$\left(e^{\lambda_1 A(t)}U(t)\right)' + p(t)e^{\lambda_1 A(t)}\varphi(U(\sigma(t))) \leq 0, \quad t \geq T. \tag{6.2.7}$$

Taking account of the hypothesis (ii), we find that

$$\begin{aligned}\varphi(U(\sigma(t))) &= \varphi\left(e^{-\lambda_1 A(\sigma(t))}e^{\lambda_1 A(\sigma(t))}U(\sigma(t))\right) \\ &\geq \varphi_1\left(e^{-\lambda_1 A(\sigma(t))}\right)\varphi_2\left(e^{\lambda_1 A(\sigma(t))}U(\sigma(t))\right), \quad t \geq T. \end{aligned} \tag{6.2.8}$$

We see from (6.2.7) and (6.2.8) that

$$\left(e^{\lambda_1 A(t)}U(t)\right)' + p(t)e^{\lambda_1 A(t)}\varphi_1\left(e^{-\lambda_1 A(\sigma(t))}\right)\varphi_2\left(e^{\lambda_1 A(\sigma(t))}U(\sigma(t))\right) \leq 0$$

for $t \geq T$, and therefore $e^{\lambda_1 A(t)}U(t)$ is an eventually positive solution of (6.2.2). Hence, the hypothesis implies that $\lim_{t\to\infty} e^{\lambda_1 A(t)}U(t) = 0$. This is a contradiction. The case where $u(x,t) < 0$ in $G \times [t_0, \infty)$ can be treated similarly, and we are also led to a contradiction. The proof is complete. □

We investigate the functional differential inequalities of the form

$$y'(t) + \tilde{p}(t)\varphi_2(y(\sigma(t))) \leq 0, \tag{6.2.9}$$

where $\tilde{p}(t) \in C([0,\infty); [0,\infty))$.

Lemma 6.2.1. *If*

$$\int_{R[\sigma]} \tilde{p}(t)\, dt = \infty, \tag{6.2.10}$$

then every eventually positive solution $y(t)$ *of* (6.2.9) *satisfies* $\lim_{t\to\infty} y(t) = 0$, *where* $R[\sigma]$ *denotes the retarded part of* $\sigma(t)$, *that is,*

$$R[\sigma] = \{t \in [0,\infty);\ \sigma(t) \leq t\}.$$

Proof. Let $y(t)$ be an eventually positive solution of (6.2.9) which does not satisfy $\lim_{t\to\infty} y(t) = 0$. Since $y'(t) \leq 0$ eventually, we find that $y(t)$ is nonincreasing. Hence, there exists $y(\infty) := \lim_{t\to\infty} y(t) \neq 0$ There exists a constant $T > 0$ such that $y(t) > 0$ and $y(\sigma(t)) > 0$ for $t \geq T$. Since

$$y'(t) \leq -\tilde{p}(t)\varphi_2(y(\sigma(t))) \leq 0, \quad t \geq T,$$

we observe that $y(t)$ is nonincreasing for $t \geq T$. Dividing (6.2.9) by $\varphi_2(y(t))$ and then integrating over $[T,t]$, we obtain

$$\int_T^t \tilde{p}(s)\frac{\varphi_2(y(\sigma(s)))}{\varphi_2(y(s))}\, ds \leq -\int_T^t \frac{y'(s)}{\varphi_2(y(s))}\, ds, \quad t \geq T. \tag{6.2.11}$$

Since $y(t)$ is nonincreasing and $\varphi_2(\xi)$ is nondecreasing, it can be shown that

$$\int_T^t \tilde{p}(s)\frac{\varphi_2\big(y(\sigma(s))\big)}{\varphi_2(y(s))}\,ds \ge \int_{R[\sigma]\cap[T,t]} \tilde{p}(s)\,ds. \tag{6.2.12}$$

It is easy to check that

$$-\int_T^t \frac{y'(s)}{\varphi_2(y(s))}\,ds = \int_{y(t)}^{y(T)} \frac{1}{\varphi_2(\xi)}\,d\xi \le \int_{y(\infty)}^{y(T)} \frac{1}{\varphi_2(\xi)}\,d\xi < \infty. \tag{6.2.13}$$

Combining (6.2.11)–(6.2.13) yields

$$\int_{R[\sigma]\cap[T,t]} \tilde{p}(s)\,ds \le \int_{y(\infty)}^{y(T)} \frac{1}{\varphi_2(\xi)}\,d\xi < \infty$$

which contradicts the hypothesis (6.2.10). This completes the proof. □

Theorem 6.2.2. *Assume that the hypotheses* (H6.1-2), (H6.1-4), (H6.2-1) *and* (i), (ii) *of Theorem* 6.2.1 *hold. If*

$$\int_{R[\sigma]} p(t)e^{\lambda_1 A(t)}\varphi_1\left(e^{-\lambda_1 A(\sigma(t))}\right)dt = \infty,$$

then the conclusion of Theorem 6.2.1 *is valid.*

Proof. The conclusion follows by combining Theorem 6.2.1 with Lemma 6.2.1. □

Theorem 6.2.3. *Assume that the hypotheses* (H6.1-2), (H6.1-4), (H6.2-1) *and* (i) *of Theorem* 6.2.1 *hold. If every eventually positive solution* $y(t)$ *of the functional differential inequality*

$$y'(t) + p(t)\varphi\big(y(\sigma(t))\big) \le 0 \tag{6.2.14}$$

satisfies $\lim_{t\to\infty} y(t) = 0$, *then every solution* u *of the boundary value problem* (6.2.1), $(\tilde{\mathrm{B}}_0)$ *is oscillatory in* Ω *or satisfies*

$$\lim_{t\to\infty} \tilde{U}(t) = 0, \tag{6.2.15}$$

where

$$\tilde{U}(t) = \frac{1}{|G|}\int_G u(x,t)\,dx \quad \left(|G| = \int_G dx\right). \tag{6.2.16}$$

Proof. Suppose that $u(x,t)$ is a nonoscillatory solution of the problem (6.2.1), $(\tilde{\mathrm{B}}_0)$ which does not satisfy (6.2.15). Without loss of generality we may assume that $u(x,t) > 0$ in $G \times [t_0,\infty)$ for some $t_0 > 0$. Then

$u(x,\sigma(t)) > 0$ and $z_i[u](x,t) \geq 0$ in $G \times [T,\infty)$ $(i = 1,2,...,\tilde{m})$ for some $T > t_0$. Integrating (6.2.1) over G, we obtain

$$\frac{d}{dt}\int_G u(x,t)\,dx - a(t)\int_G \Delta u(x,t)\,dx + \int_G c(x,t,u(x,\sigma(t)),(z_i[u](x,t))_{i=1}^{\tilde{m}})\,dx = 0. \qquad (6.2.17)$$

Using the divergence theorem [242, p.81] yields

$$\int_G \Delta u(x,t)\,dx = \int_{\partial G} \frac{\partial u}{\partial \nu}(x,t)\,dS. \qquad (6.2.18)$$

The hypothesis (H6.1-2) and Jensen's inequality imply that

$$\begin{aligned}
&\int_G c(x,t,u(x,\sigma(t)),(z_i[u](x,t))_{i=1}^{\tilde{m}})\,dx \\
&\geq \int_G p(x,t)\varphi(u(x,\sigma(t)))\,dx \\
&\geq p(t)\int_G \varphi(u(x,\sigma(t)))\,dx \\
&\geq p(t)|G|\varphi\left(\frac{1}{|G|}\int_G u(x,\sigma(t))\,dx\right), \quad t \geq T. \qquad (6.2.19)
\end{aligned}$$

Combining (6.2.17)–(6.2.19) yields

$$\begin{aligned}
\tilde{U}'(t) + p(t)\varphi(\tilde{U}(\sigma(t))) &\leq a(t)\frac{1}{|G|}\int_{\partial G}\frac{\partial u}{\partial \nu}(x,t)\,dS \\
&= -a(t)\frac{1}{|G|}\int_{\partial G} \mu \cdot u(x,t)\,dS \leq 0, \quad t \geq T
\end{aligned}$$

which shows that $\tilde{U}(t)$ is an eventually positive solution of (6.2.14) which does not satisfy $\lim_{t\to\infty}\tilde{U}(t) = 0$. This contradicts the hypothesis and completes the proof. □

Theorem 6.2.4. *Assume that the hypotheses* (H6.1-2), (H6.1-4), (H6.2-1) *and* (i) *of Theorem* 6.2.1 *are satisfied. If*

$$\int_{R[\sigma]} p(t)\,dt = \infty,$$

then the conclusion of Theorem 6.2.3 *is valid.*

Proof. The conclusion follows from Theorem 6.2.3 and Lemma 6.2.1. □

A special case of the problem (6.2.1), (B_0) is the following:

$$\frac{\partial u}{\partial t}(x,t) - \frac{\partial^2 u}{\partial x^2}(x,t) + p(t)\big(u(x,t-\sigma_0)\big)^\gamma = 0 \quad \text{in } (0,L)\times(0,\infty), \tag{6.2.20}$$

$$u(0,t) = u(L,t) = 0, \quad t > 0, \tag{6.2.21}$$

where σ_0, L are positive numbers and γ (≥ 1) is the quotient of odd integers.

Corollary 6.2.1. *If*

$$\int_{t_0}^{\infty} p(t) e^{(\pi/L)^2(1-\gamma)t}\, dt = \infty \quad \textit{for some } t_0 > 0, \tag{6.2.22}$$

then every solution u of the boundary value problem (6.2.20), (6.2.21) *is oscillatory in* $(0,L)\times(0,\infty)$ *or satisfies*

$$\lim_{t\to\infty} e^{(\pi/L)^2 t} \int_0^L u(x,t) \sin\frac{\pi}{L}x\, dx = 0. \tag{6.2.23}$$

Proof. In the case where $n = 1$ and $G = (0,L)$, we observe that $\lambda_1 = (\pi/L)^2$ and $\Phi(x) = \sin(\pi/L)x$. Since $\varphi(\xi) = \xi^\gamma$, we may choose $\varphi_1(\xi) = \varphi_2(\xi) = \xi^\gamma$. It is easy to verify that (6.2.22) implies

$$\int_{t_0}^{\infty} p(t) e^{(\pi/L)^2 t}\left(e^{-(\pi/L)^2(t-\sigma_0)}\right)^\gamma dt = \infty.$$

Hence, the conclusion follows from Theorem 6.2.2. □

Example 6.2.1. We consider the equation

$$\frac{\partial u}{\partial t}(x,t) - \frac{\partial^2 u}{\partial x^2}(x,t) + e^{-\sigma_0((\pi/L)^2+1)} u(x,t-\sigma_0) = 0, \quad (x,t)\in(0,L)\times(0,\infty). \tag{6.2.24}$$

Here $n = 1$, $G = (0,L)$, $\Omega = (0,L)\times(0,\infty)$, $a(t) = 1$, $\sigma(t) = t - \sigma_0$ and $\gamma = 1$. Since

$$\int_1^{\infty} e^{-\sigma_0((\pi/L)^2+1)}\, dt = \infty,$$

it follows from Corollary 6.2.1 that every nonoscillatory solution u of the problem (6.2.21), (6.2.24) satisfies (6.2.23). One such solution is

$$u(x,t) = \left(\sin\frac{\pi}{L}x\right) e^{-((\pi/L)^2+1)t}.$$

Example 6.2.2. We consider the problem

$$\frac{\partial u}{\partial t}(x,t) - \frac{\partial^2 u}{\partial x^2}(x,t) + e^{2t-3}\big(u(x,t-1)\big)^3 = 0 \quad \text{in } (0,L)\times(0,\infty), \tag{6.2.25}$$

$$-\frac{\partial u}{\partial x}(0,t) = \frac{\partial u}{\partial x}(L,t) = 0, \quad t > 0. \tag{6.2.26}$$

Here $n = 1$, $G = (0, L)$, $\Omega = (0, L) \times (0, \infty)$, $\gamma = 3$, $\sigma_0 = 1$ and $p(x, t) = p(t) = e^{2t-3}$. Since $\int_1^\infty e^{2t-3}\, dt = \infty$, Theorem 6.2.4 implies that every nonoscillatory solution u of (6.2.25), (6.2.26) satisfies

$$\lim_{t\to\infty} \frac{1}{L} \int_0^L u(x, t)\, dx = 0. \tag{6.2.27}$$

In fact, $u(x, t) = e^{-t}$ is such a solution.

Example 6.2.3. We consider the equation

$$\frac{\partial u}{\partial t}(x, t) - \frac{\partial^2 u}{\partial x^2}(x, t) + e^{-4} u(x, t-2) = 0, \quad (x, t) \in (0, \pi) \times (0, \infty). \tag{6.2.28}$$

Here $n = 1$, $G = (0, \pi)$, $\Omega = (0, \pi) \times (0, \infty)$, $a(t) = 1$, $L = \pi$, $\gamma = 1$ and $p(t) = e^{-4}$. Since $\int_2^\infty e^{-4}\, dt = \infty$, it follows from Theorem 6.2.4 that every solution u of the problem (6.2.26), (6.2.28) is oscillatory in $(0, \pi) \times (0, \infty)$, or satisfies

$$\lim_{t\to\infty} \frac{1}{\pi} \int_0^\pi u(x, t)\, dx = 0. \tag{6.2.29}$$

We conclude that there exists an oscillatory solution $u(x, t) = (\cos x)\, e^{-2t}$ of (6.2.26), (6.2.28) which satisfies (6.2.29).

Remark 6.2.1. In case $\sigma(t) \le t$ $(t \in [T, \infty))$ for some $T > 0$, it is readily verified that $R[\sigma] \supset [T, \infty)$.

Instead of (6.2.1) we deal with

$$\begin{aligned} &\frac{\partial u}{\partial t}(x, t) - a(t)\Delta u(x, t) \\ &\qquad - c(x, t, u(x, \sigma(t)), (z_i[u](x, t))_{i=1}^{\tilde{m}}) = 0, \quad (x, t) \in \Omega. \end{aligned} \tag{6.2.30}$$

Proceeding as in the proof of Theorems 6.2.1 and 6.2.3, we can establish the following two theorems.

Theorem 6.2.5. *Assume that the hypotheses* (H6.1-2), (H6.1-4), (H6.2-1) *and* (i), (ii) *of Theorem* 6.2.1 *hold. If*

$$y'(t) - p(t) e^{\lambda_1 A(t)} \varphi_1 \left(e^{-\lambda_1 A(\sigma(t))} \right) \varphi_2(y(\sigma(t))) \ge 0 \tag{6.2.31}$$

has no eventually positive solution, then every solution u of the boundary value problem (6.2.30), (B_0) *is oscillatory in* Ω.

Theorem 6.2.6. *Assume that the hypotheses* (H6.1-2), (H6.1-4), (H6.2-1) *and* (i) *of Theorem* 6.2.1 *hold. If*

$$y'(t) - p(t)\varphi\big(y(\sigma(t))\big) \geq 0 \tag{6.2.32}$$

has no eventually positive solution, then every solution u of Eq. (6.2.30) *satisfying the boundary condition*

$$\frac{\partial u}{\partial \nu} - \mu u = 0 \quad on \quad \partial G \times (0, \infty)$$

is oscillatory in Ω, where $\mu \in C(\partial G \times (0, \infty); [0, \infty))$.

By combining Theorems 6.2.5 and 6.2.6 with Lemma 6.1.2 we can obtain oscillation results for (6.2.30). For example, we have the following.

Theorem 6.2.7. *Assume that the hypotheses* (H6.1-2), (H6.1-4), (H6.2-1) *and* (i), (ii) *of Theorem* 6.2.1 *hold. Every solution u of the boundary value problem* (6.2.30), (B_0) *is oscillatory in Ω if*

$$\int_{A[\sigma]} p(t) e^{\lambda_1 A(t)} \varphi_1 \left(e^{-\lambda_1 A(\sigma(t))}\right) dt = \infty,$$

$$\int_{\xi_0}^{\infty} \frac{1}{\varphi_2(\xi)} d\xi < \infty \quad for\ any\ \xi_0 > 0,$$

where $A[\sigma]$ is the advanced part of $\sigma(t)$.

Applying Theorem 6.2.7 to the equation

$$\frac{\partial u}{\partial t}(x,t) - \frac{\partial^2 u}{\partial x^2}(x,t) - p(t)\big(u(x, t+\sigma_0)\big)^{\gamma} = 0, \quad (x,t) \in (0, L) \times (0, \infty), \tag{6.2.33}$$

where σ_0, L are positive numbers and γ (≥ 1) is the ratio of odd integers, we derive the following corollary.

Corollary 6.2.2. *If the condition* (6.2.22) *is satisfied, then every solution u of the boundary value problem* (6.2.21), (6.2.33) *is oscillatory in* $(0, L) \times (0, \infty)$.

Remark 6.2.2. Corollary 6.2.2 remains true if Eq. (6.2.33) is replaced by

$$\frac{\partial u}{\partial t}(x,t) - \frac{\partial^2 u}{\partial x^2}(x,t) - p_0(t) u(x,t) - p(t)\big(u(x, t+\sigma_0)\big)^{\gamma} = 0,$$

where $p_0(t) \in C([0, \infty); [0, \infty))$.

Example 6.2.4. It is known that Duffing's equation $y'' + 4y + 4y^3 = 0$ has a periodic solution $y(x)$. We find that the periodic solution $y(x)$ is oscillatory at $x = \infty$. There exists an interval (a, b) such that $y(a) = y(b) = 0$ and $y(x) > 0$ in (a, b). We consider the equation

$$\frac{\partial u}{\partial t}(x,t) - \frac{\partial^2 u}{\partial x^2}(x,t) - 2u(x,t) - e^{-2t-3}\big(u(x,t+1)\big)^3 = 0 \qquad (6.2.34)$$

for $(x,t) \in (0, L) \times (0, \infty)$, where $L = 2(b - a)$. It is readily seen that

$$2u(x,t) + e^{-2t-3}\big(u(x,t+1)\big)^3 \geq e^{-2t-3}\big(u(x,t+1)\big)^3 \quad \text{if } u(x,t) \geq 0.$$

Combining Corollary 6.2.2 with Remark 6.2.2, we conclude that every solution u of the problem (6.2.21), (6.2.34) is oscillatory in $(0, L) \times (0, \infty)$ if (6.2.22) is satisfied. Since

$$\int_0^\infty e^{-2t-3} \exp\left(-2\left(\frac{\pi}{L}\right)^2 t\right) dt < \infty,$$

the condition (6.2.22) is violated. In this case, there is a nonoscillatory solution $u(x,t) = y\big((x/2) + a\big)e^t$ of the problem (6.2.21), (6.2.34).

6.3 Forced Oscillation I

Oscillatory behavior of solutions of functional parabolic equations with forcing terms was studied by numerous authors, see, for example, Kusano and Yoshida [161], Mishev [203], Tanaka and Yoshida [258], Yoshida [302, 308, 311, 312].

We consider the parabolic equation with functional arguments and a forcing term

$$\begin{aligned} &\frac{\partial}{\partial t}\left(u(x,t) + \sum_{i=1}^{\ell} h_i(t)u(x,\rho_i(t))\right) \\ &\quad -a(t)\Delta u(x,t) - \sum_{i=1}^{k} b_i(t)\Delta u(x,\tau_i(t)) \\ &\quad +c(x,t,(z_i[u](x,t))_{i=1}^{\tilde{m}}) = f(x,t), \quad (x,t) \in \Omega := G \times (0,\infty), \quad (6.3.1) \end{aligned}$$

where G is a bounded domain in $\mathbb{R}^n$ with piecewise smooth boundary ∂G.

Assume that the hypothesis (H6.1-4) of Section 6.1 and the following hold:

(H6.3-1) $h_i(t) \in C^1([0,\infty);[0,\infty))$ $(i = 1, 2, ..., \ell)$, $a(t) \in C([0,\infty);[0,\infty))$, $b_i(t) \in C([0,\infty);[0,\infty))$ $(i = 1, 2, ..., k)$ and $f(x,t) \in C(\overline{\Omega};\mathbb{R})$;

(H6.3-2) $\rho_i(t) \in C^1([0,\infty);\mathbb{R})$, $\lim_{t\to\infty} \rho_i(t) = \infty$ $(i = 1, 2, ..., \ell)$,
$\tau_i(t) \in C([0,\infty);\mathbb{R})$, $\lim_{t\to\infty} \tau_i(t) = \infty$ $(i = 1, 2, ..., k)$;

(H6.3-3) $c(x, t, (\xi_i)_{i=1}^{\tilde{m}}) \in C(\overline{\Omega} \times \mathbb{R}^{\tilde{m}};\mathbb{R})$,

$$c(x, t, (\xi_i)_{i=1}^{\tilde{m}}) \geq \sum_{i=1}^{m} p_i(t)\varphi_i(\xi_i), \quad (x, t, (\xi_i)_{i=1}^{\tilde{m}}) \in \Omega \times [0,\infty)^{\tilde{m}},$$

$$c(x, t, (-\xi_i)_{i=1}^{\tilde{m}}) \leq -\sum_{i=1}^{m} p_i(t)\varphi_i(\xi_i), \quad (x, t, (\xi_i)_{i=1}^{\tilde{m}}) \in \Omega \times [0,\infty)^{\tilde{m}},$$

where $p_i(t) \in C([0,\infty);[0,\infty))$, $\varphi_i(\xi) \in C([0,\infty);[0,\infty))$, $\varphi_i(\xi)$ are convex in $(0,\infty)$ $(i = 1, 2, ..., m)$ and $[0,\infty)^j = [0,\infty) \times [0,\infty)^{j-1}$ $(j = 1, 2, ..., \tilde{m})$.

The boundary conditions to be considered are the following:

(B$_1$) $u = \psi$ on $\partial G \times (0,\infty)$,

(B$_2$) $\dfrac{\partial u}{\partial \nu} + \mu u = \tilde{\psi}$ on $\partial G \times (0,\infty)$,

where ψ, $\tilde{\psi} \in C(\partial G \times (0,\infty);\mathbb{R})$, $\mu \in C(\partial G \times (0,\infty);[0,\infty))$ and ν denotes the unit exterior normal vector to ∂G.

Definition 6.3.1. By a *solution* of Eq. (6.3.1) we mean a function $u(x,t) \in C^2(\overline{G} \times [t_{-1},\infty);\mathbb{R}) \cap C^1(\overline{G} \times [\hat{t}_{-1},\infty);\mathbb{R}) \cap C(\overline{G} \times [\tilde{t}_{-1},\infty);\mathbb{R})$ which satisfies (6.3.1), where

$$t_{-1} = \min\left\{0, \min_{1\leq i\leq k}\left\{\inf_{t\geq 0} \tau_i(t)\right\}\right\},$$

$$\hat{t}_{-1} = \min\left\{0, \min_{1\leq i\leq \ell}\left\{\inf_{t\geq 0} \rho_i(t)\right\}\right\},$$

$$\tilde{t}_{-1} = \min\left\{0, \min_{1\leq i\leq m}\left\{\inf_{t\geq 0} \sigma_i(t)\right\}, \min_{\substack{m_1+1\leq i\leq \tilde{m}\\ 1\leq j\leq N_i}}\left\{\inf_{t\geq 0} \sigma_{ij}(t)\right\}\right\}.$$

We use the following notation:

$$F(t) = \left(\int_G \Phi(x)\,dx\right)^{-1} \int_G f(x,t)\Phi(x)\,dx,$$

$$\Psi(t) = \left(\int_G \Phi(x)\,dx\right)^{-1} \int_{\partial G} \psi \frac{\partial \Phi(x)}{\partial \nu}\,dS,$$

$$\tilde{F}(t) = \frac{1}{|G|}\int_G f(x,t)\,dx,$$

$$\tilde{\Psi}(t) = \frac{1}{|G|}\int_{\partial G} \tilde{\psi}\,dS,$$

where $|G| = \int_G dx$ and $\Phi(x)$ is the eigenfunction of the eigenvalue problem (EVP) which appeared in Section 6.1.

Theorem 6.3.1. *Assume that the hypotheses* (H6.1-4), (H6.3-1)–(H6.3-3) *are satisfied. If the functional differential inequalities*

$$\frac{d}{dt}\left(y(t) + \sum_{i=1}^{\ell} h_i(t)y(\rho_i(t))\right) + \lambda_1 a(t)y(t)$$
$$+\lambda_1 \sum_{i=1}^{k} b_i(t)y(\tau_i(t)) + \sum_{i=1}^{m} p_i(t)\varphi_i\big(y(\sigma_i(t))\big) \leq \pm G(t) \quad (6.3.2)$$

have no eventually positive solutions, then every solution u of the boundary value problem (6.3.1), (B_1) *is oscillatory in Ω, where λ_1 is the first eigenvalue of the eigenvalue problem* (EVP) *appearing in Section* 6.1 *and*

$$G(t) = F(t) - a(t)\Psi(t) - \sum_{i=1}^{k} b_i(t)\Psi(\tau_i(t)). \quad (6.3.3)$$

Proof. Suppose to the contrary that there exists a solution $u(x,t)$ of the problem (6.3.1), (B_1) which is nonoscillatory in Ω. First we assume that $u(x,t) > 0$ in $G \times [t_0, \infty)$ for some $t_0 > 0$. The hypothesis (H6.3-3) implies that

$$c(x,t,(z_i[u](x,t))_{i=1}^{\tilde{m}}) \geq \sum_{i=1}^{m} p_i(t)\varphi_i\big(u(x,\sigma_i(t))\big) \quad \text{in } G \times [t_1, \infty)$$

for some $t_1 \geq t_0$. Hence, from (6.3.1) we see that

$$\frac{\partial}{\partial t}\left(u(x,t) + \sum_{i=1}^{\ell} h_i(t)u(x,\rho_i(t))\right)$$
$$-a(t)\Delta u(x,t) - \sum_{i=1}^{k} b_i(t)\Delta u(x,\tau_i(t))$$
$$+\sum_{i=1}^{m} p_i(t)\varphi_i\big(u(x,\sigma_i(t))\big) \leq f(x,t) \quad \text{in } G \times [t_1, \infty). \quad (6.3.4)$$

Multiplying (6.3.4) by $\left(\int_G \Phi(x)\,dx\right)^{-1}\Phi(x)$ and integrating over G, we have

$$\begin{aligned}
&\frac{d}{dt}\left(U(t)+\sum_{i=1}^{\ell} h_i(t)U(\rho_i(t))\right) - a(t)K_\Phi \int_G \Delta u(x,t)\Phi(x)\,dx \\
&\quad - \sum_{i=1}^{k} b_i(t)K_\Phi \int_G \Delta u(x,\tau_i(t))\Phi(x)\,dx \\
&\quad + \sum_{i=1}^{m} p_i(t)K_\Phi \int_G \varphi_i\big(u(x,\sigma_i(t))\big)\Phi(x)\,dx \\
&\quad \le F(t), \quad t \ge t_1, \qquad (6.3.5)
\end{aligned}$$

where $U(t)$ is defined by (6.2.3) and

$$K_\Phi = \left(\int_G \Phi(x)\,dx\right)^{-1}.$$

It follows from Green's formula that

$$\begin{aligned}
&K_\Phi \int_G \Delta u(x,t)\Phi(x)\,dx \\
&= K_\Phi \int_{\partial G}\left[\frac{\partial u}{\partial \nu}(x,t)\Phi(x) - u(x,t)\frac{\partial \Phi(x)}{\partial \nu}\right] dS + K_\Phi \int_G u(x,t)\Delta\Phi(x)\,dx \\
&= -K_\Phi \int_{\partial G} \psi\,\frac{\partial \Phi(x)}{\partial \nu}\,dS - \lambda_1 K_\Phi \int_G u(x,t)\Phi(x)\,dx \\
&= -\Psi(t) - \lambda_1 U(t), \quad t \ge t_1. \qquad (6.3.6)
\end{aligned}$$

Analogously we have

$$K_\Phi \int_G \Delta u(x,\tau_i(t))\Phi(x)\,dx = -\Psi(\tau_i(t)) - \lambda_1 U(\tau_i(t)), \quad t \ge t_2 \qquad (6.3.7)$$

for some $t_2 \ge t_1$. Applying Jensen's inequality, we obtain

$$K_\Phi \int_G \varphi_i\big(u(x,\sigma_i(t))\big)\Phi(x)\,dx \ge \varphi_i\big(U(\sigma_i(t))\big), \quad t \ge t_2. \qquad (6.3.8)$$

Combining (6.3.5)–(6.3.8) yields

$$\begin{aligned}
&\frac{d}{dt}\left(U(t)+\sum_{i=1}^{\ell} h_i(t)U(\rho_i(t))\right) + \lambda_1 a(t)U(t) \\
&\quad + \lambda_1 \sum_{i=1}^{k} b_i(t)U(\tau_i(t)) + \sum_{i=1}^{m} p_i(t)\varphi_i\big(U(\sigma_i(t))\big) \le G(t)
\end{aligned}$$

for $t \geq t_2$. Moreover, it is clear that $U(t) > 0$ on $[t_2, \infty)$. This contradicts the hypothesis. If $u(x,t) < 0$ in $G \times [t_0, \infty)$, it can be shown that

$$c(x,t,(z_i[u](x,t))_{i=1}^{\tilde{m}}) \leq -\sum_{i=1}^{m} p_i(t)\varphi_i(-u(x,\sigma_i(t))) \quad \text{in } G \times [t_1, \infty)$$

for some $t_1 \geq t_0$. Letting $v(x,t) = -u(x,t)$, we obtain

$$\frac{\partial}{\partial t}\left(v(x,t) + \sum_{i=1}^{\ell} h_i(t)v(x,\rho_i(t))\right)$$
$$-a(t)\Delta v(x,t) - \sum_{i=1}^{k} b_i(t)\Delta v(x,\tau_i(t))$$
$$+\sum_{i=1}^{m} p_i(t)\varphi_i(v(x,\sigma_i(t))) \leq -f(x,t) \quad \text{in } G \times [t_1, \infty).$$

Proceeding as in the case where $u(x,t) > 0$, we are led to a contradiction. The proof is complete. □

Theorem 6.3.2. *Assume that the hypotheses* (H6.1-4), (H6.3-1)–(H6.3-3) *are satisfied. If the functional differential inequalities*

$$\frac{d}{dt}\left(y(t) + \sum_{i=1}^{\ell} h_i(t)y(\rho_i(t))\right) + \sum_{i=1}^{m} p_i(t)\varphi_i(y(\sigma_i(t))) \leq \pm\tilde{G}(t) \quad (6.3.9)$$

have no eventually positive solutions, then every solution u of the boundary value problem (6.3.1), (B$_2$) *is oscillatory in* Ω, *where*

$$\tilde{G}(t) = \tilde{F}(t) + a(t)\tilde{\Psi}(t) + \sum_{i=1}^{k} b_i(t)\tilde{\Psi}(\tau_i(t)). \quad (6.3.10)$$

Proof. Assume on the contrary, that there exists a solution $u(x,t)$ of the problem (6.3.1), (B$_2$) such that $u(x,t) > 0$ in $G \times [t_0, \infty)$ for some $t_0 > 0$. Arguing as in the proof of Theorem 6.3.1, we observe that the inequality (6.3.4) holds for some $t_1 \geq t_0$. Dividing (6.3.4) by $|G|$ and then integrating over G, we obtain

$$\frac{d}{dt}\left(\tilde{U}(t) + \sum_{i=1}^{\ell} h_i(t)\tilde{U}(\rho_i(t))\right) - a(t)\frac{1}{|G|}\int_G \Delta u(x,t)\,dx$$
$$-\sum_{i=1}^{k} b_i(t)\frac{1}{|G|}\int_G \Delta u(x,\tau_i(t))\,dx + \sum_{i=1}^{m} p_i(t)\frac{1}{|G|}\int_G \varphi_i(u(x,\sigma_i(t)))\,dx$$
$$\leq \tilde{F}(t), \quad t \geq t_1, \quad (6.3.11)$$

where $\tilde{U}(t)$ is given by (6.2.16). It follows from the divergence theorem that

$$\begin{aligned}\frac{1}{|G|}\int_G \Delta u(x,t)\,dx &= \frac{1}{|G|}\int_{\partial G}\frac{\partial u}{\partial \nu}(x,t)\,dS\\ &= \frac{1}{|G|}\int_{\partial G}\left(-\mu\cdot u(x,t)+\tilde{\psi}\right)dS\\ &\le \tilde{\Psi}(t), \quad t\ge t_1. \end{aligned} \tag{6.3.12}$$

Analogously we obtain

$$\frac{1}{|G|}\int_G \Delta u(x,\tau_i(t))\,dx \le \tilde{\Psi}(\tau_i(t)), \quad t\ge t_2 \tag{6.3.13}$$

for some $t_2 \ge t_1$. An application of Jensen's inequality shows that

$$\frac{1}{|G|}\int_G \varphi_i(u(x,\sigma_i(t)))\,dx \ge \varphi_i\Big(\tilde{U}(\sigma_i(t))\Big), \quad t\ge t_2. \tag{6.3.14}$$

Combining (6.3.11)–(6.3.14) yields

$$\frac{d}{dt}\left(\tilde{U}(t)+\sum_{i=1}^{\ell} h_i(t)\tilde{U}(\rho_i(t))\right)+\sum_{i=1}^{m} p_i(t)\varphi_i\Big(\tilde{U}(\sigma_i(t))\Big)\le \tilde{G}(t)$$

for $t\ge t_2$, and furthermore $\tilde{U}(t)$ is positive on $[t_2,\infty)$. This contradicts the hypothesis. The case where $u(x,t)<0$ can be treated similarly, and we are led to a contradiction. The proof is complete. □

Theorem 6.3.3. *Assume that the hypotheses* (H6.1-4), (H6.3-1)–(H6.3-3) *are satisfied. Every solution* u *of the boundary value problem* (6.3.1), (B_1) *is oscillatory in* Ω *if*

$$\liminf_{t\to\infty}\int_T^t G(s)\,ds = -\infty,$$
$$\limsup_{t\to\infty}\int_T^t G(s)\,ds = \infty$$

for all large T. *Every solution* u *of the boundary value problem* (6.3.1), (B_2) *is oscillatory in* Ω *if*

$$\liminf_{t\to\infty}\int_T^t \tilde{G}(s)\,ds = -\infty,$$
$$\limsup_{t\to\infty}\int_T^t \tilde{G}(s)\,ds = \infty$$

for all large T.

Proof. Suppose that $y_\pm(t)$ are eventually positive solutions of the functional differential inequalities (6.3.2), respectively. Then, $y_\pm(t)$ are also eventually positive solutions of

$$\frac{d}{dt}\left(y(t)+\sum_{i=1}^{\ell}h_i(t)y(\rho_i(t))\right)\le \pm G(t)$$

and therefore $z_\pm(t) := y_\pm(t) + \sum_{i=1}^{\ell} h_i(t)y_\pm(\rho_i(t))$ satisfy

$$z_\pm{}'(t) \le \pm G(t).$$

Integrating the above inequality over $[T, t]$, we have

$$z_\pm(t) - z_\pm(T) \le \pm\int_T^t G(s)\,ds, \quad t \ge T$$

for large $T > 0$, which contradicts the hypothesis in view of the fact that $z_\pm(t) > 0$ eventually. Hence, (6.3.2) have no eventually positive solutions. The conclusion follows from Theorem 6.3.1. The boundary value problem (6.3.1), (B_2) can be treated similarly. □

Now we investigate the functional differential inequality of neutral type

$$\frac{d}{dt}\left(y(t)+\sum_{i=1}^{\ell}h_i(t)y(\rho_i(t))\right)+\sum_{i=1}^{m}p_i(t)\varphi_i\big(y(\sigma_i(t))\big)\le q(t), \qquad (6.3.15)$$

where $q(t) \in C([t_0,\infty);\mathbb{R})$ for some $t_0 > 0$.

Lemma 6.3.1. *Assume that:*

(i) $\sum_{i=1}^{\ell} h_i(t) \le 1$ *and* $\rho_i(t) \ge t$ $(i = 1, 2, ..., \ell)$;
(ii) $\varphi_j(\xi)$ *is nondecreasing in* $(0,\infty)$ *for some* $j \in \{1, 2, ..., m\}$;
(iii) *there exists a function* $Q(t) \in C^1([t_0,\infty);\mathbb{R})$ *such that* $Q(t)$ *is oscillatory at* $t = \infty$ *and* $Q'(t) \ge q(t)$.

If

$$\int_{t_0}^{\infty} p_j(t)\varphi_j\left(\left[\left(1-\sum_{i=1}^{\ell}h_i(\sigma_j(t))\right)[Q(\sigma_j(t))]_- + H(\sigma_j(t))\right]_+\right)dt=\infty, \qquad (6.3.16)$$

then (6.3.15) *has no eventually positive solution, where*

$$[Q(t)]_\pm = \max\{\pm Q(t),\ 0\},$$

$$H(t) = Q(t) - \sum_{i=1}^{\ell} h_i(t)Q(\rho_i(t)).$$

Proof. Let $y(t)$ be an eventually positive solution of (6.3.15). The hypothesis (H6.3-2) implies that there is a number $t_1 \geq t_0$ such that $y(t) > 0$, $y(\rho_i(t)) > 0$ $(i = 1, 2, ..., \ell)$ and $y(\sigma_i(t)) > 0$ $(i = 1, 2, ..., m)$ for $t \geq t_1$. We let

$$z(t) = y(t) + \sum_{i=1}^{\ell} h_i(t) y(\rho_i(t)) - Q(t).$$

It follows from (6.3.15) and the hypothesis (ii) that

$$z'(t) \leq -p_j(t)\varphi_j\big(y(\sigma_j(t))\big) \leq 0, \quad t \geq t_1. \tag{6.3.17}$$

Therefore, $z(t) > 0$ or $z(t) \leq 0$ on $[t_2, \infty)$ for some $t_2 \geq t_1$. If $z(t) \leq 0$ on $[t_2, \infty)$, then

$$y(t) + \sum_{i=1}^{\ell} h_i(t) y(\rho_i(t)) \leq Q(t), \quad t \geq t_2. \tag{6.3.18}$$

The left hand side of (6.3.18) is positive, whereas the right hand side of (6.3.18) is oscillatory at $t = \infty$. This is a contradiction. Hence, we conclude that $z(t) > 0$ on $[t_2, \infty)$. Since $z(t) + Q(t) > 0$ on $[t_2, \infty)$, we find that $z(t) > -Q(t)$ on $[t_2, \infty)$, and therefore

$$z(t) \geq [Q(t)]_-, \quad t \geq t_2.$$

In view of the fact that $y(t) \leq z(t) + Q(t)$ and $z(t)$ is nonincreasing, we obtain

$$\begin{aligned} y(t) &= z(t) - \sum_{i=1}^{\ell} h_i(t) y(\rho_i(t)) + Q(t) \\ &\geq z(t) - \sum_{i=1}^{\ell} h_i(t)\big(z(\rho_i(t)) + Q(\rho_i(t))\big) + Q(t) \\ &\geq \left(1 - \sum_{i=1}^{\ell} h_i(t)\right) z(t) + H(t), \quad t \geq t_2. \end{aligned} \tag{6.3.19}$$

Since $z(t) \geq [Q(t)]_-$ and $y(t) > 0$ for $t \geq t_2$, we have

$$y(t) \geq \left[\left(1 - \sum_{i=1}^{\ell} h_i(t)\right)[Q(t)]_- + H(t)\right]_+, \quad t \geq t_2. \tag{6.3.20}$$

Since $\varphi_j(\xi)$ is nondecreasing, we see from (6.3.17) and (6.3.20) that

$$z'(t) + p_j(t)\varphi_j\left(\left[\left(1 - \sum_{i=1}^{\ell} h_i(\sigma_j(t))\right)[Q(\sigma_j(t))]_- + H(\sigma_j(t))\right]_+\right) \leq 0 \tag{6.3.21}$$

for $t \geq t_3$. Integrating (6.3.21) over $[t_3, t]$ yields

$$\int_{t_3}^{t} p_j(s)\varphi_j\left(\left[\left(1-\sum_{i=1}^{\ell} h_i(\sigma_j(s))\right)[Q(\sigma_j(s))]_- + H(\sigma_j(s))\right]_+\right) ds$$
$$\leq -z(t) + z(t_3) \leq z(t_3), \quad t \geq t_3.$$

This contradicts the hypothesis (6.3.16) and completes the proof. □

Theorem 6.3.4. *Assume that the hypotheses* (H6.1-4), (H6.3-1)–(H6.3-3), (i), (ii) *of Lemma* 6.3.1 *are satisfied. Every solution u of the boundary value problem* (6.3.1), (B_1) [*resp.* (6.3.1), (B_2)] *is oscillatory in Ω if there exists a function $\Theta(t) \in C^1((0,\infty);\mathbb{R})$ such that $\Theta(t)$ is oscillatory at $t = \infty$, $\Theta'(t) = G(t)$* [*resp.* $\Theta'(t) = \tilde{G}(t)$], *and*

$$\int_{t_0}^{\infty} p_j(t)\varphi_j\left(\left[\left(1-\sum_{i=1}^{\ell} h_i(\sigma_j(t))\right)[\Theta(\sigma_j(t))]_{\mp} \pm\left(\Theta(\sigma_j(t)) - \sum_{i=1}^{\ell} h_i(\sigma_j(t))\Theta\big(\rho_i(\sigma_j(t))\big)\right)\right]_+\right) dt = \infty$$

for some $t_0 > 0$.

Proof. We limit ourselves to the boundary value problem (6.3.1), (B_1). Lemma 6.3.1 implies that

$$\frac{d}{dt}\left(y(t) + \sum_{i=1}^{\ell} h_i(t) y(\rho_i(t))\right) + \sum_{i=1}^{m} p_i(t)\varphi_i\big(y(\sigma_i(t))\big) \leq \pm G(t)$$

have no eventually positive solutions, and hence (6.3.2) have no eventually positive solutions. The conclusion follows from Theorem 6.3.1. □

Theorem 6.3.5. *Assume that the hypotheses* (H6.1-4), (H6.3-1)–(H6.3-3), (i) *of Lemma* 6.3.1 *are satisfied. Assume, moreover, that for some $j \in \{1, 2, ..., m\}$ and some $\beta > 0$*

$$\varphi_j(\xi) \geq \beta\xi \quad \text{for } \xi > 0. \tag{6.3.22}$$

Every solution u of the boundary value problem (6.3.1), (B_1) [*resp.* (6.3.1), (B_2)] *is oscillatory in Ω if there exists a function $\Theta(t) \in C^1((0,\infty);\mathbb{R})$ such that $\Theta(t)$ is oscillatory at $t = \infty$, $\Theta'(t) = G(t)$* [*resp.* $\Theta'(t) = \tilde{G}(t)$], *and the functional differential inequalities*

$$y'(t) + \beta p_j(t)\left(1 - \sum_{i=1}^{\ell} h_i(\sigma_j(t))\right) y(\sigma_j(t)) \leq \mp\tilde{\Theta}(t) \tag{6.3.23}$$

have no eventually positive solutions, where

$$\tilde{\Theta}(t) = \beta p_j(t) \left(\Theta(\sigma_j(t)) - \sum_{i=1}^{\ell} h_i(\sigma_j(t))\Theta(\rho_i(\sigma_j(t))) \right). \tag{6.3.24}$$

Proof. It is sufficient to show that the functional differential inequality

$$\frac{d}{dt}\left(y(t) + \sum_{i=1}^{\ell} h_i(t)y(\rho_i(t)) \right) + \beta p_j(t)y(\sigma_j(t)) \leq \pm G(t) \tag{6.3.25}$$

have no eventually positive solutions. Let $y(t)$ be an eventually positive solution of

$$\frac{d}{dt}\left(y(t) + \sum_{i=1}^{\ell} h_i(t)y(\rho_i(t)) \right) + \beta p_j(t)y(\sigma_j(t)) \leq G(t). \tag{6.3.26}$$

Arguing as in the proof of Lemma 6.3.1, we see that

$$z(t) := y(t) + \sum_{i=1}^{\ell} h_i(t)y(\rho_i(t)) - \Theta(t) > 0,$$

$$z'(t) \leq -p_j(t)\varphi_j\big(y(\sigma_j(t))\big) \leq -\beta p_j(t)y(\sigma_j(t)), \tag{6.3.27}$$

$$y(\sigma_j(t)) \geq \left(1 - \sum_{i=1}^{\ell} h_i(\sigma_j(t)) \right) z(\sigma_j(t)) + H_\Theta(\sigma_j(t)) \tag{6.3.28}$$

for $t \geq t_2$, where t_2 is a positive number and

$$H_\Theta(t) = \Theta(t) - \sum_{i=1}^{\ell} h_i(t)\Theta(\rho_i(t)).$$

Combining (6.3.27) with (6.3.28) yields

$$z'(t) + \beta p_j(t) \left(1 - \sum_{i=1}^{\ell} h_i(\sigma_j(t)) \right) z(\sigma_j(t)) \leq -\beta p_j(t) H_\Theta(\sigma_j(t)) \tag{6.3.29}$$

for $t \geq t_2$. Hence, $z(t)$ is an eventually positive solution of (6.3.29). Consequently, if (6.3.29) has no eventually positive solution, then (6.3.26) has no eventually positive solution. The case where (6.3.26) has the forcing term $G(t)$ replaced by $-G(t)$ can be handled similarly. We conclude that the hypothesis implies that (6.3.25) have no eventually positive solutions, and therefore (6.3.2) have no eventually positive solutions. The conclusion follows from Theorem 6.3.1. □

Theorem 6.3.6. *Assume that the hypotheses* (H6.1-4), (H6.3-1)–(H6.3-3), (i) *of Lemma* 6.3.1 *and* (6.3.22) *are satisfied. Assume, moreover, that* $\sigma_j(t) \le t$ *and* $\sigma_j(t)$ *is nondecreasing in* $(0, \infty)$. *Every solution* u *of the boundary value problem* (6.3.1), (B_1) [*resp.* (6.3.1), (B_2)] *is oscillatory in* Ω *if there exists a function* $\Theta(t) \in C^1((0,\infty);\mathbb{R})$ *such that* $\Theta(t)$ *is oscillatory at* $t = \infty$, $\Theta'(t) = G(t)$ [*resp.* $\Theta'(t) = \tilde{G}(t)$], *and there is a sequence* $\{t_n\}_{n=1}^{\infty}$ *such that:*

$$\lim_{n\to\infty} t_n = \infty,$$

$$\int_{\sigma_j(t_n)}^{t_n} p_j(t)\left(1 - \sum_{i=1}^{\ell} h_i(\sigma_j(t))\right) dt \ge \frac{1}{\beta},$$

$$\int_{\sigma_j(t_n)}^{t_n} \tilde{\Theta}(t)\, dt + \beta \int_{\sigma_j(t_n)}^{t_n} p_j(t)\left(1 - \sum_{i=1}^{\ell} h_i(\sigma_j(t))\right)\left(\int_{\sigma_j(t)}^{\sigma_j(t_n)} \tilde{\Theta}(s)\, ds\right) dt = 0,$$

where $\tilde{\Theta}(t)$ *is given by* (6.3.24).

Proof. It follows from a result of Yoshida [311, Proposition 1] that (6.3.23) have no eventually positive solutions. The conclusion follows from Theorem 6.3.5. □

Example 6.3.1. We consider the problem

$$\frac{\partial}{\partial t}\left(u(x,t) + \frac{1}{2}\, u(x, t+2\pi)\right) - \frac{\partial^2 u}{\partial x^2}(x,t) - \frac{\partial^2 u}{\partial x^2}(x, t - \tfrac{3}{2}\pi)$$
$$+ \frac{5}{2}\, u(x, t - \tfrac{\pi}{2}) = (\sin x)\sin t, \quad (x,t) \in (0,\pi) \times (0,\infty), \tag{6.3.30}$$
$$u(0,t) = u(\pi,t) = 0, \quad t > 0. \tag{6.3.31}$$

Here $n = 1$, $G = (0,\pi)$, $\Omega = (0,\pi)\times(0,\infty)$, $\ell = k = m = \tilde{m} = 1$, $h_1(t) = 1/2$, $\rho_1(t) = t + 2\pi$, $a(t) = 1$, $b_1(t) = 1$, $\tau_1(t) = t - (3/2)\pi$, $p_1(t) = 5/2$, $\sigma_1(t) = t - (1/2)\pi$, $\varphi_1(\xi) = \xi$ and $f(x,t) = (\sin x)\sin t$. It is readily seen that $\lambda_1 = 1$, $\Phi(x) = \sin x$, $\Psi(t) = 0$ and $F(t) = G(t) = (1/4)\pi \sin t$. We can choose $\Theta(t) = -(1/4)\pi\cos t$. It is easy to check that $\Theta(\sigma_1(t)) = -(1/4)\pi\sin t$, $\Theta\big(\rho_1(\sigma_1(t))\big) = -(1/4)\pi \sin t$, and that

$$\int_1^{\infty} \frac{5}{2}\left[\left(1 - \frac{1}{2}\right)\left[-\frac{1}{4}\pi\sin t\right]_{\mp} \pm \left(-\frac{1}{4}\pi \sin t - \frac{1}{2}\left(-\frac{1}{4}\pi\sin t\right)\right)\right]_{+} dt$$
$$= \frac{5}{16}\pi \int_1^{\infty} \left[[-\sin t]_{\mp} \mp \sin t\right]_{+} dt$$
$$= \frac{5}{16}\pi \int_1^{\infty} [\sin t]_{\mp}\, dt = \infty.$$

Hence, Theorem 6.3.4 implies that every solution u of the problem (6.3.30), (6.3.31) is oscillatory in $(0,\pi)\times(0,\infty)$. For example, $u(x,t)=(\sin x)\sin t$ is such a solution.

Example 6.3.2. We consider the problem

$$\frac{\partial}{\partial t}\left(u(x,t)+\frac{1}{2}\,u(x,t+\pi)\right)-\frac{\partial^2 u}{\partial x^2}(x,t)-\frac{\partial^2 u}{\partial x^2}(x,t+\pi)$$
$$+u(x,t+\tfrac{\pi}{2})=3(\cos x+1)\cos t,\quad (x,t)\in(0,\pi)\times(0,\infty),\qquad (6.3.32)$$
$$-\frac{\partial u}{\partial x}(0,t)=\frac{\partial u}{\partial x}(\pi,t)=0,\quad t>0.\qquad (6.3.33)$$

Here $n=1$, $G=(0,\pi)$, $\Omega=(0,\pi)\times(0,\infty)$, $\ell=k=m=\tilde{m}=1$, $h_1(t)=1/2$, $\rho_1(t)=t+\pi$, $a(t)=1$, $b_1(t)=1$, $\tau_1(t)=t+\pi$, $p_1(t)=1$, $\sigma_1(t)=t+(\pi/2)$, $\varphi_1(\xi)=\xi$ and $f(x,t)=3(\cos x+1)\cos t$. Since $\tilde{\Psi}(t)=0$, we easily see that

$$\tilde{F}(t)=\tilde{G}(t)=\frac{1}{\pi}\int_0^{\pi}3(\cos x+1)\cos t\,dx=3\cos t.$$

Choosing $\Theta(t)=3\sin t$, we observe that $\Theta(\sigma_1(t))=3\cos t$, $\Theta(\rho_1(\sigma_1(t)))=-3\cos t$, and that

$$\int_1^{\infty}\left[\left(1-\frac{1}{2}\right)[3\cos t]_{\mp}\pm\left(3\cos t-\frac{1}{2}\,(-3\cos t)\right)\right]_+dt$$
$$=\frac{3}{2}\int_1^{\infty}\big[[\cos t]_{\mp}\mp 3\cos t\big]_+dt$$
$$=\frac{3}{2}\int_1^{\infty}[3\cos t]_{\pm}\,dt=\infty.$$

Therefore, it follows from Theorem 6.3.4 that every solution u of the problem (6.3.32), (6.3.33) is oscillatory in $(0,\pi)\times(0,\infty)$. One such solution is $u(x,t)=2(\cos x+1)\sin t$.

Remark 6.3.1. Theorem 6.3.1 remains true if (6.3.2) are replaced by

$$\frac{d}{dt}\left(y(t)+\sum_{i=1}^{\ell}h_i(t)y(\rho_i(t))\right)+\lambda_1 a(t)y(t)\le\pm G(t)$$

or

$$\frac{d}{dt}\left(y(t)+\sum_{i=1}^{\ell}h_i(t)y(\rho_i(t))\right)+\lambda_1\sum_{i=1}^{k}b_i(t)y(\tau_i(t))\le\pm G(t).$$

Remark 6.3.2. In the case where $p_j(t) > 0$ for some $j \in \{1, 2, ..., m\}$, the hypothesis (H6.3-3) means that $\varphi_j(0) = 0$. Then we cannot apply Theorem 6.3.4 to the case where $f(x,t) = 0$ and $\psi = 0$ (or $\tilde{\psi} = 0$). However, Theorem 6.3.6 can be applied to this case.

Remark 6.3.3. When $\ell = k = m = 1$ and $m_1 = \tilde{m} = 2$, Theorem 6.3.3 was established by Mishev [203]. The case where $m = \tilde{m},\ h_i(t) = 0\ (i = 1, 2, ..., \ell),\ p_i(t) = 0\ (i = 1, 2, ..., m)$ was treated by Yoshida [308]. In case $m = \tilde{m},\ \rho_i(t) = t - \rho_i\ (\rho_i > 0)\ (i = 1, 2, ..., \ell),\ \varphi_i(\xi) = \xi\ (i = 1, 2, ..., m)$ and $f(x,t) = 0$, Theorem 6.3.3 reduces to a result of Cui [56].

Remark 6.3.4. Oscillations of (6.3.1) with $c(x, t, (z_i[u](x,t))_{i=1}^{\tilde{m}})$ replaced by $-c(x, t, (z_i[u](x,t))_{i=1}^{\tilde{m}})$ were studied by Kusano and Yoshida [161] (see also Yoshida [300, 312]).

6.4 Forced Oscillation II

In this section we consider the case where the coefficients $h_i(t)$ of Eq. (6.3.1) of Section 6.3 are nonpositive, that is, we deal with the parabolic equation with functional arguments

$$\begin{aligned}
&\frac{\partial}{\partial t}\left(u(x,t) - \sum_{i=1}^{\ell} h_i(t) u(x, \rho_i(t))\right) \\
&-a(t)\Delta u(x,t) - \sum_{i=1}^{k} b_i(t) \Delta u(x, \tau_i(t)) \\
&+c(x, t, (z_i[u](x,t))_{i=1}^{\tilde{m}}) = f(x,t), \quad (x,t) \in \Omega := G \times (0, \infty), \quad (6.4.1)
\end{aligned}$$

where $h_i(t) \geq 0$ and G is a bounded domain in $\mathbb{R}^n$ with piecewise smooth boundary ∂G.

It is assumed that the hypotheses (H6.1-4) of Section 6.1 and (H6.3-1)–(H6.3-3) of Section 6.3 are satisfied.

We study the boundary value problems (6.4.1), (B_i) $(i = 1, 2)$, where the boundary conditions (B_i) are given in Section 6.3.

By a *solution* of Eq. (6.4.1) we mean a function $u(x,t) \in C^2(\overline{G} \times [t_{-1}, \infty); \mathbb{R}) \cap C^1(\overline{G} \times [\hat{t}_{-1}, \infty); \mathbb{R}) \cap C(\overline{G} \times [\tilde{t}_{-1}, \infty); \mathbb{R})$ which satisfies (6.4.1), where $t_{-1},\ \hat{t}_{-1},\ \tilde{t}_{-1}$ are the same as those given in Definition 6.3.1 of Section 6.3.

We note that the functions $F(t),\ \Psi(t),\ \tilde{F}(t),\ \tilde{\Psi}(t)$ are those given in Section 6.3.

Theorem 6.4.1. *Assume that the hypotheses* (H6.1-4), (H6.3-1)–(H6.3-3) *hold. If the functional differential inequalities*

$$\frac{d}{dt}\left(y(t)-\sum_{i=1}^{\ell}h_i(t)y(\rho_i(t))\right)+\lambda_1 a(t)y(t)$$
$$+\lambda_1\sum_{i=1}^{k}b_i(t)y(\tau_i(t))+\sum_{i=1}^{m}p_i(t)\varphi_i\big(y(\sigma_i(t))\big)\le \pm G(t) \qquad (6.4.2)$$

have no eventually positive solutions, then every solution u of the boundary value problem (6.4.1), (B_1) *is oscillatory in* Ω, *where* $G(t)$ *is the function defined by* (6.3.3).

Proof. The conclusion follows by employing the same arguments as were used in the proof of Theorem 6.3.1. □

Theorem 6.4.2. *Assume that the hypotheses* (H6.1-4), (H6.3-1)–(H6.3-3) *are satisfied. If the functional differential inequalities*

$$\frac{d}{dt}\left(y(t)-\sum_{i=1}^{\ell}h_i(t)y(\rho_i(t))\right)+\sum_{i=1}^{m}p_i(t)\varphi_i\big(y(\sigma_i(t))\big)\le \pm\tilde{G}(t) \qquad (6.4.3)$$

have no eventually positive solutions, then every solution u of the boundary value problem (6.4.1), (B_2) *is oscillatory in* Ω, *where the function* $\tilde{G}(t)$ *is given by* (6.3.10).

Proof. The theorem follows by the same arguments as in the proof of Theorem 6.3.2. □

We derive sufficient conditions for the functional differential inequality

$$\frac{d}{dt}\left(y(t)-\sum_{i=1}^{\ell}h_i(t)y(\rho_i(t))\right)+\sum_{i=1}^{m}p_i(t)\varphi_i\big(y(\sigma_i(t))\big)\le q(t) \qquad (6.4.4)$$

to have no eventually positive solution, where $q(t)\in C([t_0,\infty);\mathbb{R})$ for some $t_0>0$.

Lemma 6.4.1. *Assume that:*

(i) $\sum_{i=1}^{\ell}h_i(t)\le 1$ *and* $\rho_i(t)\le t$ $(i=1,2,...,\ell)$;

(ii) *there is an integer* $j\in\{1,2,...,m\}$ *such that* $\varphi_j(\xi)$ *is nondecreasing,* $\varphi_j(\xi)>0$ *for* $\xi>0$, $p_j(t)>0$ *on* $[t_0,\infty)$ *and*

$$\int_{t_0}^{\infty}p_j(t)\,dt=\infty;$$

(iii) *there exists a function* $Q(t) \in C^1([t_0,\infty);\mathbb{R})$ *satisfying* $Q'(t) \ge q(t)$ *and* $\lim_{t\to\infty} Q(t) = 0$.

If the functional differential inequality

$$y'(t) + p_j(t)\varphi_j\left([y(\sigma_j(t)) + Q(\sigma_j(t))]_+\right) \le 0 \tag{6.4.5}$$

has no eventually positive solution, then (6.4.4) *has no eventually positive solution.*

Proof. Suppose that $y(t)$ is an eventually positive solution of (6.4.4). There exists a number $t_1 \ge t_0$ for which $y(t) > 0$, $y(\rho_i(t)) > 0$ $(i = 1, 2, ..., \ell)$ and $y(\sigma_i(t)) > 0$ $(i = 1, 2, ..., m)$ for $t \ge t_1$. Letting

$$z(t) = y(t) - \sum_{i=1}^{\ell} h_i(t)y(\rho_i(t)) - Q(t), \tag{6.4.6}$$

we see from (6.4.4) that

$$z'(t) \le -p_j(t)\varphi_j(y(\sigma_j(t))) < 0, \quad t \ge t_1 \tag{6.4.7}$$

and hence $z(t)$ is decreasing for $t \ge t_1$. We claim that $z_\infty := \lim_{t\to\infty} z(t) > -\infty$. If $\lim_{t\to\infty} z(t) = -\infty$, then the hypothesis (iii) implies that $\lim_{t\to\infty} \tilde{z}(t) = -\infty$, where

$$\tilde{z}(t) = y(t) - \sum_{i=1}^{\ell} h_i(t)y(\rho_i(t)). \tag{6.4.8}$$

In case $y(t)$ is bounded from above, we observe that $\tilde{z}(t)$ is bounded from below. Hence, $y(t)$ is not bounded from above. Then there exists a number $t_2 \ge t_1$ such that $\tilde{z}(t_2) < 0$ and $\max_{t_1 \le t \le t_2} y(t) = y(t_2)$. Therefore we obtain

$$\begin{aligned}
\tilde{z}(t_2) &= y(t_2) - \sum_{i=1}^{\ell} h_i(t_2)y(\rho_i(t_2)) \\
&\ge y(t_2) - \sum_{i=1}^{\ell} h_i(t_2)y(t_2) \\
&= \left(1 - \sum_{i=1}^{\ell} h_i(t_2)\right) y(t_2) \ge 0
\end{aligned}$$

which contradicts the fact that $\tilde{z}(t_2) < 0$. Hence, we observe that $\lim_{t\to\infty} z(t) = z_\infty > -\infty$. Using (6.4.7), we see that

$$\begin{aligned} 0 &\le \int_{t_1}^{t} p_j(s)\varphi_j\big(y(\sigma_j(s))\big)\,ds \\ &\le \int_{t_1}^{t} \sum_{i=1}^{m} p_i(s)\varphi_i\big(y(\sigma_i(s))\big)\,ds \\ &\le -\int_{t_1}^{t} z'(s)\,ds = z(t_1) - z(t) \le z(t_1) - z_\infty \end{aligned}$$

and therefore $p_j(t)\varphi_j\big(y(\sigma_j(t))\big) \in L^1(t_1,\infty)$. Since $y(t) > 0$ for $t \ge t_1$, we find that $k_0 := \liminf_{t\to\infty} y(t) \ge 0$. If $k_0 > 0$, then

$$y(\sigma_j(t)) \ge \frac{k_0}{2}, \quad t \ge t_3$$

for some $t_3 \ge t_1$. We easily see that

$$\int_{t_1}^{t} p_j(s)\varphi_j\big(y(\sigma_j(s))\big)\,ds \ge \int_{t_3}^{t} p_j(s)\varphi_j\big(y(\sigma_j(s))\big)\,ds \ge \varphi_j\left(\frac{k_0}{2}\right)\int_{t_3}^{t} p_j(s)\,ds \tag{6.4.9}$$

which contradicts the fact that $p_j(t)\varphi_j(y(\sigma_j(t))) \in L^1(t_1,\infty)$ in view of the hypothesis (ii). Hence, we conclude that $k_0 = 0$, that is, $\liminf_{t\to\infty} y(t) = 0$. Since $\lim_{t\to\infty} Q(t) = 0$, we obtain $\lim_{t\to\infty} \tilde{z}(t) = z_\infty$. Taking the inferior limit as $t \to \infty$ of the inequality

$$y(t) = \tilde{z}(t) + \sum_{i=1}^{\ell} h_i(t)y(\rho_i(t)) \ge \tilde{z}(t),$$

we find that $z_\infty \le 0$. Let $z_\infty < 0$. If $y(t)$ is not bounded from above, there exists a sequence $\{t_n\}_{n=1}^{\infty}$ satisfying

$$\begin{aligned} &\lim_{n\to\infty} t_n = \infty, \\ &\lim_{n\to\infty} y(t_n) = \infty, \\ &\max_{t_1\le t\le t_n} y(t) = y(t_n). \end{aligned}$$

Then we see that for all large n

$$\begin{aligned} 0 > \tilde{z}(t_n) &= y(t_n) - \sum_{i=1}^{\ell} h_i(t_n)y(\rho_i(t_n)) \\ &\ge \left(1 - \sum_{i=1}^{\ell} h_i(t_n)\right) y(t_n) \ge 0 \end{aligned}$$

which yields a contradiction. Hence, we conclude that $y(t)$ is bounded from above. Then there exists $\overline{y}_\infty := \limsup_{t\to\infty} y(t)$, and therefore there exists a sequence $\{\tilde{t}_n\}_{n=1}^\infty$ such that $\lim_{n\to\infty} \tilde{t}_n = \infty$ and $\lim_{n\to\infty} y(\tilde{t}_n) = \overline{y}_\infty$. It is readily checked that

$$\begin{aligned}\tilde{z}(\tilde{t}_n) &= y(\tilde{t}_n) - \sum_{i=1}^{\ell} h_i(\tilde{t}_n) y(\rho_i(\tilde{t}_n)) \\ &\geq y(\tilde{t}_n) - \overline{y}_\infty + \sum_{i=1}^{\ell} h_i(\tilde{t}_n)\big(\overline{y}_\infty - y(\rho_i(\tilde{t}_n))\big). \qquad (6.4.10)\end{aligned}$$

Since

$$\liminf_{n\to\infty} \big(\overline{y}_\infty - y(\rho_i(\tilde{t}_n))\big) \geq \overline{y}_\infty - \limsup_{n\to\infty} y(\rho_i(\tilde{t}_n)) \geq 0,$$

taking inferior limit of (6.4.10) as $n \to \infty$ yields

$$0 > z_\infty \geq \liminf_{n\to\infty} y(\tilde{t}_n) - \overline{y}_\infty = 0.$$

This is a contradiction, and consequently we conclude that $z_\infty = 0$. We see that $z(t) > 0$ for $t \geq t_1$ in view of the fact that $z(t)$ is decreasing. Since $y(t) \geq z(t) + Q(t)$ and $y(t) > 0$ for $t \geq t_1$, we obtain

$$y(t) \geq [z(t) + Q(t)]_+, \quad t \geq t_1$$

and therefore

$$y(\sigma_j(t)) \geq [z(\sigma_j(t)) + Q(\sigma_j(t))]_+, \quad t \geq T$$

for some $T \geq t_1$. The hypothesis (ii) implies that

$$\varphi_j\big(y(\sigma_j(t))\big) \geq \varphi_j\big([z(\sigma_j(t)) + Q(\sigma_j(t))]_+\big), \quad t \geq T. \qquad (6.4.11)$$

Combining (6.4.7) with (6.4.11) yields

$$z'(t) + p_j(t)\varphi_j\big([z(\sigma_j(t)) + Q(\sigma_j(t))]_+\big) \leq 0, \quad t \geq T$$

and consequently $z(t)$ is an eventually positive solution of (6.4.5). This contradicts the hypothesis and completes the proof. □

Lemma 6.4.2. *Assume that the hypothesis* (ii) *of Lemma* 6.4.1 *holds. If*

$$\int_{t_0}^{\infty} p_j(t)\varphi_j\big([Q(\sigma_j(t))]_+\big)\, dt = \infty, \qquad (6.4.12)$$

then (6.4.5) *has no eventually positive solution.*

Proof. Let $y(t)$ be an eventually positive solution of (6.4.5). Then there exists a number $t_1 \geq t_0$ such that $y(t) > 0$ and $y(\sigma_j(t)) > 0$ for $t \geq t_1$. Since

$$y(\sigma_j(t)) + Q(\sigma_j(t)) \geq Q(\sigma_j(t)), \quad t \geq t_1,$$

we obtain

$$[y(\sigma_j(t)) + Q(\sigma_j(t))]_+ \geq [Q(\sigma_j(t))]_+, \quad t \geq t_1$$

and therefore

$$\varphi_j\big([y(\sigma_j(t)) + Q(\sigma_j(t))]_+\big) \geq \varphi_j\big([Q(\sigma_j(t))]_+\big), \quad t \geq t_1.$$

Hence, (6.4.5) implies

$$y'(t) + p_j(t)\varphi_j\big([Q(\sigma_j(t))]_+\big) \leq 0, \quad t \geq t_1. \tag{6.4.13}$$

Integrating (6.4.13) over $[t_1, t]$ yields

$$\int_{t_1}^{t} p_j(s)\varphi_j\big([Q(\sigma_j(s))]_+\big)\, ds \leq -\int_{t_1}^{t} y'(s)\, ds \leq y(t_1)$$

which contradicts the hypothesis (6.4.12). The proof is complete. □

Theorem 6.4.3. *Assume that the hypotheses* (H6.1-4), (H6.3-1)–(H6.3-3), (i), (ii) *of Lemma* 6.4.1 *are satisfied, and assume, moreover, that there exists a function* $\Theta(t) \in C^1((0,\infty);\mathbb{R})$ *such that* $\Theta'(t) = G(t)$ [*resp.* $\Theta'(t) = \tilde{G}(t)$], $\lim_{t\to\infty} \Theta(t) = 0$. *If the functional differential inequalities*

$$y'(t) + p_j(t)\varphi_j\left([y(\sigma_j(t)) \pm \Theta(\sigma_j(t))]_+\right) \leq 0$$

have no eventually positive solutions, then every solution u of the boundary value problem (6.4.1), (B_1) [*resp.* (6.4.1), (B_2)] *is oscillatory in* Ω.

Proof. We note that Theorem 6.4.1 remains true if (6.4.2) are replaced by

$$\frac{d}{dt}\left(y(t) - \sum_{i=1}^{\ell} h_i(t)y(\rho_i(t))\right) + \sum_{i=1}^{m} p_i(t)\varphi_i\big(y(\sigma_i(t))\big) \leq \pm G(t).$$

The conclusion follows by combining Theorems 6.4.1 and 6.4.2 and Lemma 6.4.1. □

Theorem 6.4.4. *Assume that the hypotheses* (H6.1-4), (H6.3-1)–(H6.3-3), (i), (ii) *of Lemma* 6.4.1 *are satisfied. Every solution* u *of the boundary value problem* (6.4.1), (B_1) [*resp.* (6.4.1), (B_2)] *is oscillatory in* Ω *if there exists a function* $\Theta(t) \in C^1((0,\infty);\mathbb{R})$ *such that* $\Theta'(t) = G(t)$ [*resp.* $\Theta'(t) = \tilde{G}(t)$], $\lim_{t\to\infty} \Theta(t) = 0$, *and*

$$\int_{t_0}^{\infty} p_j(t)\varphi_j\big([\Theta(\sigma_j(t))]_{\pm}\big)\,dt = \infty$$

for some $t_0 > 0$.

Proof. The conclusion follows from Theorem 6.4.3 and Lemma 6.4.2. □

Lemma 6.4.3. *Assume that the hypotheses* (i), (iii) *of Lemma* 6.4.1 *are satisfied. Assume, moreover, that:*

(i) *there exist positive constants* $\tilde{\rho}_i$ *such that*

$$\rho_i'(t) \geq \tilde{\rho}_i \quad (i = 1, 2, ..., \ell);$$

(ii) *there is an integer* $j \in \{1, 2, ..., m\}$ *for which*

$$p_j(t) \geq \tilde{p}_j \quad \textit{for some } \tilde{p}_j > 0,$$
$$0 \leq \sigma_j'(t) \leq \tilde{\sigma}_j \quad \textit{for some } \tilde{\sigma}_j > 0$$

and (6.3.22) *holds for some* $\beta > 0$.

If the functional differential inequality

$$y'(t) + \beta p_j(t)y(\sigma_j(t)) \leq -\beta p_j(t)Q(\sigma_j(t)) \tag{6.4.14}$$

has no eventually positive solution, then (6.4.4) *has no eventually positive solution.*

Proof. Let $y(t)$ be an eventually positive solution of (6.4.4). Proceeding as in the proof of Lemma 6.4.1, the function $z(t)$ defined by (6.4.6) is decreasing on $[t_1, \infty)$ for some $t_1 \geq t_0$, and that there exists $z_\infty := \lim_{t\to\infty} z(t) > -\infty$. Since $\lim_{t\to\infty} Q(t) = 0$, we obtain

$$\lim_{t\to\infty} \tilde{z}(t) = z_\infty > -\infty, \tag{6.4.15}$$

where $\tilde{z}(t)$ is given by (6.4.8). The hypothesis (ii) and (6.4.4) imply that

$$\begin{aligned}
\tilde{p}_j\beta \int_{t_1}^{t} |y(\sigma_j(s))|\,ds &\leq \int_{t_1}^{t} p_j(s)\varphi_j\big(y(\sigma_j(s))\big)\,ds \\
&\leq \int_{t_1}^{t} \sum_{i=1}^{m} p_i(s)\varphi_i\big(y(\sigma_i(s))\big)\,ds \\
&\leq -\int_{t_1}^{t} z'(s)\,ds = z(t_1) - z(t)
\end{aligned}$$

and therefore $y(\sigma_j(t)) \in L^1(t_1, \infty)$. It is easy to see that

$$\int_{t_1}^{t} |y(\sigma_j(s))|\, ds \geq \int_{t_1}^{t} |y(\sigma_j(s))| \frac{\sigma_j'(s)}{\tilde{\sigma}_j}\, ds = \frac{1}{\tilde{\sigma}_j} \int_{\sigma_j(t_1)}^{\sigma_j(t)} |y(s)|\, ds.$$

Letting $t \to \infty$, we find that $y(t) \in L^1(t_2, \infty)$ for some $t_2 \geq t_1$. It follows from the hypothesis (i) that

$$\begin{aligned}\int_{t_2}^{t} \left| \sum_{i=1}^{\ell} h_i(s) y(\rho_i(s)) \right| ds &\leq \sum_{i=1}^{\ell} \int_{t_2}^{t} |y(\rho_i(s))| \frac{\rho_i'(s)}{\tilde{\rho}_i}\, ds \\ &= \sum_{i=1}^{\ell} \frac{1}{\tilde{\rho}_i} \int_{\rho_i(t_2)}^{\rho_i(t)} |y(s)|\, ds\end{aligned}$$

and letting $t \to \infty$ yields

$$\sum_{i=1}^{\ell} h_i(t) y(\rho_i(t)) \in L^1(t_2, \infty).$$

Hence, we observe that $\tilde{z}(t) \in L^1(t_2, \infty)$, which, together with (6.4.15), implies that $\lim_{t\to\infty} \tilde{z}(t) = 0$. Consequently, we conclude that $\lim_{t\to\infty} z(t) = 0$, and therefore $z(t) > 0$ for $t \geq t_1$. Since $y(t) \geq z(t) + Q(t)$, (6.3.22) and (6.4.4) imply that

$$\begin{aligned}0 &\geq z'(t) + \beta p_j(t) y(\sigma_j(t)) \\ &\geq z'(t) + \beta p_j(t) \big(z(\sigma_j(t)) + Q(\sigma_j(t)) \big), \quad t \geq t_1.\end{aligned}$$

Therefore, $z(t)$ is an eventually positive solution of (6.4.14). This contradicts the hypothesis and completes the proof. □

Theorem 6.4.5. *Assume that the hypotheses* (H6.1-4), (H6.3-1)–(H6.3-3), (i) *of Lemma* 6.4.1, *and* (i), (ii) *of Lemma* 6.4.3 *are satisfied. Assume, moreover, that* $\sigma_j(t) \leq t$, $\sigma_j(t)$ *is nondecreasing in* $(0, \infty)$, *and there exists a function* $\Theta(t) \in C^1((0,\infty); \mathbb{R})$ *such that* $\Theta'(t) = G(t)$ [*resp.* $\Theta'(t) = \tilde{G}(t)$], $\lim_{t\to\infty} \Theta(t) = 0$. *If the functional differential inequalities*

$$y'(t) + \beta p_j(t) y(\sigma_j(t)) \leq \mp \beta p_j(t) \Theta(\sigma_j(t))$$

have no eventually positive solutions, then every solution u of the boundary value problem (6.4.1), (B_1) [*resp.* (6.4.1), (B_2)] *is oscillatory in* Ω.

Proof. The conclusion follows by combining Theorems 6.4.1 and 6.4.2 and Lemma 6.4.3. □

Theorem 6.4.6. *Assume that the hypotheses* (H6.1-4), (H6.3-1)–(H6.3-3), (i) *of Lemma* 6.4.1, *and* (i), (ii) *of Lemma* 6.4.3 *are satisfied. Assume, moreover, that $\sigma_j(t) \le t$ and $\sigma_j(t)$ is nondecreasing in $(0,\infty)$. Every solution u of the boundary value problem* (6.4.1), (B$_1$) [*resp.* (6.4.1), (B$_2$)] *is oscillatory in Ω if there exists a function $\Theta(t) \in C^1((0,\infty);\mathbb{R})$ such that $\Theta'(t) = G(t)$ [resp. $\Theta'(t) = \tilde{G}(t)$], $\lim_{t\to\infty} \Theta(t) = 0$, and there is a sequence $\{t_n\}_{n=1}^{\infty}$ such that:*

$$\lim_{n\to\infty} t_n = \infty,$$

$$\int_{\sigma_j(t_n)}^{t_n} p_j(t)\,dt \ge \frac{1}{\beta},$$

$$\int_{\sigma_j(t_n)}^{t_n} p_j(t)\Theta(\sigma_j(t))\,dt + \beta \int_{\sigma_j(t_n)}^{t_n} p_j(t)\left(\int_{\sigma_j(t)}^{\sigma_j(t_n)} p_j(s)\Theta(\sigma_j(s))\,ds\right)dt = 0.$$

Proof. The conclusion follows from Theorem 6.4.5 and a result of Yoshida [311, Proposition 1]. □

Example 6.4.1. We consider the problem

$$\frac{\partial}{\partial t}\left(u(x,t) - \frac{1}{e^{2\pi}}\,u(x,t-2\pi)\right) - \frac{\partial^2 u}{\partial x^2}(x,t) - \frac{1}{e^{\pi}}\frac{\partial^2 u}{\partial x^2}(x,t-\pi)$$
$$+\frac{1}{e^{2\pi}}e^t\,u(x,t-2\pi) = (\sin x)\sin t, \quad (x,t) \in (0,\pi)\times(0,\infty), \tag{6.4.16}$$

$$u(0,t) = u(\pi,t) = 0, \quad t > 0. \tag{6.4.17}$$

Here $n = 1$, $G = (0,\pi)$, $\Omega = (0,\pi)\times(0,\infty)$, $\ell = k = m = \tilde{m} = 1$, $h_1(t) = e^{-2\pi}$, $\rho_1(t) = t - 2\pi$, $a(t) = 1$, $b_1(t) = e^{-\pi}$, $\tau_1(t) = t - \pi$, $p_1(t) = e^{-2\pi}e^t$, $\sigma_1(t) = t - 2\pi$, $\varphi_1(\xi) = \xi$ and $f(x,t) = (\sin x)\sin t$. It is easy to see that $\lambda_1 = 1$, $\Phi(x) = \sin x$, $\Psi(t) = 0$ and $F(t) = G(t) = (1/4)\pi\sin t$. Choosing $\Theta(t) = -(1/4)\pi\cos t$, we find that $\Theta(\sigma_1(t)) = -(1/4)\pi\cos t$, and that

$$\int_1^{\infty}\frac{1}{e^{2\pi}}e^t\left[-\frac{1}{4}\pi\cos t\right]_{\pm} dt \ge \frac{\pi}{4e^{2\pi}}\int_1^{\infty}[-\cos t]_{\pm}\,dt = \infty.$$

Hence, it follows from Theorem 6.4.4 that every solution u of the problem (6.4.16), (6.4.17) is oscillatory in $(0,\pi)\times(0,\infty)$. One such solution is $u(x,t) = (\sin x)e^{-t}\sin t$.

Example 6.4.2. We consider the problem

$$\frac{\partial}{\partial t}\left(u(x,t) - \frac{1}{e^{\pi}}\,u(x,t-\pi)\right) - 2\frac{\partial^2 u}{\partial x^2}(x,t) - e^{\pi}\frac{\partial^2 u}{\partial x^2}(x,t-\pi)$$
$$+u(x,t-2\pi) = -2(\cos x)e^{-t}\sin t, \quad (x,t)\in(0,\pi)\times(0,\infty), \tag{6.4.18}$$

$$-\frac{\partial u}{\partial x}(0,t) = \frac{\partial u}{\partial x}(\pi,t) = 0, \quad t > 0. \tag{6.4.19}$$

Here $n = 1$, $G = (0, \pi)$, $\Omega = (0, \pi) \times (0, \infty)$, $\ell = k = m = \tilde{m} = 1$, $h_1(t) = e^{-\pi}$, $\rho_1(t) = t - \pi$, $a(t) = 2$, $b_1(t) = e^{\pi}$, $\tau_1(t) = t - \pi$, $p_1(t) = 1$, $\sigma_1(t) = t - 2\pi$, $\varphi_1(\xi) = \xi$ and $f(x,t) = -2(\cos x)e^{-t} \sin t$. It is easily checked that $\tilde{\Psi}(t) = 0$ and $\tilde{F}(t) = \tilde{G}(t) = 0$. We can choose $\Theta(t) = 0$. Since

$$\int_{\sigma_1(t_n)}^{t_n} p_1(t)\, dt = \int_{t_n - 2\pi}^{t_n} dt = 2\pi \geq 1$$

for any sequence $\{t_n\}$, Theorem 6.4.6 implies that every solution u of the problem (6.4.18), (6.4.19) is oscillatory in $(0, \pi) \times (0, \infty)$. In fact, $u(x,t) = (\cos x)e^{-t} \cos t$ is such a solution.

Remark 6.4.1. Lemma 6.4.3 holds true if the functional differential inequality (6.4.14) is replaced by

$$y'(t) + \beta \tilde{p}_j y(\sigma_j(t)) \leq -\beta \tilde{p}_j Q(\sigma_j(t)).$$

Remark 6.4.2. The case where $m = \tilde{m}$, $\psi = \tilde{\psi} = 0$, $f(x,t) = 0$, $\rho_i(t) = t - \rho_i$ $(\rho_i > 0)$ $(i = 1, 2, ..., \ell)$, $\tau_i(t) = t - \tau_i$ $(\tau_i > 0)$ $(i = 1, 2, ..., k)$, $\sigma_i(t) = t - \sigma_i$ $(\sigma_i > 0)$ $(i = 1, 2, ..., m)$, $\varphi_i(\xi) = \xi$ $(i = 1, 2, ..., m)$ was studied by Mishev and Bainov [209]. When $m = \tilde{m}$, $\psi = \tilde{\psi} = 0$, $f(x,t) = 0$, $\rho_i(t) = t - \rho_i$ $(\rho_i > 0)$ $(i = 1, 2, ..., \ell)$ and (6.3.22) holds, the boundary value problems (6.4.1), (B_i) $(i = 1, 2)$ were treated by Cui [55].

6.5 Impulsive Parabolic Equations

Since the pioneering work [75] on impulsive partial differential equations was published in 1991, there has been an increasing interest in studying oscillation properties of impulsive partial differential equations. In particular, we mention the paper [22] by Bainov and Minchev where oscilllation results for impulsive parabolic equations were first obtained. We refer the reader to [23,24,57,61,63,82,83,175] for impulsive parabolic equations, and to [169] for impulsive parabolic systems.

Let G be a bounded domain in $\mathbb{R}^n$ with piecewise smooth boundary ∂G, and $\{t_k\}_{k=0}^{\infty}$ be a sequence of real numbers such that

$$0 = t_0 < t_1 < \cdots < t_k < \cdots$$

and $\lim_{k \to \infty} t_k = \infty$.

We use the following notation:

$$\Omega = G \times (0, \infty),$$
$$J_{\text{imp}} = \{t_k;\ k = 1, 2, ...\},$$
$$\Omega_{\text{imp}} = \{(x,t) \in \Omega;\ t \in J_{\text{imp}}\},$$
$$\overline{\Omega}_{\text{imp}} = \{(x,t) \in \overline{\Omega};\ t \in J_{\text{imp}}\}.$$

Definition 6.5.1. The function space $C_{\text{imp}}([0,\infty);\mathbb{R})$ is defined to be the set of all real-valued functions $g(t)$ on $[0,\infty)$ such that:

(i) each $g(t)$ is continuous in (t_k, t_{k+1}) for $k = 0, 1, 2, ...$;
(ii) for $k = 1, 2, ...$ there exist the limits

$$g(t_k^+) = \lim_{\substack{s \to t_k \\ s > t_k}} g(s), \qquad g(t_k^-) = \lim_{\substack{s \to t_k \\ s < t_k}} g(s);$$

(iii) $g(t_k) = g(t_k^-)$, that is, $g(t)$ is left continuous at t_k.

Definition 6.5.2. The function space $C_{\text{imp}}([0,\infty);[0,\infty))$ is defined to be the set of all nonnegative-valued functions $g(t)$ on $[0,\infty)$ which satisfy the conditions (i)–(iii) of Definition 6.5.1.

Definition 6.5.3. The function space $C_{\text{imp}}(\overline{\Omega};\mathbb{R})$ is defined to be the set of all real-valued functions $u(x,t)$ on $\overline{\Omega}$ such that:

(i) each $u(x,t)$ is continuous in $\overline{\Omega} \setminus \overline{\Omega}_{\text{imp}}$;
(ii) for $(x,t) \in \overline{\Omega}_{\text{imp}}$ there exist the limits

$$u(x,t^+) = \lim_{\substack{(y,s)\to(x,t) \\ s > t}} u(y,s), \quad u(x,t^-) = \lim_{\substack{(y,s)\to(x,t) \\ s < t}} u(y,s);$$

(iii) $u(x,t) = u(x,t^-)$ for $(x,t) \in \overline{\Omega}_{\text{imp}}$.

Definition 6.5.4. The function space $C_{\text{imp}}(\overline{\Omega};[0,\infty))$ is defined to be the set of all nonnegative-valued functions $u(x,t)$ on $\overline{\Omega}$ which satisfy the conditions (i)–(iii) of Definition 6.5.3.

Definition 6.5.5. A function $y(t) \in C_{\text{imp}}([0,\infty);\mathbb{R})$ is said to be *eventually positive* (*eventually negative*) if there exists a number $T \geq 0$ such that $y(t) > 0$ ($y(t) < 0$) for $t \geq T$ (*cf.* Definition 1.2.2 of Section 1.2).

We deal with the impulsive parabolic equation

$$\frac{\partial u}{\partial t}(x,t) - a(t)\Delta u(x,t) + p(x,t)\varphi\big(u(x,t-\sigma)\big) = 0, \ (x,t) \in \Omega \setminus \Omega_{\text{imp}}, \quad (6.5.1)$$

$$u(x,t^+) - u(x,t^-) = h(x,t,u(x,t)), \ (x,t) \in \overline{\Omega}_{\text{imp}}. \quad (6.5.2)$$

It is assumed that:

(H6.5-1) $\sigma \geq 0$ is a constant and $t_{k+\ell} = t_k + \sigma$ $(k = 0, 1, 2, ...)$ for some $\ell \in \mathbb{N} \cup \{0\}$;

(H6.5-2) $a(t) \in C_{\rm imp}([0,\infty); [0,\infty))$;

(H6.5-3) $p(x,t) \in C_{\rm imp}(\overline{\Omega}; [0,\infty))$;

(H6.5-4) $\varphi(\xi) \in C(\mathbb{R}; \mathbb{R})$, $\varphi(-\xi) = -\varphi(\xi)$ for $\xi \geq 0$, $\varphi(\xi) \geq K\xi$ $(\xi \geq 0)$ for some $K > 0$;

(H6.5-5) $h(x,t,\xi) \in C(\overline{\Omega}_{\rm imp} \times \mathbb{R}, \mathbb{R})$, $h(x, t_k, -\xi) = -h(x, t_k, \xi)$ and

$$h(x, t_k, \xi) \leq M_k \xi \quad \text{for } x \in \overline{G},\ \xi \geq 0,\ k = 1, 2, ...,$$

where $M_k \geq 0$ are constants.

We consider two kinds of boundary conditions:

$$(\mathrm{B}_0) \qquad u = 0 \quad \text{on } \partial G \times \big([0,\infty) \setminus J_{\rm imp}\big),$$

$$(\tilde{\mathrm{B}}_0) \qquad \frac{\partial u}{\partial \nu} + \mu u = 0 \quad \text{on } \partial G \times \big([0,\infty) \setminus J_{\rm imp}\big),$$

where $\mu \in C_{\rm imp}(\partial G \times [0,\infty); [0,\infty))$ and ν denotes the unit exterior normal vector to ∂G. We note that $C_{\rm imp}(\partial G \times [0,\infty); [0,\infty))$ is defined to be the same as $C_{\rm imp}(\overline{\Omega}; \mathbb{R})$ (*cf.* Definition 6.5.4).

Definition 6.5.6. A real-valued function $u(x,t)$ on $\overline{G} \times [-\sigma, \infty)$ is said to be a *solution* of the problem (6.5.1), (6.5.2), (B_0) [or (6.5.1), (6.5.2), $(\tilde{\mathrm{B}}_0)$] if the following conditions are satisfied:

(i) $u(x,t) \in C_{{\rm imp}(-\sigma)}(\overline{G} \times [-\sigma,\infty); \mathbb{R}) \cap C^1(\overline{\Omega} \setminus \overline{\Omega}_{\rm imp})$ and there exist the continuous derivatives $\frac{\partial u}{\partial t}(x,t)$, $\frac{\partial^2 u}{\partial x_i^2}(x,t)$ $(i = 1, 2, ...n)$ for $(x,t) \in \Omega \setminus \Omega_{\rm imp}$;

(ii) $u(x,t)$ satisfies (6.5.1), (6.5.2), (B_0) [or (6.5.1), (6.5.2), $(\tilde{\mathrm{B}}_0)$],

where $C_{{\rm imp}(-\sigma)}(\overline{G} \times [-\sigma,\infty); \mathbb{R})$ is defined to be the space $C_{\rm imp}(\overline{\Omega}; \mathbb{R})$ with $\overline{\Omega}$ replaced by $\overline{G} \times [-\sigma, \infty)$ and $\overline{\Omega}_{\rm imp}$ by

$$\begin{aligned}&\overline{G} \times [-\sigma,\infty)_{{\rm imp}(-\sigma)} \\ &\quad = \{(x,t) \in \overline{G} \times [-\sigma,\infty);\ t \in \{-t_{\ell-1}, -t_{\ell-2}, ..., -t_1, 0\} \cup J_{\rm imp}\},\end{aligned}$$

respectively.

Definition 6.5.7. A solution u of (6.5.1) is said to be *nonoscillatory* in Ω if it is either eventually positive or eventually negative (*cf.* Definition 6.5.5). Otherwise, it is called *oscillatory* in Ω.

Let λ_1 (> 0) be the first eigenvalue of the eigenvalue problem (EVP) and $\Phi(x)$ be the corresponding eigenfunction which is chosen so that $\Phi(x) > 0$ in G (see Section 6.1).

Lemma 6.5.1. *Assume that the hypotheses* (H6.5-1)–(H6.5-5) *hold. If* u *is an eventually positive solution of the problem* (6.5.1), (6.5.2), (B_0), *then we obtain*

$$U'(t) + \lambda_1 a(t)U(t) + KP(t)U(t-\sigma) \leq 0, \quad t \neq t_k,$$
$$U(t_k^+) \leq (1 + M_k)U(t_k)$$

eventually, where

$$P(t) = \min_{x \in \overline{G}} p(x,t),$$
$$U(t) = \int_G u(x,t)\Phi(x)\,dx.$$

Proof. Let $u(x,t) > 0$ in $G \times [T,\infty)$ for some $T \geq 0$ and let $t \geq T + \sigma$. Multiplying (6.5.1) by $\Phi(x)$ and then integrating over G, we obtain

$$\frac{d}{dt}\int_G u(x,t)\Phi(x)\,dx - a(t)\int_G \Delta u(x,t)\Phi(x)\,dx + \int_G p(x,t)\varphi\big(u(x,t-\sigma)\big)\Phi(x)\,dx = 0, \quad t \neq t_k. \tag{6.5.3}$$

It follows from Green's formula that

$$\int_G \Delta u(x,t)\Phi(x)\,dx = -\lambda_1 \int_G u(x,t)\Phi(x)\,dx, \quad t \neq t_k \tag{6.5.4}$$

(see, for example, (6.2.5)). The hypothesis (H6.5-4) implies

$$\int_G p(x,t)\varphi\big(u(x,t-\sigma)\big)\Phi(x)\,dx \geq P(t)K\int_G u(x,t-\sigma)\Phi(x)\,dx, \quad t \neq t_k. \tag{6.5.5}$$

Combining (6.5.3)–(6.5.5) yields

$$U'(t) + \lambda_1 a(t)U(t) + KP(t)U(t-\sigma) \leq 0, \quad t \neq t_k$$

which is the desired inequality. For $t = t_k$ we observe that

$$U(t_k^+) - U(t_k^-) \leq M_k \int_G u(x,t_k)\Phi(x)\,dx = M_k U(t_k)$$

and therefore

$$U(t_k^+) \leq (1 + M_k)U(t_k).$$

This completes the proof. □

Theorem 6.5.1. *Assume that the hypotheses* (H6.5-1)–(H6.5-5) *hold. If every eventually positive solution* $y(t)$ *of*

$$y'(t) + \lambda_1 a(t)y(t) + KP(t)y(t-\sigma) \leq 0, \quad t \neq t_k, \tag{6.5.6}$$

$$y(t_k^+) \leq (1+M_k)y(t_k) \tag{6.5.7}$$

satisfies the condition $\lim_{t\to\infty} y(t) = 0$, *then every solution* u *of the problem* (6.5.1), (6.5.2), (B_0) *is oscillatory in* Ω, *or satisfies*

$$\lim_{t\to\infty} \int_G u(x,t)\Phi(x)\,dx = 0. \tag{6.5.8}$$

Proof. Suppose to the contrary that there is a nonoscillatory solution $u(x,t)$ of the problem (6.5.1), (6.5.2), (B_0) which does not satisfy (6.5.8). Let $u(x,t)$ have no zero in $G \times [T,\infty)$ for some $T > 0$. Without loss of generality we may assume that $u(x,t) > 0$ in $G \times [T,\infty)$. From Lemma 6.5.1 we see that $U(t)$ is a positive solution of (6.5.6), (6.5.7) for $t \geq T+\sigma$. Since $U(t)$ does not satisfy the condition $\lim_{t\to\infty} U(t) = 0$, we are led to a contradiction. The proof is complete. □

Now we consider the impulsive differential inequality

$$y'(t) + Q(t)y(t-\sigma) \leq 0, \quad t \neq t_k, \tag{6.5.9}$$

$$y(t_k^+) \leq (1+L_k)y(t_k), \tag{6.5.10}$$

where $Q(t) \in C_{\text{imp}}([0,\infty);[0,\infty))$ and $L_k \geq 0$ are constants.

Lemma 6.5.2. *If*

$$\int_T^\infty Q(t)\,dt = \infty \quad \textit{for each } T \geq 0, \tag{6.5.11}$$

$$\sum_{k=1}^{\infty} L_k < \infty, \tag{6.5.12}$$

then every eventually positive solution $y(t)$ *of* (6.5.9), (6.5.10) *satisfies the condition* $\lim_{t\to\infty} y(t) = 0$.

Proof. Let $y(t)$ be an eventually positive solution of (6.5.9), (6.5.10). Then there exists a number $T > 0$ such that $y(t) > 0$ for $t \geq T$. Since $y'(t) \leq 0$ for $t \neq t_k$, $t \geq T+\sigma$, we see that

$$y(t-\sigma) \geq \frac{1}{\prod\limits_{t-\sigma\leq t_k<t} (1+L_k)}\, y(t)$$

for $t \geq T + 2\sigma$. Hence we obtain

$$y'(t) + \frac{Q(t)}{\prod\limits_{t-\sigma \leq t_k < t} (1 + L_k)} y(t) \leq 0, \quad t \neq t_k, \ t \geq T + 2\sigma.$$

A direct computation yields

$$y(t) \leq y(T+2\sigma) \prod_{T+2\sigma \leq t_k < t} (1 + L_k) \exp\left(- \int_{T+2\sigma}^{t} \frac{Q(s)}{\prod\limits_{s-\sigma \leq t_k < s} (1 + L_k)} ds \right)$$

and therefore

$$y(t) \leq y(T+2\sigma) \prod_{k=1}^{\infty} (1+L_k) \exp\left(- \frac{1}{\prod\limits_{k=1}^{\infty} (1 + L_k)} \int_{T+2\sigma}^{t} Q(s)\, ds \right) \quad (6.5.13)$$

(*cf.* Lakshmikantham, Bainov and Simeonov [163, Theorem 1.4.1]). Using the inequality $1+L_k \leq \exp L_k$ $(L_k \geq 0)$, we note that $\prod_{k=1}^{\infty}(1+L_k) < \infty$ by (6.5.12). Taking the limit as $t \to \infty$ in (6.5.13), we observe, using (6.5.11), that $\lim_{t\to\infty} y(t) = 0$. □

Theorem 6.5.2. *Assume that the hypotheses* (H6.5-1)–(H6.5-5) *hold. If*

$$\int_{T}^{\infty} P(t)\, dt = \infty \quad \textit{for each } T \geq 0, \quad (6.5.14)$$

$$\sum_{k=1}^{\infty} M_k < \infty, \quad (6.5.15)$$

then every solution u of the problem (6.5.1), (6.5.2), (B$_0$) *is oscillatory in* Ω, *or satisfies the condition* (6.5.8).

Proof. The conclusion follows by combining Theorem 6.5.1 with Lemma 6.5.2. □

Next we consider the boundary condition ($\tilde{\mathrm{B}}_0$).

Lemma 6.5.3. *Assume that the hypotheses* (H6.5-1)–(H6.5-5) *hold. If* $u \in C^2(\Omega \setminus \Omega_{\mathrm{imp}}) \cap C^1(\overline{\Omega} \setminus \overline{\Omega}_{\mathrm{imp}})$ *is an eventually positive solution of the problem* (6.5.1), (6.5.2), ($\tilde{\mathrm{B}}_0$), *then we obtain*

$$\tilde{U}'(t) + KP(t)\tilde{U}(t-\sigma) \leq 0, \quad t \neq t_k,$$
$$\tilde{U}(t_k^+) \leq (1 + M_k)\tilde{U}(t_k)$$

eventually, where

$$\tilde{U}(t) = \int_G u(x,t)\, dx.$$

Proof. Suppose that $u(x,t) > 0$ in $G \times [T,\infty)$ for some $T \geq 0$ and let $t \geq T + \sigma$. Integrating (6.5.1) over G yields

$$\frac{d}{dt}\int_G u(x,t)\,dx - a(t)\int_G \Delta u(x,t)\,dx$$
$$+\int_G p(x,t)\varphi\big(u(x,t-\sigma)\big)\,dx = 0, \quad t \neq t_k. \tag{6.5.16}$$

Using Green's formula, we have

$$\int_G \Delta u(x,t)\,dx = \int_{\partial G}\frac{\partial u}{\partial \nu}(x,t)\,dS = -\int_{\partial G}\gamma \cdot u(x,t)\,dS \leq 0, \quad t \neq t_k. \tag{6.5.17}$$

It follows from the hypothesis (H6.5-4) that

$$\int_G p(x,t)\varphi\big(u(x,t-\sigma)\big)\,dx \geq KP(t)\int_G u(x,t-\sigma)\,dx, \quad t \neq t_k. \tag{6.5.18}$$

Combining (6.5.16)–(6.5.18), we obtain

$$\tilde{U}'(t) + KP(t)\tilde{U}(t-\sigma) \leq 0, \quad t \neq t_k$$

which is the desired inequality. For $t = t_k$ we see that

$$\tilde{U}(t_k^+) - \tilde{U}(t_k^-) \leq M_k \int_G u(x,t_k)\,dx = M_k\tilde{U}(t_k)$$

and hence

$$\tilde{U}(t_k^+) \leq (1+M_k)\tilde{U}(t_k).$$

The proof is complete. □

Theorem 6.5.3. *Assume that the hypotheses* (H6.5-1)–(H6.5-5) *hold. If every eventually positive solution $y(t)$ of*

$$y'(t) + KP(t)y(t-\sigma) \leq 0, \quad t \neq t_k,$$
$$y(t_k^+) \leq (1+M_k)y(t_k)$$

satisfies the condition $\lim_{t\to\infty} y(t) = 0$, *then every solution u of the problem* (6.5.1), (6.5.2), ($\tilde{\text{B}}_0$) *is oscillatory in* Ω, *or satisfies*

$$\lim_{t\to\infty}\int_G u(x,t)\,dx = 0. \tag{6.5.19}$$

Proof. The proof is quite similar to that of Theorem 6.5.1. □

Theorem 6.5.4. *Assume that the hypotheses* (H6.5-1)–(H6.5-5) *hold. If the conditions* (6.5.14), (6.5.15) *hold, then every solution u of the problem* (6.5.1), (6.5.2), ($\tilde{\text{B}}_0$) *is oscillatory in* Ω, *or satisfies the condition* (6.5.19).

Proof. The conclusion follows by combining Lemma 6.5.2 with Theorem 6.5.3. □

6.6 Parabolic Systems I

In 1970 Domšlak [65] introduced the concept of H-oscillation to study the oscillatory character of vector differential equations, where H is a unit vector in $\mathbb{R}^N$. Several authors investigated H-oscillation of vector differential equations. We refer the reader to [65,66,139] for vector ordinary differential equations, and to [222,223] for vector partial differential equations without functional arguments.

We consider the vector parabolic equation with functional arguments

$$\frac{\partial}{\partial t}\left(U(x,t)+\sum_{i=1}^{\ell} h_i(t)U(x,\rho_i(t))\right)-a(t)\Delta U(x,t)-\sum_{i=1}^{k} b_i(t)\Delta U(x,\tau_i(t))$$
$$+\sum_{i=1}^{m} c_i\big(x,t,U(x,\sigma_i(t))\big)U(x,\sigma_i(t))=F(x,t) \tag{6.6.1}$$

for $(x,t)\in\Omega:=G\times(0,\infty)$, where G is a bounded domain in $\mathbb{R}^n$ with piecewise smooth boundary ∂G.

We assume that:

(H6.6-1) $h_i(t)\in C^1([0,\infty);\mathbb{R})$ $(i=1,2,...,\ell)$,
$b_i(t)\in C([0,\infty);[0,\infty))$ $(i=1,2,...,k)$,
$a(t)\in C([0,\infty);[0,\infty))$ and $F(x,t)\in C(\overline{\Omega};\mathbb{R}^N)$;

(H6.6-2) $\rho_i(t)\in C^1([0,\infty);\mathbb{R})$, $\lim_{t\to\infty}\rho_i(t)=\infty$ $(i=1,2,...,\ell)$,
$\tau_i(t)\in C([0,\infty);\mathbb{R})$, $\lim_{t\to\infty}\tau_i(t)=\infty$ $(i=1,2,...,k)$,
$\sigma_i(t)\in C([0,\infty);\mathbb{R})$, $\lim_{t\to\infty}\sigma_i(t)=\infty$ $(i=1,2,...,m)$;

(H6.6-3) $c_i(x,t,\Xi)\in C(\overline{\Omega}\times\mathbb{R}^N;\mathbb{R})$ $(i=1,2,...,m)$,
$c_i(x,t,\Xi)\ge p_i(t)\psi_i(|\Xi|)$ for $(x,t)\in\Omega$, $\Xi\in\mathbb{R}^N$,
$p_i(t)\in C([0,\infty);[0,\infty))$, $\psi_i(\xi)\in C([0,\infty);[0,\infty))$ and $\psi_i(\xi)$ are nondecreasing for $\xi\ge 0$.

The following two kinds of boundary conditions are considered:

(B$_1$) $\quad U=\Psi \quad$ on $\quad \partial G\times(0,\infty)$,

(B$_2$) $\quad \dfrac{\partial U}{\partial\nu}+\mu U=\tilde{\Psi} \quad$ on $\quad \partial G\times(0,\infty)$,

where $\Psi,\ \tilde{\Psi}\in C(\partial G\times(0,\infty);\mathbb{R}^N)$, $\mu\in C(\partial G\times(0,\infty);[0,\infty))$ and ν denotes the unit exterior normal vector to ∂G.

Definition 6.6.1. By a *solution* of Eq. (6.6.1) we mean a function $U(x,t)$ $\in C^2(\overline{G}\times[t_{-1},\infty);\mathbb{R}^N)\cap C^1(\overline{G}\times[\hat{t}_{-1},\infty);\mathbb{R}^N)\cap C(\overline{G}\times[\tilde{t}_{-1},\infty);\mathbb{R}^N)$ which satisfies (6.6.1), where

$$t_{-1} = \min\left\{0, \min_{1\le i\le k}\left\{\inf_{t\ge 0}\tau_i(t)\right\}\right\},$$

$$\hat{t}_{-1} = \min\left\{0, \min_{1\le i\le m}\left\{\inf_{t\ge 0}\rho_i(t)\right\}\right\},$$

$$\tilde{t}_{-1} = \min\left\{0, \min_{1\le i\le m}\left\{\inf_{t\ge 0}\sigma_i(t)\right\}\right\}.$$

Definition 6.6.2. Let H be a unit vector in $\mathbb{R}^N$. A solution $U(x,t)$ of (6.6.1) is said to be *H-oscillatory* in Ω if the scalar product $\langle U(x,t), H\rangle$ has a zero in $G \times (t, \infty)$ for any $t > 0$.

We give two examples of $c_i(x,t,\Xi)$ which satisfy the hypothesis (H6.6-3) (*cf.* [224]). Let $M_i(x,t) \in C^1(\Omega)$ be symmetric, positive definite matrix functions, and $\lambda_i(x,t)$ be the smallest eigenvalue of $M_i(x,t)$. Then it can be shown that

$$\begin{aligned} c_i(x,t,\Xi) &= (|\Xi|^\gamma \Xi)^T M_i(x,t)\,(|\Xi|^\gamma \Xi) \\ &\ge p_i(t)|\Xi|^{2\gamma+2} \quad \text{for } (x,t)\in\Omega,\ \Xi\in\mathbb{R}^N, \end{aligned}$$

where T denotes the transpose, $\gamma \ge -1$ and $p_i(t) = \min_{x\in\overline{G}} \lambda_i(x,t)$. Another example is the following

$$c_i(x,t,\Xi) = \left(\Xi^T M_i(x,t)\Xi\right)^\delta \quad (\delta \ge 0).$$

It is easily seen that

$$c_i(x,t,\Xi) \ge \left(p_i(t)|\Xi|^2\right)^\delta = (p_i(t))^\delta\,|\Xi|^{2\delta}.$$

We use the following notation:

$$u_H(x,t) = \langle U(x,t), H\rangle,$$
$$f_H(x,t) = \langle F(x,t), H\rangle,$$

where H is a unit vector in $\mathbb{R}^N$ and $\langle U, V\rangle$ denotes the scalar product of $U, V \in \mathbb{R}^N$.

Theorem 6.6.1. *Assume that* (H6.6-1)–(H6.6-3) *hold. Let* $U(x,t)$ *be a solution of* (6.6.1). *If* $u_H(x,t)$ *is eventually positive, then* $u_H(x,t)$ *satisfies the scalar parabolic differential inequality*

$$\frac{\partial}{\partial t}\left(u_H(x,t) + \sum_{i=1}^{\ell} h_i(t)u_H(x,\rho_i(t))\right) - a(t)\Delta u_H(x,t)$$
$$- \sum_{i=1}^{k} b_i(t)\Delta u_H(x,\tau_i(t)) + \sum_{i=1}^{m} p_i(t)\varphi_i(u_H(x,\sigma_i(t))) \le f_H(x,t), \quad (6.6.2)$$

where $\varphi_i(\xi) = \xi\psi_i(|\xi|)$. *If* $u_H(x,t)$ *is eventually negative, then* $v_H(x,t) := -u_H(x,t)$ *satisfies the scalar parabolic differential inequality*

$$\frac{\partial}{\partial t}\left(v_H(x,t) + \sum_{i=1}^{\ell} h_i(t) v_H(x,\rho_i(t))\right) - a(t)\Delta v_H(x,t)$$
$$- \sum_{i=1}^{k} b_i(t)\Delta v_H(x,\tau_i(t)) + \sum_{i=1}^{m} p_i(t)\varphi_i\big(v_H(x,\sigma_i(t))\big) \le -f_H(x,t). \quad (6.6.3)$$

Proof. Let $u_H(x,t)$ be eventually positive. The scalar product of (6.6.1) and the unit vector H yields

$$\frac{\partial}{\partial t}\left(\langle U(x,t), H\rangle + \sum_{i=1}^{\ell} h_i(t)\langle U(x,\rho_i(t)), H\rangle\right)$$
$$-a(t)\langle \Delta U(x,t), H\rangle - \sum_{i=1}^{k} b_i(t)\langle \Delta U(x,\tau_i(t)), H\rangle$$
$$+\sum_{i=1}^{m} c_i\big(x,t,U(x,\sigma_i(t))\big)\langle U(x,\sigma_i(t)), H\rangle = \langle F(x,t), H\rangle. \quad (6.6.4)$$

The hypothesis (H6.6-3) implies that

$$c_i\big(x,t,U(x,\sigma_i(t))\big)\langle U(x,\sigma_i(t)), H\rangle$$
$$\ge p_i(t)\psi_i\left(|U(x,\sigma_i(t))|\right)\langle U(x,\sigma_i(t)), H\rangle. \quad (6.6.5)$$

It follows from Schwarz's inequality that

$$\langle U(x,\sigma_i(t)), H\rangle \le |U(x,\sigma_i(t))||H| = |U(x,\sigma_i(t))|. \quad (6.6.6)$$

Combining (6.6.5) with (6.6.6), we obtain

$$c_i\big(x,t,U(x,\sigma_i(t))\big)\langle U(x,\sigma_i(t)), H\rangle$$
$$\ge p_i(t)\psi_i\big(|U(x,\sigma_i(t))|\big)\langle U(x,\sigma_i(t)), H\rangle$$
$$\ge p_i(t)\psi_i\big(\langle U(x,\sigma_i(t)), H\rangle\big)\langle U(x,\sigma_i(t)), H\rangle$$
$$= p_i(t)\psi_i\big(u_H(x,\sigma_i(t))\big)u_H(x,\sigma_i(t))$$
$$= p_i(t)\varphi_i\big(u_H(x,\sigma_i(t))\big). \quad (6.6.7)$$

From (6.6.4) and (6.6.7) we conclude that $u_H(x,t)$ satisfies the differential inequality (6.6.2). If $u_H(x,t)$ is eventually negative, we find that

$$c_i\big(x,t,U(x,\sigma_i(t))\big)\langle U(x,\sigma_i(t)), H\rangle$$
$$\le p_i(t)\psi_i\big(|U(x,\sigma_i(t))|\big)\langle U(x,\sigma_i(t)), H\rangle$$
$$\le p_i(t)\psi_i\big(-\langle U(x,\sigma_i(t)), H\rangle\big)\langle U(x,\sigma_i(t)), H\rangle$$
$$= -p_i(t)\psi_i\big(-u_H(x,\sigma_i(t))\big)\big(-u_H(x,\sigma_i(t))\big)$$
$$= -p_i(t)\varphi_i\big(-u_H(x,\sigma_i(t))\big). \quad (6.6.8)$$

We multiply (6.6.4) by -1 and taking into account (6.6.8) to conclude that $v_H(x,t) = -u_H(x,t)$ satisfies the differential inequality (6.6.3). □

Associated with the boundary conditions (B_i) $(i = 1, 2)$, we consider the following scalar boundary conditions:

$(\tilde{\mathrm{B}}_1)$ $\quad u = \psi_H \quad \text{on} \quad \partial G \times (0, \infty),$

$(\tilde{\mathrm{B}}_2)$ $\quad \dfrac{\partial u}{\partial \nu} + \mu u = \tilde{\psi}_H \quad \text{on} \quad \partial G \times (0, \infty),$

where

$$\psi_H = \langle \Psi, H \rangle,$$
$$\tilde{\psi}_H = \langle \tilde{\Psi}, H \rangle.$$

Theorem 6.6.2. *Assume that* (H6.6-1)–(H6.6-3) *are satisfied. If the scalar parabolic differential inequalities*

$$\frac{\partial}{\partial t}\left(u(x,t) + \sum_{i=1}^{\ell} h_i(t) u(x, \rho_i(t))\right) - a(t)\Delta u(x,t)$$
$$- \sum_{i=1}^{k} b_i(t) \Delta u(x, \tau_i(t)) + \sum_{i=1}^{m} p_i(t) \varphi_i\big(u(x, \sigma_i(t))\big) \le \pm f_H(x,t) \quad (6.6.9)$$

have no eventually positive solutions which satisfy the boundary conditions $(\tilde{\mathrm{B}}_i)$ $(i = 1, 2)$, *then every solution* $U(x,t)$ *of the boundary value problems* (6.6.1), (B_i) $(i = 1, 2)$ *is* H*-oscillatory in* Ω, *respectively.*

Proof. Suppose to the contrary that there is a solution $U(x,t)$ of the problem (6.6.1), (B_1) which is not H-oscillatory in Ω. If $u_H(x,t)$ is eventually positive, then Theorem 6.6.1 implies that $u_H(x,t)$ satisfies (6.6.9) with $+f_H(x,t)$. It is obvious that $u_H(x,t)$ satisfies the boundary condition $(\tilde{\mathrm{B}}_1)$. This contradicts the hypothesis. If $u_H(x,t)$ is eventually negative, then $v_H(x,t) = -u_H(x,t)$ is an eventually positive solution of (6.6.9) with $-f_H(x,t)$ which satisfies the boundary condition $(\tilde{\mathrm{B}}_1)$. This contradicts the hypothesis. In case we treat the boundary condition (B_2), we are also led to a contradiction. This completes the proof. □

The following notation will be used:

$$F_H(t) = \left(\int_G \Phi(x)\,dx \right)^{-1} \int_G f_H(x,t)\Phi(x)\,dx,$$

$$\Psi_H(t) = \left(\int_G \Phi(x)\,dx \right)^{-1} \int_{\partial G} \psi_H \frac{\partial \Phi}{\partial \nu}(x)\,dS,$$

$$\tilde{F}_H(t) = \frac{1}{|G|} \int_G f_H(x,t)\,dx,$$

$$\tilde{\Psi}_H(t) = \frac{1}{|G|} \int_{\partial G} \tilde{\psi}_H\,dS,$$

where $|G| = \int_G dx$.

Theorem 6.6.3. *Assume that* (H6.6-1)–(H6.6-3) *hold, and the following hypothesis* (H6.6-4) *holds*:

(H6.6-4) $\varphi_i(\xi) = \xi\psi_i(|\xi|)$ $(i = 1, 2, ..., m)$ *are convex in* $(0, \infty)$.

If the functional differential inequalities

$$\frac{d}{dt}\left(y(t) + \sum_{i=1}^{\ell} h_i(t) y(\rho_i(t)) \right) + \lambda_1 a(t) y(t)$$
$$+ \lambda_1 \sum_{i=1}^{k} b_i(t) y(\tau_i(t)) + \sum_{i=1}^{m} p_i(t)\varphi_i\big(y(\sigma_i(t))\big) \le \pm G_H(t) \qquad (6.6.10)$$

have no eventually positive solutions, then every solution $U(x,t)$ *of the boundary value problem* (6.6.1), (B_1) *is* H*-oscillatory in* Ω, *where*

$$G_H(t) = F_H(t) - a(t)\Psi_H(t) - \sum_{i=1}^{k} b_i(t)\Psi_H(\tau_i(t)).$$

Proof. Suppose to the contrary that there is a solution $U(x,t)$ of the problem (6.6.1), (B_1) which is not H-oscillatory in Ω. We consider the case where $u_H(x,t) > 0$ in $G \times [t_0, \infty)$ for some $t_0 > 0$. We observe that $\varphi_i(\xi) \in C(\mathbb{R};\mathbb{R})$ $(i = 1, 2, ..., m)$, $\varphi_i(-\xi) = -\varphi_i(\xi)$, $\varphi_i(\xi) > 0$ for $\xi > 0$, and $\varphi_i(\xi)$ are nondecreasing in $(0, \infty)$. We easily see that

$$U_H(t) := \left(\int_G \Phi(x)\,dx \right)^{-1} \int_G u_H(x,t)\Phi(x)\,dx$$

is an eventually positive solution of the differential inequality (6.6.10) with $+G_H(t)$ (*cf.* [258, Theorem 1]). This contradicts the hypothesis. The case where $u_H(x,t) < 0$ in $G \times [t_0, \infty)$ can be handled similarly, and we are also led to a contradiction. This completes the proof. $\square$

Theorem 6.6.4. *Assume that* (H6.6-1)–(H6.6-4) *hold. If the functional differential inequalities*

$$\frac{d}{dt}\left(y(t)+\sum_{i=1}^{\ell}h_i(t)y(\rho_i(t))\right)+\sum_{i=1}^{m}p_i(t)\varphi_i\big(y(\sigma_i(t))\big)\leq \pm\tilde{G}_H(t) \quad (6.6.11)$$

have no eventually positive solutions, then every solution $U(x,t)$ *of the boundary value problem* (6.6.1), (B_2) *is* H*-oscillatory in* Ω, *where*

$$\tilde{G}_H(t)=\tilde{F}_H(t)+a(t)\tilde{\Psi}_H(t)+\sum_{i=1}^{k}b_i(t)\tilde{\Psi}_H(\tau_i(t)).$$

Proof. The proof is quite similar to that of Theorem 6.6.3, and hence will be omitted. □

Now we consider each of the following cases:

(H6.6-5) $h_i(t)\geq 0$ $(i=1,2,...,\ell)$ and $\sum_{i=1}^{\ell}h_i(t)\leq 1$;

(H6.6-6) $h_i(t)\leq 0$ $(i=1,2,...,\ell)$ and $\sum_{i=1}^{\ell}h_i(t)\geq -1$.

Theorem 6.6.5. *Assume that* (H6.6-1)–(H6.6-5) *hold, and assume, moreover, that*

(i) $\rho_i(t)\geq t$ $(i=1,2,...,\ell)$;
(ii) *there exists a function* $\Theta_H(t)\in C^1([t_0,\infty);\mathbb{R})$ *such that* $\Theta_H(t)$ *is oscillatory at* $t=\infty$ *and* ${\Theta_H}'(t)=G_H(t)$ [*resp.* ${\Theta_H}'(t)=\tilde{G}_H(t)$], *where* t_0 *is some positive number.*

If

$$\int_{t_0}^{\infty}p_j(s)\varphi_j\left(\left[\left(1-\sum_{i=1}^{\ell}h_i(\sigma_j(s))\right)[\Theta_H(\sigma_j(s))]_{\mp}+\hat{\Theta}_H(\sigma_j(s))\right]_{+}\right)ds=\infty \quad (6.6.12)$$

for some $j\in\{1,2,...,m\}$, *then every solution* $U(x,t)$ *of the boundary value problem* (6.6.1), (B_1) [*resp.* (1), (B_2)] *is* H*-oscillatory in* Ω, *where*

$$[\Theta_H(t)]_{\pm}=\max\{\pm\Theta_H(t),0\},$$
$$\hat{\Theta}_H(t)=\Theta_H(t)-\sum_{i=1}^{\ell}h_i(t)\Theta_H(\rho_i(t)).$$

Proof. We observe that (6.6.10) have no eventually positive solutions in view of Lemma 6.3.1 of Section 6.3. Hence, the conclusion follows from Theorem 6.6.3. □

Example 6.6.1. We consider the problem

$$\begin{aligned}&\frac{\partial}{\partial t}\left(U(x,t)+\frac{1}{2}U(x,t+2\pi)\right)-\frac{\partial^2 U}{\partial x^2}(x,t)\\&\quad-\frac{\partial^2 U}{\partial x^2}(x,t-\tfrac{3}{2}\pi)+\frac{5}{2}U(x,t-\tfrac{\pi}{2})\\&\quad=\begin{pmatrix}(\sin x)\sin t\\(2+e^{2\pi}+e^{-(3/2)\pi}+\frac{5}{2}e^{-(1/2)\pi})(\sin x)e^t\end{pmatrix},\end{aligned}\tag{6.6.13}$$

$$U(0,t)=U(\pi,t)=\begin{pmatrix}0\\0\end{pmatrix}\tag{6.6.14}$$

for $(x,t)\in(0,\pi)\times(0,\infty)$. Here $n=1$, $G=(0,\pi)$, $\Omega=(0,\pi)\times(0,\infty)$, $\ell=k=m=1$, $N=2$, $h_1(t)=1/2$, $\rho_1(t)=t+2\pi$, $a(t)=b_1(t)=1$, $\tau_1(t)=t-(3/2)\pi$, $\sigma_1(t)=t-(1/2)\pi$, $c_1(x,t,\Xi)=1$, $p_1(t)=5/2$, $\psi_1(\xi)=1$, $\varphi_1(\xi)=\xi$, $\Psi=\begin{pmatrix}0\\0\end{pmatrix}$ and

$$F(x,t)=\begin{pmatrix}(\sin x)\sin t\\(2+e^{2\pi}+e^{-(3/2)\pi}+\frac{5}{2}e^{-(1/2)\pi})(\sin x)e^t\end{pmatrix}.$$

It is easy to see that $\lambda_1=1$, $\Phi(x)=\sin x$. Letting $H=e_1=\begin{pmatrix}1\\0\end{pmatrix}$, we find that $f_{e_1}(x,t)=(\sin x)\sin t$, $F_{e_1}(t)=(\pi/4)\sin t$, $\psi_{e_1}=0$, $\Psi_{e_1}(t)=0$ and $G_{e_1}(t)=F_{e_1}(t)=(\pi/4)\sin t$. Choosing $\Theta_{e_1}(t)=-(\pi/4)\cos t$, we observe that $\hat{\Theta}_{e_1}(t)=-(\pi/8)\cos t$ and

$$\begin{aligned}&\int_{t_0}^{\infty}\frac{5}{2}\left[\left(1-\frac{1}{2}\right)\left[-\frac{\pi}{4}\cos\left(t-\frac{1}{2}\pi\right)\right]_{\mp}-\frac{\pi}{8}\cos t\right]_{+}ds\\&=\frac{5}{16}\pi\int_{t_0}^{\infty}\left[-[\sin t]_{\mp}-\cos t\right]_{+}ds\\&=\infty.\end{aligned}$$

Hence, Theorem 6.6.5 implies that every solution $U(x,t)$ of the problem (6.6.13), (6.6.14) is e_1-oscillatory in $(0,\pi)\times(0,\infty)$. One such solution is

$$U(x,t)=\begin{pmatrix}(\sin x)\sin t\\(\sin x)e^t\end{pmatrix}.$$

We note that the above solution $U(x,t)$ is not e_2-oscillatory in $(0,\pi)\times(0,\infty)$, where $e_2=\begin{pmatrix}0\\1\end{pmatrix}$.

Theorem 6.6.6. *Assume that* (H6.6-1)–(H6.6-4), (H6.6-6) *hold, and assume, moreover, that*

(i) $\rho_i(t) \le t \ (i = 1, 2, ..., \ell)$;

(ii) *there exists an integer* $j \in \{1, 2, ..., m\}$ *such that* $\varphi_j(\xi) > 0$ *for* $\xi > 0$, $p_j(t) > 0$ *on* $[t_0, \infty)$ *and*

$$\int_{t_0}^{\infty} p_j(t)\, dt = \infty$$

for some $t_0 > 0$;

(iii) *there is a function* $\Theta_H(t) \in C^1([t_0, \infty); \mathbb{R})$ *such that* ${\Theta_H}'(t) = G_H(t)$ *[resp.* ${\Theta_H}'(t) = \tilde{G}_H(t)$*]*, $\lim_{t\to\infty} \Theta_H(t) = 0$ *and*

$$\int_{t_0}^{\infty} p_j(t)\varphi_j\left([\Theta_H(\sigma_j(t))]_{\pm}\right) dt = \infty.$$

Then every solution $U(x,t)$ *of the boundary value problem* (6.6.1), (B$_1$)*[resp.* (6.6.1), (B$_2$)*] is* H*-oscillatory in* Ω.

Proof. Combining Theorems 6.6.3, 6.6.4 and Lemmas 6.4.1, 6.4.2, we observe that the conclusion follows. □

Theorem 6.6.7. *Assume that* (H6.6-1)–(H6.6-4), (H6.6-6), (i) *of Theorem* 6.6.6 *hold, and assume that:*

(i) *there exist constants* $\tilde{p}_i$ *such that*

$$\rho_i'(t) \ge \tilde{\rho}_i > 0 \quad (i = 1, 2, ..., \ell);$$

(ii) *there is an integer* $j \in \{1, 2, ..., m\}$ *for which*

$$p_j(t) \ge \tilde{p}_j > 0 \quad \textit{for some positive constant } \tilde{p}_j,$$
$$0 \le \sigma_j'(t) \le \tilde{\sigma}_j \quad \textit{for some positive constant } \tilde{\sigma}_j;$$

(iii) *for some* $\beta > 0$

$$\varphi_j(\xi) \ge \beta\xi \quad \textit{for } \xi > 0.$$

Assume, moreover, that $\sigma_j(t) \le t$ *and* $\sigma_j(t)$ *is nondecreasing in* $(0, \infty)$. *Every solution* $U(x,t)$ *of the boundary value problem* (6.6.1), (B$_1$)*[resp.* (6.6.1), (B$_2$)*] is* H*-oscillatory in* Ω *if there exists a function* $\Theta_H(t) \in C^1((0,\infty); \mathbb{R})$ *such that* ${\Theta_H}'(t) = G_H(t)$ *[resp.* ${\Theta_H}'(t) = \tilde{G}_H(t)$*]*,

$\lim_{t\to\infty} \Theta_H(t) = 0$ *and there is a sequence* $\{t_n\}_{n=1}^{\infty}$ *such that:*

$$\lim_{n\to\infty} t_n = \infty,$$

$$\int_{\sigma_j(t_n)}^{t_n} p_j(t)\, dt \geq \frac{1}{\beta},$$

$$\int_{\sigma_j(t_n)}^{t_n} p_j(t)\Theta_H(\sigma_j(t))\, dt + \beta \int_{\sigma_j(t_n)}^{t_n} p_j(t) \left(\int_{\sigma_j(t)}^{\sigma_j(t_n)} p_j(s)\Theta_H(\sigma_j(s))\, ds \right) dt = 0.$$

Proof. The conclusion follows by combining Theorems 6.6.3, 6.6.4 and a result of Yoshida [318, Theorem 9]. □

Example 6.6.2. We consider the problem

$$\frac{\partial}{\partial t}\left(U(x,t) - \frac{1}{e^{\pi}} U(x, t-\pi) \right) - 2\frac{\partial^2 U}{\partial x^2}(x,t) - e^{\pi}\frac{\partial^2 U}{\partial x^2}(x, t-\pi) + U(x, t-2\pi) = \begin{pmatrix} -2(\cos x)e^{-t}\sin t \\ 2(1+e^{2\pi})(\cos x)e^{-t} \end{pmatrix}, \quad (x,t) \in (0,\pi)\times(0,\infty), \tag{6.6.15}$$

$$-\frac{\partial U}{\partial x}(0,t) = \frac{\partial U}{\partial x}(\pi, t) = \begin{pmatrix} 0 \\ 0 \end{pmatrix}, \quad t > 0. \tag{6.6.16}$$

Here $n = 1$, $G = (0,\pi)$, $\Omega = (0,\pi)\times(0,\infty)$, $\ell = k = m = 1$, $N = 2$, $h_1(t) = -e^{-\pi}$, $\rho_1(t) = t - \pi$, $a(t) = 2$, $b_1(t) = e^{\pi}$, $\tau_1(t) = t - \pi$, $\sigma_1(t) = t - 2\pi$, $c_1(x,t,\Xi) = 1$, $p_1(t) = 1$, $\psi_1(\xi) = 1$, $\varphi_1(\xi) = \xi$, $\tilde{\Psi} = \begin{pmatrix} 0 \\ 0 \end{pmatrix}$ and

$$F(x,t) = \begin{pmatrix} -2(\cos x)e^{-t}\sin t \\ 2(1+e^{2\pi})(\cos x)e^{-t} \end{pmatrix}.$$

Letting $H = e_1$, we see that $\tilde{\psi}_{e_1} = 0$, $\tilde{\Psi}_{e_1}(t) = 0$, $\tilde{G}_{e_1}(t) = \tilde{F}_{e_1}(t) = 0$. Therefore, we can choose $\Theta_{e_1}(t) = 0$. Since

$$\int_{\sigma_1(t_n)}^{t_n} p_1(t)\, dt = \int_{t_n - 2\pi}^{t_n} dt = 2\pi \geq 1$$

for any sequence $\{t_n\}$, it follows from Theorem 6.6.7 that every solution $U(x,t)$ of the problem (6.6.15), (6.6.16) is e_1-oscillatory in $(0,\pi)\times(0,\infty)$. For example,

$$U(x,t) = \begin{pmatrix} (\cos x)e^{-t}\cos t \\ (\cos x)e^{-t} \end{pmatrix}$$

is such a solution. We note that the above solution $U(x,t)$ is H-oscillatory in $(0,\pi)\times(0,\infty)$ for any unit vector $H \in \mathbb{R}^2$.

6.7 Parabolic Systems II

In 1990 Gopalsamy [90] investigated oscillations of solutions of certain parabolic system of neutral type, and Liu [181] obtained oscillation results for systems of nonlinear delay parabolic equations. For recent results we refer the reader to Li and Cui [170], Li and Meng [177], Li, Cui and Debnath [172] and the references cited therein.

We are concerned with the oscillation of the system of parabolic equations with functional arguments

$$\begin{aligned}&\frac{\partial}{\partial t}\Big(U(x,t)+\sum_{i=1}^{\ell}H_i(t)U(x,\rho_i(t))\Big)-A(t)\Delta U(x,t)\\&-\sum_{i=1}^{K}B_i(t)\Delta U(x,\tau_i(t))+\sum_{i=1}^{m}P_i(x,t)\varphi_i\big(U(x,\sigma_i(t))\big)\\&=F(x,t),\quad (x,t)\in\Omega:=G\times(0,\infty),\end{aligned}\tag{6.7.1}$$

where G is a bounded domain in $\mathbb{R}^n$ with piecewise smooth boundary ∂G, Δ is the Laplacian in $\mathbb{R}^n$, and

$$\begin{aligned}&H_i(t)=\big(h_{ijk}(t)\big)_{j,k=1}^{M},\ A(t)=\big(a_{jk}(t)\big)_{j,k=1}^{M},\ B_i(t)=\big(b_{ijk}(t)\big)_{j,k=1}^{M},\\&P_i(x,t)=\big(p_{ijk}(x,t)\big)_{j,k=1}^{M},\\&U(x,t)=\big(u_1(x,t),...,u_M(x,t)\big)^T,\\&U(x,\rho_i(t))=\big(u_1(x,\rho_i(t)),...,u_M(x,\rho_i(t))\big)^T,\\&U(x,\tau_i(t))=\big(u_1(x,\tau_{i1}(t)),...,u_M(x,\tau_{iM}(t))\big)^T,\\&\varphi_i\big(U(x,\sigma_i(t))\big)=\big(\varphi_{i1}(u_1(x,\sigma_i(t))),...,\varphi_{iM}(u_M(x,\sigma_i(t)))\big)^T,\\&F(x,t)=\big(f_1(x,t),...,f_M(x,t)\big)^T,\end{aligned}$$

the superscript T denoting the transpose.

We easily see that (6.7.1) is equivalent to the following system:

$$\begin{aligned}&\frac{\partial}{\partial t}\Big(u_j(x,t)+\sum_{i=1}^{\ell}\sum_{k=1}^{M}h_{ijk}(t)u_k(x,\rho_i(t))\Big)\\&-\sum_{k=1}^{M}a_{jk}(t)\Delta u_k(x,t)-\sum_{i=1}^{K}\sum_{k=1}^{M}b_{ijk}(t)\Delta u_k(x,\tau_{ik}(t))\\&+\sum_{i=1}^{m}\sum_{k=1}^{M}p_{ijk}(x,t)\varphi_{ik}\big(u_k(x,\sigma_i(t))\big)=f_j(x,t)\quad(j=1,2,...,M)\end{aligned}\tag{6.7.2}$$

for $(x,t) \in \Omega$.

The boundary condition to be considered is the following:

(BC) $\alpha(\hat{x})\dfrac{\partial u_j}{\partial \nu} + \mu(\hat{x})\, u_j = \alpha(\hat{x})\tilde{\psi}_j + \mu(\hat{x})\psi_j$ on $\partial G \times (0,\infty)$ $(j = 1,2,...,M)$,

where ψ_j, $\tilde{\psi}_j \in C(\partial G \times (0,\infty); \mathbb{R})$, $\alpha(\hat{x})$, $\mu(\hat{x}) \in C(\partial G; [0,\infty))$, $\alpha(\hat{x})^2 + \mu(\hat{x})^2 \neq 0$, and ν denotes the unit exterior normal vector to ∂G. If $\alpha(\hat{x}) \equiv 0$ on ∂G, then $\mu(\hat{x}) \neq 0$ on ∂G, and therefore the boundary condition (BC) reduces to

$$u_j = \psi_j \quad \text{on} \quad \partial G \times (0,\infty) \quad (j = 1,2,...,M).$$

In the case where $\alpha(\hat{x}) = 1$ on ∂G, then (BC) can be written as

$$\frac{\partial u_j}{\partial \nu} + \mu(\hat{x})\, u_j = \tilde{\psi}_j + \mu(\hat{x})\psi_j \quad \text{on} \quad \partial G \times (0,\infty) \quad (j = 1,2,...,M).$$

Furthermore, if $\mu(\hat{x}) \equiv 0$ on ∂G, then (BC) can be written in the form

$$\frac{\partial u_j}{\partial \nu} = \tilde{\psi}_j \quad \text{on} \quad \partial G \times (0,\infty) \quad (j = 1,2,...,M).$$

It is assumed that:

(H6.7-1) $h_{ijk}(t) \in C^1([0,\infty); [0,\infty))$ $(i = 1,2,...,\ell;\ j,k = 1,2,...,M)$,
$a_{jk}(t) \in C([0,\infty); [0,\infty))$ $(j,k = 1,2,...,M)$,
$b_{ijk}(t) \in C([0,\infty); [0,\infty))$ $(i = 1,2,...,K;\ j,k = 1,2,...,M)$,
$p_{ijk}(x,t) \in C(\overline{G} \times [0,\infty); [0,\infty))$ $(i = 1,2,...,m;\ j,k = 1,2,...,M)$,
$f_j(x,t) \in C(\overline{G} \times [0,\infty); \mathbb{R})$ $(j = 1,2,...,M)$;

(H6.7-2) $\rho_i(t) \in C^1([0,\infty); \mathbb{R})$, $\lim_{t\to\infty} \rho_i(t) = \infty$ $(i = 1,2,...,\ell)$,
$\tau_{ik}(t) \in C([0,\infty); \mathbb{R})$, $\lim_{t\to\infty} \tau_{ik}(t) = \infty$ $(i = 1,2,...,K; k = 1,2,...,M)$,
$\sigma_i(t) \in C([0,\infty); \mathbb{R})$, $\lim_{t\to\infty} \sigma_i(t) = \infty$ $(i = 1,2,...,m)$;

(H6.7-3) $\varphi_{ik}(\xi) \in C(\mathbb{R}; \mathbb{R})$, $\varphi_{ik}(\xi) \geq 0$ for $\xi \geq 0$, $\varphi_{ik}(-\xi) = -\varphi_{ik}(\xi)$ for $\xi > 0$ and $\varphi_{ik}(\xi)$ are convex in $(0,\infty)$ $(i = 1,2,...,m;\ k = 1,2,...,M)$;

(H6.7-4) $h_{ikk}(t) - \sum_{\substack{j=1 \\ j\neq k}}^{M} h_{ijk}(t) \geq 0$ $(i = 1,2,...,\ell;\ k = 1,2,...,M)$;

(H6.7-5) $a_{kk}(t) - \sum_{\substack{j=1 \\ j\neq k}}^{M} a_{jk}(t) \geq 0$ $(k = 1,2,...,M)$;

(H6.7-6) $b_{ikk}(t) - \sum_{\substack{j=1 \\ j \neq k}}^{M} b_{ijk}(t) \geq 0 \quad (i = 1, 2, ..., K; \; k = 1, 2, ..., M);$

(H6.7-7) $p_{ikk}(x,t) - \sum_{\substack{j=1 \\ j \neq k}}^{M} p_{ijk}(x,t) \geq 0 \quad (i = 1, 2, ..., m; \; k = 1, 2, ..., M);$

(H6.7-8) $\rho_i(t) \geq t \; (i = 1, 2, ..., \ell);$

(H6.7-9) $\hat{\varphi}_i(\xi) := \min_{1 \leq k \leq M} \varphi_{ik}(\xi)$ is nondecreasing and convex in $(0, \infty)$ $(i = 1, 2, ..., m)$.

Definition 6.7.1. By a *solution* of system (6.7.2) we mean a vector function $(u_1(x,t), ..., u_M(x,t))$ such that $u_j(x,t) \in C^2(\overline{G} \times [t_{-1}, \infty); \mathbb{R}) \cap C^1(\overline{G} \times [\hat{t}_{-1}, \infty); \mathbb{R}) \cap C(\overline{G} \times [\tilde{t}_{-1}, \infty); \mathbb{R})$, and $u_j(x,t)$ $(j = 1, 2, ..., M)$ satisfy (6.7.2) in Ω, where

$$t_{-1} = \min \left\{ 0, \; \min_{\substack{1 \leq i \leq K \\ 1 \leq k \leq M}} \left\{ \inf_{t \geq 0} \tau_{ik}(t) \right\} \right\},$$

$$\hat{t}_{-1} = \min \left\{ 0, \; \min_{1 \leq i \leq \ell} \left\{ \inf_{t \geq 0} \rho_i(t) \right\} \right\},$$

$$\tilde{t}_{-1} = \min \left\{ 0, \; \min_{1 \leq i \leq m} \left\{ \inf_{t \geq 0} \sigma_i(t) \right\} \right\}.$$

Definition 6.7.2. A solution $(u_1(x,t), ..., u_M(x,t))$ of system (6.7.2) is said to be *weakly oscillatory* in Ω if at least one of its components is oscillatory in Ω (*cf.* Ladde, Lakshmikantham and Zhang [162, Definition 6.2.1]).

It is known that the smallest eigenvalue λ_1 of the eigenvalue problem

$$-\Delta w = \lambda w \quad \text{in } G,$$

$$\alpha(\hat{x}) \frac{\partial w}{\partial \nu} + \mu(\hat{x}) \, w = 0 \quad \text{on } \partial G$$

is nonnegative, and the corresponding eigenfunction $\Phi(x)$ can be chosen so that $\Phi(x) > 0$ in G (see Ye and Li [293, Theorem 3.3.22]). In case $\mu(\hat{x}) \equiv 0$ on ∂G, then we can choose $\lambda_1 = 0$ and $\Phi(x) = 1$. If $\mu(\hat{x}) \not\equiv 0$ on ∂G, then there exist $\lambda_1 > 0$ and the eigenfunction $\Phi(x) > 0$ in G.

We use the notation:

$$\Gamma_1 = \{ \, \hat{x} \in \partial G; \; \alpha(\hat{x}) = 0 \},$$

$$\Gamma_2 = \{ \, \hat{x} \in \partial G; \; \alpha(\hat{x}) \neq 0 \}.$$

Lemma 6.7.1. *If u_j satisfy the boundary condition* (BC) *and $\Phi(x)$ is the eigenfunction corresponding to the smallest eigenvalue $\lambda_1 \geq 0$, then we obtain*

$$K_\Phi \int_{\partial G} \left[\frac{\partial u_j}{\partial \nu} \Phi(x) - u_j \frac{\partial \Phi(x)}{\partial \nu} \right] dS = \Psi_j(t), \tag{6.7.3}$$

where

$$K_\Phi = \left(\int_G \Phi(x)\, dx \right)^{-1},$$

$$\Psi_j(t) = K_\Phi \left(- \int_{\Gamma_1} \psi_j \frac{\partial \Phi(x)}{\partial \nu} dS + \int_{\Gamma_2} \left(\tilde{\psi}_j + \frac{\mu(\hat{x})}{\alpha(\hat{x})} \psi_j \right) \Phi(x)\, dS \right).$$

Proof. It is obvious that

$$\int_{\partial G} \left[\frac{\partial u_j}{\partial \nu} \Phi(x) - u_j \frac{\partial \Phi(x)}{\partial \nu} \right] dS = \int_{\Gamma_1} \left[\frac{\partial u_j}{\partial \nu} \Phi(x) - u_j \frac{\partial \Phi(x)}{\partial \nu} \right] dS + \int_{\Gamma_2} \left[\frac{\partial u_j}{\partial \nu} \Phi(x) - u_j \frac{\partial \Phi(x)}{\partial \nu} \right] dS. \tag{6.7.4}$$

Since $u_j = \psi_j$ on Γ_1 and $\Phi(x) = 0$ on Γ_1, it can be shown that

$$\int_{\Gamma_1} \left[\frac{\partial u_j}{\partial \nu} \Phi(x) - u_j \frac{\partial \Phi(x)}{\partial \nu} \right] dS = - \int_{\Gamma_1} \psi_j \frac{\partial \Phi(x)}{\partial \nu} dS. \tag{6.7.5}$$

From the boundary condition (BC) it follows that

$$\frac{\partial u_j}{\partial \nu} = \tilde{\psi}_j + \frac{\mu(\hat{x})}{\alpha(\hat{x})} \psi_j - \frac{\mu(\hat{x})}{\alpha(\hat{x})} u_j \quad \text{on } \Gamma_2,$$

$$\frac{\partial \Phi(x)}{\partial \nu} = - \frac{\mu(\hat{x})}{\alpha(\hat{x})} \Phi(x) \quad \text{on } \Gamma_2.$$

Therefore we find that

$$\int_{\Gamma_2} \left[\frac{\partial u_j}{\partial \nu} \Phi(x) - u_j \frac{\partial \Phi(x)}{\partial \nu} \right] dS = \int_{\Gamma_2} \left(\tilde{\psi}_j + \frac{\mu(\hat{x})}{\alpha(\hat{x})} \psi_j \right) \Phi(x)\, dS. \tag{6.7.6}$$

Combining (6.7.4)–(6.7.6) yields the desired identity (6.7.3). □

We note that if $\alpha(\hat{x}) = 0$ on ∂G, then $\Gamma_2 = \emptyset$ and

$$K_\Phi \int_{\partial G} \left[\frac{\partial u_j}{\partial \nu} \Phi(x) - u_j \frac{\partial \Phi(x)}{\partial \nu} \right] dS = -K_\Phi \int_{\partial G} \psi_j \frac{\partial \Phi(x)}{\partial \nu} dS.$$

In case $\alpha(\hat{x}) = 1$ $(\hat{x} \in \partial G)$ and $\mu(\hat{x}) = 0$ $(\hat{x} \in \partial G)$, then $\Gamma_1 = \emptyset$ and

$$K_\Phi \int_{\partial G} \left[\frac{\partial u_j}{\partial \nu} \Phi(x) - u_j \frac{\partial \Phi(x)}{\partial \nu} \right] dS = \frac{1}{|G|} \int_{\partial G} \tilde{\psi}_j\, dS,$$

where $|G| = \int_G dx$ denotes the volume of G.

We use the notation:

$$F_j(t) = K_\Phi \int_G f_j(x,t)\Phi(x)\,dx \quad (j = 1,2,...,M),$$

$$\left[\Theta(t)\right]_\pm = \max\{\pm\Theta(t), 0\},$$

$$\Gamma = \left\{ \left((-1)^{\alpha_1}, (-1)^{\alpha_2}, ..., (-1)^{\alpha_M}\right);\ \alpha_j = 0,1\ (j = 1,2,...,M) \right\}.$$

We note that $\#\Gamma = 2^M$ and $-\gamma \in \Gamma$ for $\gamma \in \Gamma$, and hence $\Gamma = \left\{\pm\gamma;\ \gamma \in \tilde{\Gamma}\right\}$ for some $\tilde{\Gamma} \subset \Gamma$ with $\#\tilde{\Gamma} = 2^{M-1}$. For example, we let $M = 2$. Then we observe that

$$\Gamma = \{(1,1),(1,-1),(-1,1),(-1,-1)\}$$

and

$$\Gamma = \{\pm(1,1), \pm(1,-1)\} = \{\pm\gamma;\ \gamma \in \tilde{\Gamma}\},$$

where

$$\tilde{\Gamma} = \{(1,1),(1,-1)\}.$$

Theorem 6.7.1. *Assume that the hypotheses* (H6.7-1)–(H6.7-9) *hold. If the following conditions are satisfied:*

(H6.7-10) $\displaystyle\sum_{i=1}^{\ell}\sum_{j=1}^{M} \tilde{h}_{ij}(t) \le 1$ *on* $[t_0,\infty)$ *for some* $t_0 > 0$;

(H6.7-11) *there exist functions* $\Theta_\gamma(t) \in C^1([t_0,\infty);\mathbb{R})$ $(\gamma \in \Gamma)$ *such that* $\Theta_\gamma(t)$ *are oscillatory at* $t = \infty$ *and* $\Theta'_\gamma(t) = \gamma \cdot \left(G_j(t)\right)_{j=1}^{M}$ *(the dot* $\cdot$ *denotes the scalar product),*

and if for some $j_0 \in \{1,2,...,m\}$

$$\int_{t_0}^{\infty} \hat{p}_{j_0}(t)\hat{\varphi}_{j_0}\left(\left[\left(1 - \sum_{i=1}^{\ell}\sum_{j=1}^{M}\tilde{h}_{ij}(\sigma_{j_0}(t))\right)[\Theta_\gamma(\sigma_{j_0}(t))]_- \right.\right.$$
$$\left.\left. + \hat{\Theta}_\gamma(\sigma_{j_0}(t))\right]_+\right)dt = \infty, \tag{6.7.7}$$

then every solution $\left(u_1(x,t),...,u_M(x,t)\right)$ *of the boundary value problem*

(6.7.2), (BC) *is weakly oscillatory in* Ω, *where*

$$\tilde{h}_{ij}(t) = \max_{1 \le k \le M} h_{ijk}(t),$$

$$G_j(t) = \frac{1}{M}\left(F_j(t) + \sum_{k=1}^{M} a_{jk}(t)\Psi_k(t) + \sum_{i=1}^{K}\sum_{k=1}^{M} b_{ijk}(t)\Psi_k(\tau_{ik}(t))\right),$$

$$\hat{p}_i(t) = \min_{x \in \overline{G}}\left\{\min_{1 \le k \le M}\left(p_{ikk}(x,t) - \sum_{\substack{j=1 \\ j \ne k}}^{M} p_{ijk}(x,t)\right)\right\},$$

$$\hat{\Theta}_\gamma(t) = \Theta_\gamma(t) - \sum_{i=1}^{\ell}\sum_{j=1}^{M} \tilde{h}_{ij}(t)\Theta_\gamma(\rho_i(t)).$$

Proof. Suppose that there is a solution $\big(u_1(x,t), ..., u_M(x,t)\big)$ of the problem (6.7.2), (BC) which is not weakly oscillatory in Ω. Then, each component $u_j(x,t)$ is nonoscillatory in Ω. We easily see that there is a number $t_1 \ge t_0$ such that $|u_j(x,t)| > 0$ for $x \in G,\ t \ge t_1$ $(j = 1, 2, ..., M)$. Letting

$$w_j(x,t) = \delta_j u_j(x,t),$$

where $\delta_j = \text{sgn}\ u_j(x,t)$, we find that $w_j(x,t) = |u_j(x,t)| > 0$ in $G \times [t_1, \infty)$. There is a number $t_2 \ge t_1$ such that $w_j(x,t) > 0$, $w_j(x, \rho_i(t)) > 0$, $w_j(x, \tau_{ij}(t)) > 0$, $w_j(x, \sigma_i(t)) > 0$ in $G \times [t_2, \infty)$. Multiplying (6.7.2) by δ_j, we obtain

$$\begin{aligned}
&\frac{\partial}{\partial t}\left(w_j(x,t) + \sum_{i=1}^{\ell}\sum_{k=1}^{M} \delta_j h_{ijk}(t)\delta_k w_k(x, \rho_i(t))\right) \\
&-\delta_j \sum_{k=1}^{M} a_{jk}(t)\Delta u_k(x,t) - \delta_j \sum_{i=1}^{K}\sum_{k=1}^{M} b_{ijk}(t)\Delta u_k(x, \tau_{ik}(t)) \\
&+\sum_{i=1}^{m}\left(p_{ijj}(x,t)\varphi_{ij}\big(w_j(x,\sigma_i(t))\big) + \sum_{\substack{k=1 \\ k \ne j}}^{M} \delta_j p_{ijk}(x,t)\delta_k \varphi_{ik}\big(w_k(x,\sigma_i(t))\big)\right) \\
&= \delta_j f_j(x,t), \ (j = 1, 2, ..., M). \qquad (6.7.8)
\end{aligned}$$

It is easy to check that

$$\sum_{j=1}^{M}\left(p_{ijj}(x,t)\varphi_{ij}\big(w_j(x,\sigma_i(t))\big)+\sum_{\substack{k=1\\k\neq j}}^{M}\delta_j p_{ijk}(x,t)\delta_k\varphi_{ik}\big(w_k(x,\sigma_i(t))\big)\right)$$

$$\geq\sum_{j=1}^{M}\left(p_{ijj}(x,t)\varphi_{ij}\big(w_j(x,\sigma_i(t))\big)-\sum_{\substack{k=1\\k\neq j}}^{M}p_{ijk}(x,t)\varphi_{ik}\big(w_k(x,\sigma_i(t))\big)\right)$$

$$=\sum_{j=1}^{M}p_{ijj}(x,t)\varphi_{ij}\big(w_j(x,\sigma_i(t))\big)-\sum_{k=1}^{M}\sum_{\substack{j=1\\j\neq k}}^{M}p_{ijk}(x,t)\varphi_{ik}\big(w_k(x,\sigma_i(t))\big)$$

$$=\sum_{k=1}^{M}\left(p_{ikk}(x,t)-\sum_{\substack{j=1\\j\neq k}}^{M}p_{ijk}(x,t)\right)\varphi_{ik}\big(w_k(x,\sigma_i(t))\big)\ \text{ in } G\times[t_2,\infty). \quad (6.7.9)$$

The hypotheses (H6.7-7) and (H6.7-9) imply that

$$\sum_{i=1}^{m}\sum_{k=1}^{M}\left(p_{ikk}(x,t)-\sum_{\substack{j=1\\j\neq k}}^{M}p_{ijk}(x,t)\right)\varphi_{ik}\big(w_k(x,\sigma_i(t))\big)$$

$$\geq\sum_{i=1}^{m}\min_{1\leq k\leq M}\left(p_{ikk}(x,t)-\sum_{\substack{j=1\\j\neq k}}^{M}p_{ijk}(x,t)\right)\sum_{k=1}^{M}\varphi_{ik}\big(w_k(x,\sigma_i(t))\big)$$

$$\geq\sum_{i=1}^{m}\min_{x\in\overline{G}}\left\{\min_{1\leq k\leq M}\left(p_{ikk}(x,t)-\sum_{\substack{j=1\\j\neq k}}^{M}p_{ijk}(x,t)\right)\right\}\sum_{k=1}^{M}\varphi_{ik}\big(w_k(x,\sigma_i(t))\big)$$

$$=\sum_{i=1}^{m}\hat{p}_i(t)\sum_{k=1}^{M}\varphi_{ik}\big(w_k(x,\sigma_i(t))\big)$$

$$\geq\sum_{i=1}^{m}\hat{p}_i(t)\sum_{k=1}^{M}\hat{\varphi}_i\big(w_k(x,\sigma_i(t))\big)\quad \text{in } G\times[t_2,\infty). \quad (6.7.10)$$

Combining (6.7.9) with (6.7.10) yields the following

$$\sum_{i=1}^{m}\sum_{j=1}^{M}\left(p_{ijj}(x,t)\varphi_{ij}\big(w_j(x,\sigma_i(t))\big)+\sum_{\substack{k=1\\k\neq j}}^{M}\delta_j p_{ijk}(x,t)\delta_k\varphi_{ik}\big(w_k(x,\sigma_i(t))\big)\right)$$
$$\geq\sum_{i=1}^{m}\hat{p}_i(t)\sum_{k=1}^{M}\hat{\varphi}_i\big(w_k(x,\sigma_i(t))\big)\quad\text{in } G\times[t_2,\infty). \tag{6.7.11}$$

Summing (6.7.8) over j ($j=1,2,...,M$) and taking account of (6.7.11), we obtain

$$\frac{\partial}{\partial t}\left(\sum_{j=1}^{M}w_j(x,t)+\sum_{i=1}^{\ell}\sum_{j,k=1}^{M}\delta_j\delta_k h_{ijk}(t)w_k(x,\rho_i(t))\right)$$
$$-\sum_{j,k=1}^{M}\delta_j\delta_k a_{jk}(t)\Delta w_k(x,t)-\sum_{i=1}^{K}\sum_{j,k=1}^{M}\delta_j\delta_k b_{ijk}(t)\Delta w_k(x,\tau_{ik}(t))$$
$$+\sum_{i=1}^{m}\hat{p}_i(t)\sum_{k=1}^{M}\hat{\varphi}_i\big(w_k(x,\sigma_i(t))\big)\leq\sum_{j=1}^{M}\delta_j f_j(x,t)\quad\text{in } G\times[t_2,\infty). \tag{6.7.12}$$

Multiplying (6.7.12) by $K_\Phi\Phi(x)$ and then integrating over G, we have

$$\frac{\partial}{\partial t}\left(\sum_{j=1}^{M}W_j(t)+\sum_{i=1}^{\ell}\sum_{j,k=1}^{M}\delta_j\delta_k h_{ijk}(t)W_k(\rho_i(t))\right)$$
$$-\sum_{j,k=1}^{M}\delta_j\delta_k a_{jk}(t)K_\Phi\int_G\Delta w_k(x,t)\Phi(x)\,dx$$
$$-\sum_{i=1}^{K}\sum_{j,k=1}^{M}\delta_j\delta_k b_{ijk}(t)K_\Phi\int_G\Delta w_k(x,\tau_{ik}(t))\Phi(x)\,dx$$
$$+\sum_{i=1}^{m}\hat{p}_i(t)\sum_{k=1}^{M}K_\Phi\int_G\hat{\varphi}_i\big(w_k(x,\sigma_i(t))\big)\Phi(x)\,dx\leq\sum_{j=1}^{M}\delta_j F_j(t) \tag{6.7.13}$$

for $t\geq t_2$, where

$$W_j(t)=K_\Phi\int_G w_j(x,t)\Phi(x)\,dx\quad(j=1,2,...,M).$$

It follows from Green's formula and Lemma 6.7.1 that

$$\begin{aligned}
&K_\Phi \int_G \Delta w_k(x,t)\Phi(x)\,dx \\
&= K_\Phi \int_{\partial G} \left[\frac{\partial w_k(x,t)}{\partial \nu}\Phi(x) - w_k(x,t)\frac{\partial \Phi(x)}{\partial \nu} \right] dS \\
&\quad + K_\Phi \int_G w_k(x,t)\Delta\Phi(x)\,dx \\
&= \delta_k \Psi_k(t) - \lambda_1 W_k(t)
\end{aligned}$$

and therefore

$$\begin{aligned}
&-\sum_{j,k=1}^{M} \delta_j \delta_k a_{jk}(t) K_\Phi \int_G \Delta w_k(x,t)\Phi(x)\,dx \\
&= -\sum_{j=1}^{M} \delta_j \left(\sum_{k=1}^{M} a_{jk}(t)\Psi_k(t) \right) + \lambda_1 \sum_{j,k=1}^{M} \delta_j \delta_k a_{jk}(t) W_k(t) \\
&\ge -\sum_{j=1}^{M} \delta_j \left(\sum_{k=1}^{M} a_{jk}(t)\Psi_k(t) \right) + \lambda_1 \sum_{k=1}^{M} \left(a_{kk}(t) - \sum_{\substack{j=1 \\ j\neq k}}^{M} a_{jk}(t) \right) W_k(t) \\
&\ge -\sum_{j=1}^{M} \delta_j \left(\sum_{k=1}^{M} a_{jk}(t)\Psi_k(t) \right), \quad t \ge t_2 \qquad (6.7.14)
\end{aligned}$$

in view of the hypothesis (H6.7-5). Analogously we have

$$\begin{aligned}
&-\sum_{i=1}^{K} \sum_{j,k=1}^{M} \delta_j \delta_k b_{ijk}(t) K_\Phi \int_G \Delta w_k(x,\tau_{ik}(t))\Phi(x)\,dx \\
&\qquad \ge -\sum_{j=1}^{M} \delta_j \left(\sum_{i=1}^{K} \sum_{k=1}^{M} b_{ijk}(t)\Psi_k(\tau_{ik}(t)) \right), \quad t \ge t_2. \qquad (6.7.15)
\end{aligned}$$

An application of Jensen's inequality shows that

$$K_\Phi \int_G \hat{\varphi}_i\big(w_k(x,\sigma_i(t))\big)\Phi(x)\,dx \ge \hat{\varphi}_i\big(W_k(\sigma_i(t))\big), \quad t \ge t_2. \qquad (6.7.16)$$

Combining (6.7.13)–(6.7.16) yields

$$\begin{aligned}
&\frac{d}{dt}\left(\sum_{j=1}^{M} W_j(t) + \sum_{i=1}^{\ell} \sum_{j,k=1}^{M} \delta_j \delta_k h_{ijk}(t) W_k(\rho_i(t)) \right) \\
&\qquad + \sum_{i=1}^{m} \hat{p}_i(t) \sum_{k=1}^{M} \hat{\varphi}_i\big(W_k(\sigma_i(t))\big) \\
&\le \sum_{j=1}^{M} \delta_j \left(F_j(t) + \sum_{k=1}^{M} a_{jk}(t)\Psi_k(t) + \sum_{i=1}^{K} \sum_{k=1}^{M} b_{ijk}(t)\Psi_k(\tau_{ik}(t)) \right) \qquad (6.7.17)
\end{aligned}$$

for $t \geq t_2$. We use Jensen's inequality to obtain

$$\sum_{k=1}^{M} \hat{\varphi}_i\big(W_k(\sigma_i(t))\big) \geq M\hat{\varphi}_i\left(\frac{\sum\limits_{k=1}^{M} W_k(\sigma_i(t))}{M}\right), \quad t \geq t_2. \tag{6.7.18}$$

From (6.7.17) and (6.7.18) we see that

$$\frac{d}{dt}\left(V(t) + \frac{1}{M}\sum_{i=1}^{\ell}\sum_{j,k=1}^{M} \delta_j\delta_k h_{ijk}(t)W_k(\rho_i(t))\right) + \sum_{i=1}^{m} \hat{p}_i(t)\hat{\varphi}_i\big(V(\sigma_i(t))\big)$$

$$\leq \sum_{j=1}^{M} \delta_j G_j(t), \quad t \geq t_2, \tag{6.7.19}$$

where

$$V(t) = \frac{\sum\limits_{k=1}^{M} W_k(t)}{M}.$$

There exists $\gamma \in \Gamma$ such that $\sum\limits_{j=1}^{M} \delta_j G_j(t) = \gamma \cdot \big(G_j(t)\big)_{j=1}^{M}$. Letting

$$Y(t) = V(t) + \frac{1}{M}\sum_{i=1}^{\ell}\sum_{j,k=1}^{M} \delta_j\delta_k h_{ijk}(t)W_k(\rho_i(t)) - \Theta_\gamma(t),$$

we find that

$$Y'(t) \leq -\sum_{i=1}^{m} \hat{p}_i(t)\hat{\varphi}_i(V(\sigma_i(t))) \leq 0, \quad t \geq t_2. \tag{6.7.20}$$

Therefore, $Y(t) > 0$ or $Y(t) \leq 0$ on $[t_3, \infty)$ for some $t_3 \geq t_2$. If $Y(t) \leq 0$ on $[t_3, \infty)$, then

$$V(t) + \frac{1}{M}\sum_{i=1}^{\ell}\sum_{j,k=1}^{M} \delta_j\delta_k h_{ijk}(t)W_k(\rho_i(t)) \leq \Theta_\gamma(t), \quad t \geq t_3$$

and therefore

$$V(t) + \frac{1}{M}\sum_{i=1}^{\ell}\sum_{k=1}^{M}\left(h_{ikk}(t) - \sum_{\substack{j=1 \\ j\neq k}}^{M} h_{ijk}(t)\right)W_k(\rho_i(t)) \leq \Theta_\gamma(t), \quad t \geq t_3. \tag{6.7.21}$$

The left hand side of (6.7.21) is positive in view of the hypothesis (H6.7-4), whereas the right hand side of (6.7.21) is oscillatory at $t = \infty$. This is a contradiction. Hence, we conclude that $Y(t) > 0$ on $[t_3, \infty)$. Since $Y(t) + \Theta_\gamma(t) \geq 0$ on $[t_3, \infty)$, we see that

$$Y(t) \geq [\Theta_\gamma(t)]_-, \quad t \geq t_3.$$

In view of the fact that $V(t) \leq Y(t) + \Theta_\gamma(t)$ and $Y(t)$ is nonincreasing, we obtain

$$\begin{aligned}
V(t) &= Y(t) - \frac{1}{M}\sum_{i=1}^{\ell}\sum_{j,k=1}^{M} \delta_j\delta_k h_{ijk}(t)W_k(\rho_i(t)) + \Theta_\gamma(t) \\
&\geq Y(t) - \sum_{i=1}^{\ell}\sum_{j=1}^{M} \tilde{h}_{ij}(t)V(\rho_i(t)) + \Theta_\gamma(t) \\
&\geq Y(t) - \sum_{i=1}^{\ell}\sum_{j=1}^{M} \tilde{h}_{ij}(t)\big(Y(\rho_i(t)) + \Theta_\gamma(\rho_i(t))\big) + \Theta_\gamma(t) \\
&\geq \left(1 - \sum_{i=1}^{\ell}\sum_{j=1}^{M} \tilde{h}_{ij}(t)\right) Y(t) + \hat{\Theta}_\gamma(t), \quad t \geq t_3. \qquad (6.7.22)
\end{aligned}$$

Since $Y(t) \geq [\Theta_\gamma(t)]_-$ and $V(t) > 0$ for $t \geq t_3$, from (6.7.22) we have

$$V(t) \geq \left[\left(1 - \sum_{i=1}^{\ell}\sum_{j=1}^{M} \tilde{h}_{ij}(t)\right)[\Theta_\gamma(t)]_- + \hat{\Theta}_\gamma(t)\right]_+, \quad t \geq t_3. \qquad (6.7.23)$$

Since $\hat{\varphi}_{j_0}(\xi)$ is nondecreasing, we see from (6.7.20) and (6.7.23) that

$$\begin{aligned}
&\hat{p}_{j_0}(t)\hat{\varphi}_{j_0}\left(\left[\left(1 - \sum_{i=1}^{\ell}\sum_{j=1}^{M} \tilde{h}_{ij}(\sigma_{j_0}(t))\right)[\Theta_\gamma(\sigma_{j_0}(t))]_- + \hat{\Theta}_\gamma(\sigma_{j_0}(t))\right]_+\right) \\
&\leq -Y'(t), \quad t \geq t_4 \qquad (6.7.24)
\end{aligned}$$

for some $t_4 \geq t_3$. Integrating (6.7.24) over $[t_4, t]$ yields

$$\begin{aligned}
&\int_{t_4}^{t} \hat{p}_{j_0}(s)\hat{\varphi}_{j_0}\left(\left[\left(1 - \sum_{i=1}^{\ell}\sum_{j=1}^{M} \tilde{h}_{ij}(\sigma_{j_0}(s))\right)[\Theta_\gamma(\sigma_{j_0}(s))]_- + \hat{\Theta}_\gamma(\sigma_{j_0}(s))\right]_+\right) ds \\
&\leq -Y(t) + Y(t_4) \leq Y(t_4), \quad t \geq t_4.
\end{aligned}$$

This contradicts the hypothesis (6.7.7) and completes the proof. □

Remark 6.7.1. Since $\Gamma = \left\{\pm\gamma;\ \gamma \in \tilde{\Gamma}\right\}$ for some $\tilde{\Gamma} \subset \Gamma$ with $\#\tilde{\Gamma} = 2^{M-1}$, Theorem 6.7.1 holds true if the hypothesis (H6.7-11) and the condition (6.7.7) are replaced by

($\tilde{\text{H}}$6.7-11) there exist functions $\Theta_\gamma(t) \in C^1([t_0,\infty);\mathbb{R})$ $(\gamma \in \tilde{\Gamma})$ such that $\Theta_\gamma(t)$ are oscillatory at $t=\infty$ and $\Theta'_\gamma(t) = \gamma \cdot \big(G_j(t)\big)_{j=1}^M$ (the dot $\cdot$ denotes the scalar product)

and

$$\int_{t_0}^{\infty} \hat{p}_{j_0}(t)\hat{\varphi}_{j_0}\left(\left[\left(1-\sum_{i=1}^{\ell}\sum_{j=1}^{M}\tilde{h}_{ij}(\sigma_{j_0}(t))\right)[\Theta_\gamma(\sigma_{j_0}(t))]_\mp \pm\hat{\Theta}_\gamma(\sigma_{j_0}(t))\right]_+\right)dt = \infty,$$

respectively.

Theorem 6.7.2. *Assume that the hypotheses* (H6.7-1)–(H6.7-9) *hold. Every solution* $\big(u_1(x,t),...,u_M(x,t)\big)$ *of the boundary value problem* (6.7.2), (BC) *is weakly oscillatory in* Ω *if for any* $\gamma \in \Gamma$

$$\liminf_{t\to\infty} \int_T^t \gamma \cdot \big(G_j(s)\big)_{j=1}^M \, ds = -\infty \tag{6.7.25}$$

for all large T.

Proof. Suppose that there is a solution $\big(u_1(x,t),...,u_M(x,t)\big)$ of the problem (6.7.2), (BC) which is not weakly oscillatory in Ω. Proceeding as in the proof of Theorem 6.7.1, we observe that (6.7.19) holds, and therefore

$$\frac{d}{dt}\left(V(t)+\frac{1}{M}\sum_{i=1}^{\ell}\sum_{j,k=1}^{M}\delta_j\delta_k h_{ijk}(t)W_k(\rho_i(t))\right) \le \sum_{j=1}^{M}\delta_j G_j(t), \quad t \ge t_2 \tag{6.7.26}$$

for some $t_2 > 0$. Integrating (6.7.26) over $[t_2, t]$ yields

$$\begin{aligned}\tilde{V}(t)-\tilde{V}(t_0) &\le \int_{t_0}^{t}\sum_{j=1}^{M}\delta_j G_j(s)\,ds \\ &= \int_{t_0}^{t}\gamma\cdot\big(G_j(s)\big)_{j=1}^M\,ds \end{aligned}\tag{6.7.27}$$

for some $\gamma \in \Gamma$, where

$$\tilde{V}(t) = V(t)+\frac{1}{M}\sum_{i=1}^{\ell}\sum_{j,k=1}^{M}\delta_j\delta_k h_{ijk}(t)W_k(\rho_i(t)).$$

We note that

$$\tilde{V}(t) \geq V(t) + \frac{1}{M}\sum_{i=1}^{\ell}\sum_{k=1}^{M}\left(h_{ikk}(t) - \sum_{\substack{j=1\\ j\neq k}}^{M} h_{ijk}(t)\right) W_k(\rho_i(t))$$
$$> 0, \quad t > t_3$$

for some $t_3 \geq t_2$ (*cf.* (6.7.21)). The left hand side of (6.7.27) is bounded from below, whereas the right hand side of (6.7.27) is not bounded from below by (6.7.25). This is a contradiction and the proof is complete. □

Example 6.7.1. We consider the parabolic system

$$\begin{aligned}
&\frac{\partial}{\partial t}\Big(u_1(x,t) + \frac{1}{2}u_1(x,t+2\pi) + \frac{1}{4}u_2(x,t+2\pi)\Big) \\
&\quad -\frac{\partial^2 u_1}{\partial x^2}(x,t) - \frac{1}{8}e^{-2\pi}\frac{\partial^2 u_2}{\partial x^2}(x,t) - \frac{\partial^2 u_1}{\partial x^2}(x,t-\pi) - \frac{1}{16}\frac{\partial^2 u_2}{\partial x^2}(x,t+2\pi) \\
&\quad +\frac{1}{2}u_1(x,t-\tfrac{\pi}{2}) + \frac{1}{16}e^{-(5/2)\pi}u_2(x,t-\tfrac{\pi}{2}) = (\sin x)\cos t, \qquad (6.7.28)\\
&\frac{\partial}{\partial t}\Big(u_2(x,t) + \frac{1}{4}u_1(x,t+2\pi) + \frac{1}{4}u_2(x,t+2\pi)\Big) \\
&\quad -\frac{\partial^2 u_1}{\partial x^2}(x,t) - \frac{\partial^2 u_2}{\partial x^2}(x,t) - \frac{\partial^2 u_1}{\partial x^2}(x,t-\pi) - \frac{1}{4}\frac{\partial^2 u_2}{\partial x^2}(x,t+2\pi) \\
&\quad +\frac{1}{4}u_1(x,t-\tfrac{\pi}{2}) + e^{-(\pi/2)}u_2(x,t-\tfrac{\pi}{2}) = (\sin x)\, e^{-t} \qquad (6.7.29)
\end{aligned}$$

for $(x,t) \in (0,\pi) \times (0,\infty)$, and the boundary condition

$$u_j(0,t) = u_j(\pi,t) = 0, \quad t > 0 \ (j = 1,2). \qquad (6.7.30)$$

Here $G = (0,\pi)$, $n = 1$, $M = 2$, $l = K = m = 1$, $h_{111}(t) = 1/2$, $h_{112}(t) = h_{121}(t) = h_{122}(t) = 1/4$, $\rho_1(t) = t + 2\pi$, $a_{11}(t) = 1$, $a_{12}(t) = (1/8)e^{-2\pi}$, $a_{21}(t) = a_{22}(t) = 1$, $b_{111}(t) = 1$, $b_{112}(t) = 1/16$, $b_{121}(t) = 1$, $b_{122}(t) = 1/4$, $\tau_{11}(t) = t - \pi$, $\tau_{12}(t) = t + 2\pi$, $p_{111}(x,t) = 1/2$, $p_{112}(t) = (1/16)e^{-(5/2)\pi}$, $p_{121}(t) = 1/4$, $p_{122}(t) = e^{-(\pi/2)}$, $\sigma_1(t) = t - (\pi/2)$, $\varphi_{11}(\xi) = \varphi_{12}(\xi) = \xi$, $f_1(x,t) = (\sin x)\cos t$, $f_2(x,t) = (\sin x)\, e^{-t}$, $\psi_j = 0$ $(j = 1,2)$, $\alpha(\hat{x}) = 0$ and $\Gamma_2 = \emptyset$. We see that

$$\tilde{h}_{11}(t) = \frac{1}{2}, \ \tilde{h}_{12}(t) = \frac{1}{4},$$
$$\hat{p}_1(t) = e^{-(\pi/2)}\left(1 - \frac{1}{16}e^{-2\pi}\right).$$

It is easily seen that $\lambda_1 = 1$, $\Phi(x) = \sin x$, $\Psi_j(t) = 0$ $(j = 1,2)$, $G_1(t) = (1/2)F_1(t) = (\pi/8)\cos t$ and $G_2(t) = (1/2)F_2(t) = (\pi/8)e^{-t}$. Since we can choose

$$\Theta_{(1,1)}(t) = \frac{\pi}{8}\left(\sin t - e^{-t}\right),$$
$$\Theta_{(1,-1)}(t) = \frac{\pi}{8}\left(\sin t + e^{-t}\right),$$

we obtain

$$\hat{\Theta}_{(1,1)}(t) = \frac{\pi}{8}\left(\frac{1}{4}\sin t - e^{-t} + \frac{3}{4}e^{-t-2\pi}\right),$$
$$\hat{\Theta}_{(1,-1)}(t) = \frac{\pi}{8}\left(\frac{1}{4}\sin t + e^{-t} - \frac{3}{4}e^{-t-2\pi}\right)$$

and therefore

$$\int_{t_0}^{\infty} \frac{\pi}{8}e^{-(\pi/2)}\left(1 - \frac{1}{16}e^{-2\pi}\right)\left[\left(1 - \frac{3}{4}\right)\left[\sin\left(t - \frac{\pi}{2}\right) - e^{-t+(\pi/2)}\right]_{\mp} \pm \left(\frac{1}{4}\sin\left(t - \frac{\pi}{2}\right) - e^{-t+(\pi/2)} + \frac{3}{4}e^{-t-(3/2)\pi}\right)\right]_{+} dt$$
$$\geq \frac{\pi}{8}e^{-(\pi/2)}\left(1 - \frac{1}{16}e^{-2\pi}\right)\int_{t_0}^{\infty}\left[-\frac{1}{4}\cos t - e^{-t+(\pi/2)} + \frac{3}{4}e^{-t-(3/2)\pi}\right]_{\pm} dt$$
$$= \infty$$

and

$$\int_{t_0}^{\infty} \frac{\pi}{8}e^{-(\pi/2)}\left(1 - \frac{1}{16}e^{-2\pi}\right)\left[\left(1 - \frac{3}{4}\right)\left[\sin\left(t - \frac{\pi}{2}\right) + e^{-t+(\pi/2)}\right]_{\mp} \pm \left(\frac{1}{4}\sin\left(t - \frac{\pi}{2}\right) + e^{-t+(\pi/2)} - \frac{3}{4}e^{-t-(3/2)\pi}\right)\right]_{+} dt$$
$$\geq \frac{\pi}{8}e^{-(\pi/2)}\left(1 - \frac{1}{16}e^{-2\pi}\right)\int_{t_0}^{\infty}\left[-\frac{1}{4}\cos t + e^{-t+(\pi/2)} - \frac{3}{4}e^{-t-(3/2)\pi}\right]_{\pm} dt$$
$$= \infty.$$

Hence, it follows from Theorem 6.7.1 and Remark 6.7.1 that every solution $(u_1(x,t), u_2(x,t))$ of the problem (6.7.28)–(6.7.30) is weakly oscillatory in $(0,\pi)\times(0,\infty)$. One such solution is

$$U(x,t) = \begin{pmatrix} (\sin x)\sin t \\ (\sin x)\, e^{-t} \end{pmatrix}.$$

Example 6.7.2. We consider the parabolic system

$$\frac{\partial}{\partial t}\Big(u_1(x,t)+\frac{1}{2}u_1(x,t+2\pi)+\frac{1}{2}u_2(x,t+2\pi)\Big)$$
$$-\frac{\partial^2 u_1}{\partial x^2}(x,t)-\frac{1}{8}e^{-2\pi}\frac{\partial^2 u_2}{\partial x^2}(x,t)$$
$$-\frac{\partial^2 u_1}{\partial x^2}(x,t-\pi)-\frac{1}{4}e^{-\pi}\frac{\partial^2 u_2}{\partial x^2}(x,t+\pi)$$
$$+\frac{5}{2}u_1(x,t-\tfrac{\pi}{2})+\frac{1}{8}e^{-(5/2)\pi}u_2(x,t-\tfrac{\pi}{2})=(\sin x)\sin t, \tag{6.7.31}$$
$$\frac{\partial}{\partial t}\Big(u_2(x,t)+\frac{1}{3}u_1(x,t+2\pi)+\frac{1}{3}u_2(x,t+2\pi)\Big)$$
$$-\frac{\partial^2 u_1}{\partial x^2}(x,t)-\frac{\partial^2 u_2}{\partial x^2}(x,t)$$
$$-\frac{\partial^2 u_1}{\partial x^2}(x,t-\pi)-e^{-\pi}\frac{\partial^2 u_2}{\partial x^2}(x,t+\pi)$$
$$+\frac{1}{3}u_1(x,t-\tfrac{\pi}{2})+\frac{1}{3}e^{-(5/2)\pi}u_2(x,t-\tfrac{\pi}{2})=(\sin x)\,e^{-t-2\pi} \tag{6.7.32}$$

for $(x,t)\in(0,\pi)\times(0,\infty)$, and the boundary condition

$$-\frac{\partial u_1}{\partial x}(0,t)+\frac{1}{2}u_1(0,t)=\frac{\partial u_1}{\partial x}(\pi,t)+\frac{1}{2}u_1(\pi,t)=-\cos t,\quad t>0, \tag{6.7.33}$$
$$-\frac{\partial u_2}{\partial x}(0,t)+\frac{1}{2}u_2(0,t)=\frac{\partial u_2}{\partial x}(\pi,t)+\frac{1}{2}u_2(\pi,t)=-e^{-t},\quad t>0. \tag{6.7.34}$$

Here $G=(0,\pi)$, $n=1$, $M=2$, $l=K=m=1$, $h_{111}(t)=h_{112}(t)=1/2$, $h_{121}(t)=h_{122}(t)=1/3$, $\rho_1(t)=t+2\pi$, $a_{11}(t)=1$, $a_{12}(t)=(1/8)e^{-2\pi}$, $a_{21}(t)=a_{22}(t)=1$, $b_{111}(t)=1$, $b_{112}(t)=(1/4)e^{-\pi}$, $b_{121}(t)=1$, $b_{122}(t)=e^{-\pi}$, $\tau_{11}(t)=t-\pi$, $\tau_{12}(t)=t+\pi$, $p_{111}(x,t)=5/2$, $p_{112}(x,t)=(1/8)e^{-(5/2)\pi}$, $p_{121}(x,t)=1/3$, $p_{122}(x,t)=(1/3)e^{-(5/2)\pi}$, $\sigma_1(t)=t-(\pi/2)$, $\varphi_{11}(\xi)=\varphi_{12}(\xi)=\xi$, $f_1(x,t)=(\sin x)\sin t$, $f_2(x,t)=(\sin x)\,e^{-t-2\pi}$, $\alpha(\hat{x})=1$, $\mu(\hat{x})=1/2$, $\Gamma_1=\emptyset$, $\tilde{\psi}_1+(1/2)\psi_1=-\cos t$ and $\tilde{\psi}_2+(1/2)\psi_2=-e^{-t}$. We find that

$$\tilde{h}_{11}(t)=\frac{1}{2},\ \tilde{h}_{12}(t)=\frac{1}{3},\ \hat{p}_1(t)=\frac{5}{24}e^{-(5/2)\pi}.$$

We observe that $\lambda_1=1/4$ and $\Phi(x)=\cos(x/2)+\sin(x/2)$ and that

$$G_1(t)=\frac{1}{8}\left(\frac{8}{3}\sin t-\frac{3}{4}e^{-t-2\pi}\right),$$
$$G_2(t)=\frac{1}{8}\left(-2e^{-t}+\frac{2}{3}e^{-t-2\pi}\right).$$

Choosing

$$\Theta_{(1,1)}(t) = \frac{1}{8}\left(-\frac{8}{3}\cos t + 2e^{-t} + \frac{1}{12}e^{-t-2\pi}\right),$$

$$\Theta_{(1,-1)}(t) = \frac{1}{8}\left(-\frac{8}{3}\cos t - 2e^{-t} + \frac{17}{12}e^{-t-2\pi}\right),$$

we obtain

$$\hat{\Theta}_{(1,1)}(t) = \frac{1}{8}\left(-\frac{4}{9}\cos t + \left(1-\frac{5}{6}e^{-2\pi}\right)\left(2+\frac{1}{12}e^{-2\pi}\right)e^{-t}\right),$$

$$\hat{\Theta}_{(1,-1)}(t) = \frac{1}{8}\left(-\frac{4}{9}\cos t - \left(1-\frac{5}{6}e^{-2\pi}\right)\left(2-\frac{17}{12}e^{-2\pi}\right)e^{-t}\right)$$

and therefore

$$\int_{t_0}^{\infty} \frac{5}{192}e^{-(5/2)\pi}\left[\frac{1}{6}\left[-\frac{8}{3}\cos\left(t-\frac{\pi}{2}\right)+2e^{-(t-(\pi/2))-2\pi}+\frac{1}{12}e^{-(t-(\pi/2))-2\pi}\right]_{\mp} \pm\left(-\frac{4}{9}\cos\left(t-\frac{\pi}{2}\right)+\left(1-\frac{5}{6}e^{-2\pi}\right)\left(2+\frac{1}{12}e^{-2\pi}\right)e^{-(t-(\pi/2))}\right)\right]_{+} dt = \infty,$$

$$\int_{t_0}^{\infty} \frac{5}{192}e^{-(5/2)\pi}\left[\frac{1}{6}\left[-\frac{8}{3}\cos\left(t-\frac{\pi}{2}\right)-2e^{-(t-(\pi/2))-2\pi}+\frac{17}{12}e^{-(t-(\pi/2))-2\pi}\right]_{\mp} \pm\left(-\frac{4}{9}\cos\left(t-\frac{\pi}{2}\right)-\left(1-\frac{5}{6}e^{-2\pi}\right)\left(2-\frac{17}{12}e^{-2\pi}\right)e^{-(t-(\pi/2))}\right)\right]_{+} dt = \infty.$$

Hence, from Theorems 6.7.1 and Remark 6.7.1 it follows that every solution $(u_1(x,t), u_2(x,t))$ of the problem (6.7.31)–(6.7.34) is weakly oscillatory in $(0,\pi)\times(0,\infty)$. In fact,

$$U(x,t) = \begin{pmatrix} (\sin x)\cos t \\ (\sin x)\, e^{-t} \end{pmatrix}$$

is such a solution.

6.8 Notes

Lemma 6.1.2 is due to Kitamura and Kusano [131]. Theorems 6.1.1 – 6.1.4 are slight generalizations of the results obtained by Bykov and Kultaev [40].

Lemma 6.2.1 is based on Kitamura and Kusano [131]. The oscillation results in Section 6.2 are extracted from Yoshida [300]. The parabolic equation (6.2.1) has no neutral term, but more general equations which have neutral terms were investigated by Shoukaku and Yoshida [244] and Wang and Feng [280].

The results in Section 6.3 are based on Tanaka and Yoshida [258] and Yoshida [317].

The results in Section 6.4 are extracted from Yoshida [311, 318]. Oscillation results in Section 6.4 were extended to more general equations which have continuous distributed arguments, see, for example, Fu and Zhuang [85], Shoukaku [243].

The results in Section 6.5 are from Bainov and Minchev [23]. For forced oscillations of impulsive parabolic equations or systems we refer to Bainov and Minchev [24] and Li [169]. For the monographs on the theory of impulsive differential equations see Benchohra, Henderson and Ntouyas [30] and Lakshmikantham, Bainov and Simeonov [163].

All results in Section 6.6 are taken from Minchev and Yoshida [199].

A general boundary condition which includes some boundary conditions is considered in Section 6.7, and the results in Section 6.7 are based on Shoukaku and Yoshida [245].

Delay parabolic equations with positive and negative coefficients were studied by Kreith and Ladas [143], Kubiaczyk and Saker [147] and Yoshida [311].

As was shown in Sections 6.1–6.7, there is a close connection between oscillation of functional partial differential equations and that of functional differential equations. We refer the reader to Bainov and Mishev [25] for oscillation theory of functional partial differential equations, and to Agarwal, Bochner and Li [3], Agarwal, Grace and O'Regan [4–6], Bainov and Mishev [25, 26], Erbe, Kong and Zhang [76], Györi and Ladas [94], Ladde, Lakshmikantham and Zhang [163] for oscillation theory of functional differential equations. In particular we mention Agarwal, Bochner, Grace and O'Regan [2], Cheng [50], Zhang and Zhou [325] dealing with the oscillation theory of difference equations.

Chapter 7

Functional Hyperbolic Equations

7.1 Hyperbolic Equations with Delays

We derive sufficient conditions for every solution of certain boundary value problems for delay hyperbolic equations to have a zero in some bounded domain. The results are based on the conditions for the non-existence of positive solutions of differential inequalities in some bounded interval.

We study the hyperbolic equation with delays

$$\frac{\partial^2 u}{\partial t^2}(x,t) - \Delta u(x,t) - \sum_{i=1}^{k} b_i(t)\Delta u(x,t-\tau_i)$$
$$+c(x,t,u(x,t),(u(x,t-\sigma_i))_{i=1}^m) = f(x,t),\ (x,t) \in \Omega := G\times(0,\infty),\ (7.1.1)$$

where G is a bounded domain in $\mathbb{R}^n$ with piecewise smooth boundary ∂G and Δ is the Laplacian in $\mathbb{R}^n$.

It is assumed that:

(H7.1-1) $b_i(t) \in C([0,\infty);[0,\infty))$ $(i=1,2,...,k)$ and $f(x,t) \in C(\overline{\Omega};\mathbb{R})$;
(H7.1-2) τ_i $(i=1,2,...,k)$, σ_i $(i=1,2,...,m)$ are nonnegative constants;
(H7.1-3) $c(x,t,\xi,(\xi_i)_{i=1}^m) \in C(\overline{\Omega}\times\mathbb{R}\times\mathbb{R}^m;\mathbb{R})$,
$c(x,t,\xi,(\xi_i)_{i=1}^m) \geq K_0^2\xi$ for $(x,t,\xi,(\xi_i)_{i=1}^m) \in \Omega\times[0,\infty)^{m+1}$,
$c(x,t,\xi,(\xi_i)_{i=1}^m) \leq K_0^2\xi$ for $(x,t,\xi,(\xi_i)_{i=1}^m) \in \Omega\times(-\infty,0]^{m+1}$,
where K_0 is a nonnegative constant.

We consider the boundary value problems (7.1.1), (B_i) $(i=1,2)$, where the boundary conditions (B_i) are given in Section 6.3.

By a *solution* of Eq. (7.1.1) we mean a function $u(x,t) \in C^2(\overline{G}\times[t_{-1},\infty);\mathbb{R}) \cap C(\overline{G}\times[\tilde{t}_{-1},\infty);\mathbb{R})$ which satisfies (7.1.1), where

$$t_{-1} = -\max_{1\leq i\leq k}\{\tau_i\},$$

$$\tilde{t}_{-1} = -\max_{1\leq i\leq m}\{\sigma_i\}.$$

We note that the functions $F(t)$, $\Psi(t)$, $\tilde{F}(t)$, $\tilde{\Psi}(t)$ are those given in Section 6.3.

Theorem 7.1.1. *Assume that the hypotheses* (H7.1-1)–(H7.1-3) *are satisfied. If there is a number $s \geq T$ such that*

$$y''(t) + (\lambda_1 + K_0^2)\, y(t) \leq \pm G(t) \tag{7.1.2}$$

has no positive solution in $[s, s+(\pi/L))$, then every solution u of the boundary value problem (7.1.1), (B_1) *has a zero in $G \times (s-T, s+(\pi/L))$, where λ_1 is the first eigenvalue of the eigenvalue problem* (EVP) *appearing in Section* 6.1, *and*

$$L = \sqrt{\lambda_1 + K_0^2},$$

$$T = \max\left\{\max_{1\leq i\leq k}\{\tau_i\},\ \max_{1\leq i\leq m}\{\sigma_i\}\right\},$$

$$G(t) = F(t) - \Psi(t) - \sum_{i=1}^{k} b_i(t)\Psi(t-\tau_i).$$

Proof. Suppose to the contrary that there exists a solution $u(x,t)$ of the problem (7.1.1), (B_1) which has no zero in $G \times (s-T, s+(\pi/L))$. Let $u(x,t) > 0$ in $G \times (s-T, s+(\pi/L))$. Multiplying (7.1.1) by $\left(\int_G \Phi(x)\,dx\right)^{-1}\Phi(x)$ and then integrating over G, we find that

$$\begin{aligned} U''(t) &- K_\Phi \int_G \Delta u(x,t)\Phi(x)\,dx \\ &- \sum_{i=1}^{k} b_i(t) K_\Phi \int_G \Delta u(x,t-\tau_i)\Phi(x)\,dx \\ &+ K_\Phi \int_G c(x,t,u(x,t),(u(x,t-\sigma_i))_{i=1}^{m})\Phi(x)\,dx = F(t), \end{aligned} \tag{7.1.3}$$

where $K_\Phi = \left(\int_G \Phi(x)\,dx\right)^{-1}$, $U(t)$ is given by (6.2.3), and $\Phi(x)$ is the eigenfunction corresponding to λ_1. Proceeding as in the proof of Theorem 6.3.1, we obtain

$$K_\Phi \int_G \Delta u(x,t)\Phi(x)\,dx = -\Psi(t) - \lambda_1 U(t), \tag{7.1.4}$$

$$K_\Phi \int_G \Delta u(x,t-\tau_i)\Phi(x)\,dx = -\Psi(t-\tau_i) - \lambda_1 U(t-\tau_i). \tag{7.1.5}$$

From the definition of T it follows that

$$u(x,t) > 0 \quad \text{in } G \times [s, s+(\pi/L)),$$

$$u(x,t-\sigma_i) \geq 0 \quad \text{in } G \times [s, s+(\pi/L)) \ (i = 1,2,\ldots,m).$$

Hence, the hypothesis (H7.1-3) implies

$$K_\Phi \int_G c(x,t,u(x,t),(u(x,t-\sigma_i))_{i=1}^m)\Phi(x)\,dx \geq K_0^2 U(t), \quad t \in [s, s+(\pi/L)). \tag{7.1.6}$$

Combining (7.1.3)–(7.1.6), we see that $U(t)$ satisfies

$$U''(t) + \left(\lambda_1 + K_0^2\right) U(t) + \lambda_1 \sum_{i=1}^{k} b_i(t) U(t-\tau_i) \leq G(t), \quad t \in [s, s+(\pi/L)). \tag{7.1.7}$$

Since $U(t) > 0$ in $(s-T, s+(\pi/L))$, we observe that $U(t-\tau_i) \geq 0$ in $[s, s+(\pi/L))$, and therefore (7.1.7) reduces to

$$U''(t) + \left(\lambda_1 + K_0^2\right) U(t) \leq G(t), \quad t \in [s, s+(\pi/L)). \tag{7.1.8}$$

Consequently, $U(t) > 0$ in $[s, s+(\pi/L))$ and $U(t)$ satisfies (7.1.8). This contradicts the hypothesis. In the case where $u(x,t) < 0$ in $G \times (s-T, s+(\pi/L))$, we conclude that $V(t) := -U(t) > 0$ in $[s, s+(\pi/L))$ and $V(t)$ satisfies

$$V''(t) + \left(\lambda_1 + K_0^2\right) V(t) \leq -G(t), \quad t \in [s, s+(\pi/L)).$$

This also contradicts the hypothesis and completes the proof. □

Corollary 7.1.1. *Assume that the hypotheses* (H7.1-1)–(H7.1-3) *hold. If there is a number* $s \geq T$ *such that*

$$\int_s^{s+(\pi/L)} G(t) \sin L(t-s)\,dt = 0,$$

then every solution u *of the boundary value problem* (7.1.1), (B_1) *has a zero in* $G \times (s-T, s+(\pi/L))$.

Proof. It follows from Lemma 3.1.1 and Remark 3.1.4 that (7.1.2) has no positive solution in $[s, s+(\pi/L))$. The conclusion follows from Theorem 7.1.1. □

Corollary 7.1.2. *Assume that the hypotheses* (H7.1-1)–(H7.1-3) *hold. If*

$$H(s) := \int_s^{s+(\pi/L)} G(t) \sin L(t-s)\,dt$$

is oscillatory at $s = \infty$, *then every solution* u *of the boundary value problem* (7.1.1), (B_1) *is oscillatory in* Ω.

Proof. For any $t > 0$ there exists a number s for which $s - T > t$ and $H(s) = 0$. Corollary 7.1.1 implies that every solution u of the problem (7.1.1), (B_1) has a zero in $G \times (s-T, s+(\pi/L)) \subset G \times (t, \infty)$ for any $t > 0$, that is, u is oscillatory in Ω. □

Theorem 7.1.2. *Assume that the hypotheses* (H7.1-1)–(H7.1-3) *are satisfied and let* $K_0 > 0$. *If there is a number* $s \geq T$ *such that*

$$y''(t) + K_0^2 y(t) \leq \pm \tilde{G}(t) \tag{7.1.9}$$

has no positive solution in $[s, s + (\pi/K_0))$, *then every solution* u *of the boundary value problem* (7.1.1), (B_2) *has a zero in* $G \times (s - T, s + (\pi/K_0))$, *where*

$$\tilde{G}(t) = \tilde{F}(t) + \tilde{\Psi}(t) + \sum_{i=1}^{k} b_i(t)\tilde{\Psi}(t - \tau_i).$$

Proof. Let $u(x,t)$ be a solution of the problem (7.1.1), (B_2) which has no zero in $G \times (s - T, s + (\pi/K_0))$. First we assume that $u(x,t) > 0$ in $G \times (s - T, s + (\pi/K_0))$. Dividing (7.1.1) by $|G|$ and then integrating over G, we obtain

$$\begin{aligned}\tilde{U}''(t) &- \frac{1}{|G|}\int_G \Delta u(x,t)\,dx - \sum_{i=1}^{k} b_i(t)\frac{1}{|G|}\int_G \Delta u(x, t-\tau_i)\,dx \\ &+ \frac{1}{|G|}\int_G c(x,t,u(x,t),(u(x,t-\sigma_i))_{i=1}^m)\,dx = \tilde{F}(t),\end{aligned} \tag{7.1.10}$$

where $\tilde{U}(t)$ is defined by (6.2.16). From the definition of T we see that

$$\begin{aligned}&u(x, t-\tau_i) \geq 0 \quad \text{in } G \times [s, s + (\pi/K_0)) \ (i = 1, 2, ..., k),\\ &u(x, t-\sigma_i) \geq 0 \quad \text{in } G \times [s, s + (\pi/K_0)) \ (i = 1, 2, ..., m).\end{aligned}$$

The divergence theorem implies

$$\begin{aligned}\frac{1}{|G|}\int_G \Delta u(x,t)\,dx &= \frac{1}{|G|}\int_{\partial G} \frac{\partial u}{\partial \nu}(x,t)\,dS \\ &= \frac{1}{|G|}\int_{\partial G} (\tilde{\psi} - \mu \cdot u(x,t))\,dS \leq \tilde{\Psi}(t), \quad t \in (s - T, s + (\pi/K_0)),\end{aligned} \tag{7.1.11}$$

$$\frac{1}{|G|}\int_G \Delta u(x, t-\tau_i)\,dx \leq \tilde{\Psi}(t - \tau_i), \quad t \in [s, s + (\pi/K_0)). \tag{7.1.12}$$

It follows from the hypothesis (H7.1-3) that

$$\frac{1}{|G|}\int_G c(x,t,u(x,t),(u(x,t-\sigma_i))_{i=1}^m)\,dx \geq K_0^2 \tilde{U}(t), \quad t \in [s, s + (\pi/K_0)). \tag{7.1.13}$$

Combining (7.1.10)–(7.1.13) yields

$$\tilde{U}''(t) + K_0^2\tilde{U}(t) \le \tilde{G}(t), \quad t \in [s, s + (\pi/K_0)). \tag{7.1.14}$$

Consequently, $\tilde{U}(t) > 0$ in $[s, s + (\pi/K_0))$ and $\tilde{U}(t)$ satisfies (7.1.14). This contradicts the hypothesis. In the case where $u(x,t) < 0$ in $G \times (s - T, s + (\pi/K_0))$, we are also led to a contradiction. The proof is complete. □

Corollary 7.1.3. *Assume that the hypotheses* (H7.1-1)–(H7.1-3) *hold and let* $K_0 > 0$. *If there is a number* $s \ge T$ *such that*

$$\int_s^{s+(\pi/K_0)} \tilde{G}(t) \sin K_0(t - s)\, dt = 0,$$

then every solution u *of the boundary value problem* (7.1.1), (B_2) *has a zero in* $G \times (s - T, s + (\pi/K_0))$.

Proof. Lemma 3.1.1 and Remark 3.1.4 imply that (7.1.9) has no positive solution in $[s, s + (\pi/K_0))$. The conclusion follows from Theorem 7.1.2. □

Corollary 7.1.4. *Assume that the hypotheses* (H7.1-1)–(H7.1-3) *hold and let* $K_0 > 0$. *If*

$$\tilde{H}(s) := \int_s^{s+(\pi/K_0)} \tilde{G}(t) \sin K_0(t - s)\, dt$$

is oscillatory at $s = \infty$, *then every solution* u *of the boundary value problem* (7.1.1), (B_2) *is oscillatory in* Ω.

Proof. Arguing as in the proof of Corollary 7.1.2, we conclude that every solution u of (7.1.1), (B_2) is oscillatory in Ω. □

Example 7.1.1. We consider the problem

$$\frac{\partial^2 u}{\partial t^2}(x,t) - \frac{\partial^2 u}{\partial x^2}(x,t) - \frac{\partial^2 u}{\partial x^2}(x, t - \tfrac{\pi}{2})$$
$$+u(x,t) + u(x, t - \tfrac{3}{2}\pi) = (\sin x)\sin t, \quad (x,t) \in (0,\pi) \times (0,\infty), \tag{7.1.15}$$
$$u(0,t) = u(\pi,t) = 0, \quad t > 0. \tag{7.1.16}$$

Here $n = 1$, $G = (0,\pi)$, $\Omega = (0,\pi) \times (0,\infty)$, $k = m = 1$, $b_1(t) = 1$, $\tau_1 = \pi/2$, $\sigma_1 = (3/2)\pi$, $K_0 = 1$, $T = (3/2)\pi$ and $f(x,t) = (\sin x)\sin t$. It is clear that $\lambda_1 = 1$, $\Phi(x) = \sin x$, $\Psi(t) = 0$, $F(t) = G(t) = (1/4)\pi \sin t$ and $L = \sqrt{2}$. Since

$$\begin{aligned} H(s) &= \int_s^{s+(\pi/\sqrt{2})} \frac{1}{4}\pi(\sin t)\sin\sqrt{2}(t - s)\, dt \\ &= \frac{\pi}{\sqrt{2}}\left(\cos\frac{\pi}{2\sqrt{2}}\right)\sin\left(s + \frac{\pi}{2\sqrt{2}}\right), \end{aligned}$$

we observe that $H(s) = 0$ for $s = s_n = \big(1 - (1/(2\sqrt{2}))\big)\pi + n\pi$ $(> (3/2)\pi)$ $(n = 1, 2, ...)$. It follows from Corollary 7.1.1 that every solution u of the problem (7.1.15), (7.1.16) has a zero in $(0, \pi) \times (s_n - (3/2)\pi, s_n + (\pi/\sqrt{2}))$. One such solution is $u(x,t) = (\sin x) \sin t$.

Example 7.1.2. We consider the problem

$$\frac{\partial^2 u}{\partial t^2}(x,t) - \frac{\partial^2 u}{\partial x^2}(x,t) - \frac{\partial^2 u}{\partial x^2}(x, t-\pi)$$
$$+4\,u(x,t) = 3\big(\sin\left(\tfrac{x}{2} + \tfrac{\pi}{4}\right)\big) \sin t, \quad (x,t) \in (0,\pi) \times (0,\infty), \tag{7.1.17}$$
$$-\frac{\partial u}{\partial x}(0,t) + \frac{1}{2}u(0,t) = \frac{\partial u}{\partial x}(\pi,t) + \frac{1}{2}u(\pi,t) = 0, \quad t > 0. \tag{7.1.18}$$

Here $n = 1$, $G = (0,\pi)$, $\Omega = (0,\pi) \times (0,\infty)$, $k = 1$, $b_1(t) = 1$, $\tau_1 = \pi$, $K_0 = 2$, $\mu = 1/2$, $T = \pi$ and $f(x,t) = 3\big(\sin((x/2) + (\pi/4))\big) \sin t$. It is readily seen that $\tilde{\Psi}(t) = 0$, $\tilde{F}(t) = \tilde{G}(t) = (6\sqrt{2}/\pi) \sin t$ and

$$\tilde{H}(s) = \int_s^{s+\pi/2} \frac{6\sqrt{2}}{\pi} (\sin t) \sin 2(t-s)\, dt = \frac{8}{\pi} \sin\left(s + \tfrac{\pi}{4}\right).$$

Since $\tilde{H}(s) = 0$ for $s = s_n = (3/4)\pi + n\pi$ $(> \pi)$ $(n = 1, 2, ...)$, we see from Corollary 7.1.3 that every solution u of the problem (7.1.17), (7.1.18) has a zero in $(0,\pi) \times (s_n - \pi, s_n + (\pi/2))$. In fact, $u(x,t) = \big(\sin((x/2) + (\pi/4))\big) \sin t$ is such a solution.

Remark 7.1.1. In the case where

$$c(x,t,\xi,(\xi_i)_{i=1}^m) = K_0^2 \xi + \sum_{i=1}^m K_i(x,t) \xi_i^{\gamma_i}, \tag{7.1.19}$$

where $K_i(x,t) \in C(\Omega; [0,\infty))$ and γ_i are the quotients of odd integers $(i = 1, 2, ...m)$, we observe that (7.1.19) satisfies the hypothesis (H7.1-3).

Remark 7.1.2. Oscillation results in this section can be extended to the more general hyperbolic equation with functional arguments

$$\frac{\partial^2 u}{\partial t^2}(x,t) + \alpha \frac{\partial u}{\partial t}(x,t) - \Delta u(x,t) - \sum_{i=1}^k b_i(t) \Delta u(x, \tau_i(t))$$
$$+c(x,t,u(x,t),(u(x,\sigma_i(t)))_{i=1}^m) = f(x,t), \quad (x,t) \in G \times (0,\infty)$$

(see Yoshida [309]).

7.2 Equations with Forcing Terms I

In 1984 oscillations of delay hyperbolic equations were studied by Mishev and Bainov [207]. We refer to Lalli, Yu and Cui [164], Mishev [205], Mishev and Bainov [210] for functional hyperbolic equations without forcing terms, and to Mishev [202], Mishev and Bainov [207,208], Tanaka [256] for functional hyperbolic equations with forcing terms.

We investigate forced oscillations of the hyperbolic equation with functional arguments

$$\begin{aligned} &\frac{\partial^2}{\partial t^2}\left(u(x,t)+\sum_{i=1}^{\ell} h_i(t)u(x,\rho_i(t))\right) \\ &-a(t)\Delta u(x,t)-\sum_{i=1}^{k} b_i(t)\Delta u(x,\tau_i(t)) \\ &+c(x,t,(z_i[u](x,t))_{i=1}^{\tilde{m}})=f(x,t), \quad (x,t)\in\Omega:=G\times(0,\infty), \quad (7.2.1) \end{aligned}$$

where G is a bounded domain in $\mathbb{R}^n$ with piecewise smooth boundary ∂G.

It is assumed that the hypothesis (H6.1-4) of Section 6.1, the hypothesis (H6.3-3) of Section 6.3 and the following hold:

(H7.2-1) $h_i(t)\in C^2([0,\infty);[0,\infty))$ $(i=1,2,...,\ell)$, $a(t)\in C([0,\infty);[0,\infty))$, $b_i(t)\in C([0,\infty);[0,\infty))$ $(i=1,2,...,k)$ and $f(x,t)\in C(\overline{\Omega};\mathbb{R})$;

(H7.2-2) $\rho_i(t)\in C^2([0,\infty);\mathbb{R})$, $\lim_{t\to\infty}\rho_i(t)=\infty$ $(i=1,2,...,\ell)$, $\tau_i(t)\in C([0,\infty);\mathbb{R})$, $\lim_{t\to\infty}\tau_i(t)=\infty$ $(i=1,2,...,k)$.

We consider the boundary value problems (7.2.1), (B_i) $(i=1,2)$, where the boundary conditions (B_i) are given in Section 6.3.

By a *solution* of Eq. (7.2.1) we mean a function $u(x,t)\in C^2(\overline{G}\times[t_{-1},\infty);\mathbb{R})\cap C(\overline{G}\times[\tilde{t}_{-1},\infty);\mathbb{R})$ which satisfies (7.2.1), where

$$t_{-1}=\min\left\{0,\ \min_{1\le i\le \ell}\left\{\inf_{t\ge 0}\rho_i(t)\right\},\ \min_{1\le i\le k}\left\{\inf_{t\ge 0}\tau_i(t)\right\}\right\},$$

$$\tilde{t}_{-1}=\min\left\{0,\ \min_{1\le i\le m}\left\{\inf_{t\ge 0}\sigma_i(t)\right\},\ \min_{\substack{m_1+1\le i\le \tilde{m}\\ 1\le j\le N_i}}\left\{\inf_{t\ge 0}\sigma_{ij}(t)\right\}\right\}.$$

We note that the functions $F(t)$, $\Psi(t)$, $\tilde{F}(t)$, $\tilde{\Psi}(t)$, $G(t)$, $\tilde{G}(t)$ are those appearing in Section 6.3.

Theorem 7.2.1. *Assume that the hypotheses* (H6.1-4), (H6.3-3), (H7.2-1) *and* (H7.2-2) *hold. If the functional differential inequalities*

$$\frac{d^2}{dt^2}\left(y(t)+\sum_{i=1}^{\ell}h_i(t)y(\rho_i(t))\right)+\lambda_1 a(t)y(t)$$
$$+\lambda_1\sum_{i=1}^{k}b_i(t)y(\tau_i(t))+\sum_{i=1}^{m}p_i(t)\varphi_i\big(y(\sigma_i(t))\big)\leq \pm G(t) \quad (7.2.2)$$

have no eventually positive solutions, then every solution u of the boundary value problem (7.2.1), (B$_1$) *is oscillatory in Ω, where λ_1 is the first eigenvalue of the eigenvalue problem* (EVP) *appearing in Section* 6.1.

Proof. The proof follows by using exactly the same arguments as in Theorem 6.3.1. □

Theorem 7.2.2. *Assume that the hypotheses* (H6.1-4), (H6.3-3), (H7.2-1) *and* (H7.2-2) *hold. If the functional differential inequalities*

$$\frac{d^2}{dt^2}\left(y(t)+\sum_{i=1}^{\ell}h_i(t)y(\rho_i(t))\right)+\sum_{i=1}^{m}p_i(t)\varphi_i\big(y(\sigma_i(t))\big)\leq \pm\tilde{G}(t) \quad (7.2.3)$$

have no eventually positive solutions, then every solution u of the boundary value problem (7.2.1), (B$_2$) *is oscillatory in Ω.*

Proof. The proof is quite similar to that of Theorem 6.3.2. □

Theorem 7.2.3. *Assume that the hypotheses* (H6.1-4), (H6.3-3), (H7.2-1) *and* (H7.2-2) *are satisfied. Every solution u of the boundary value problem* (7.2.1), (B$_1$) *is oscillatory in Ω if*

$$\liminf_{t\to\infty}\int_T^t\left(1-\frac{s}{t}\right)G(s)\,ds=-\infty,$$
$$\limsup_{t\to\infty}\int_T^t\left(1-\frac{s}{t}\right)G(s)\,ds=\infty$$

for all large T. Every solution u of the boundary value problem (7.2.1), (B$_2$) *is oscillatory in Ω if*

$$\liminf_{t\to\infty}\int_T^t\left(1-\frac{s}{t}\right)\tilde{G}(s)\,ds=-\infty,$$
$$\limsup_{t\to\infty}\int_T^t\left(1-\frac{s}{t}\right)\tilde{G}(s)\,ds=\infty$$

for all large T.

Proof. If the functional differential inequalities (7.2.2) have eventually positive solutions, then the functional differential inequalities

$$\frac{d^2}{dt^2}\left(y(t)+\sum_{i=1}^{\ell}h_i(t)y(\rho_i(t))\right)\le \pm G(t) \tag{7.2.4}$$

have also eventually positive solutions. In case (7.2.4) have no eventually positive solutions, the first statement follows from Theorem 7.2.1. Let $y_\pm(t)$ be solutions of (7.2.4) for which $y_\pm(t)>0$ on $[t_0,\infty)$ for some $t_0>0$. Integrating (7.2.4) over $[t_0,t]$ twice, we obtain

$$z_\pm(t)\le c_1+c_2(t-t_0)\pm\int_{t_0}^{t}(t-s)G(s)\,ds,$$

where c_1 and c_2 are constants and

$$z_\pm(t)=y_\pm(t)+\sum_{i=1}^{\ell}h_i(t)y_\pm(\rho_i(t)).$$

Dividing the above inequality by t yields

$$\frac{z_\pm(t)}{t}\le\frac{c_1}{t}+c_2\left(1-\frac{t_0}{t}\right)\pm\int_{t_0}^{t}\left(1-\frac{s}{t}\right)G(s)\,ds. \tag{7.2.5}$$

The hypotheses imply that the functions

$$\pm\int_{t_0}^{t}\left(1-\frac{s}{t}\right)G(s)\,ds$$

are unbounded from below. The right hand side of (7.2.5) is unbounded from below, whereas the left hand side of (7.2.5) is positive for large t. This is a contradiction. Hence, (7.2.4) have no eventually positive solutions, and the proof of first statement is complete. Similarly, the second statement follows from Theorem 7.2.2. □

Now we study the functional differential inequality of neutral type

$$\frac{d^2}{dt^2}\left(y(t)+\sum_{i=1}^{\ell}h_i(t)y(\rho_i(t))\right)+\sum_{i=1}^{m}p_i(t)\varphi_i(y(\sigma_i(t)))\le q(t), \tag{7.2.6}$$

where $q(t)\in C([t_0,\infty);\mathbb{R})$ for some $t_0>0$.

Lemma 7.2.1. *Assume that:*

(i) $\sum\limits_{i=1}^{\ell}h_i(t)\le 1$ *and* $\rho_i(t)\le t\ (i=1,2,...,\ell)$;

(ii) *there is an integer* $j\in\{1,2,...,m\}$ *such that* $\varphi_j(\xi)$ *is nondecreasing for* $\xi>0$;

(iii) *there exists a function* $Q(t) \in C^2([t_0, \infty); \mathbb{R})$ *satisfying* $Q''(t) \geq q(t)$ *and* $Q(t)$ *is oscillatory at* $t = \infty$.

If the functional differential inequality

$$y''(t) + p_j(t)\varphi_j\left(\left[\left(1 - \sum_{i=1}^{\ell} h_i(\sigma_j(t))\right) y(\sigma_j(t)) + H(\sigma_j(t))\right]_+\right) \leq 0 \tag{7.2.7}$$

has no eventually positive solution, then (7.2.6) *has no eventually positive solution, where*

$$H(t) = Q(t) - \sum_{i=1}^{\ell} h_i(t) Q(\rho_i(t)).$$

Proof. Suppose that there exists a solution $y(t)$ of (7.2.6) which is eventually positive. Then we see that $y(t) > 0$, $y(\rho_i(t)) > 0$ $(i = 1, 2, ..., \ell)$ and $y(\sigma_i(t)) > 0$ $(i = 1, 2, ..., m)$ for $t \geq t_1$, where t_1 is some number such that $t_1 \geq t_0$. Letting

$$z(t) = y(t) + \sum_{i=1}^{\ell} h_i(t) y(\rho_i(t)) - Q(t),$$

we see from (7.2.6) that

$$z''(t) \leq -p_j(t)\varphi_j\big(y(\sigma_j(t))\big) \leq 0, \quad t \geq t_1. \tag{7.2.8}$$

Therefore, $z(t) > 0$ or $z(t) \leq 0$ on $[t_2, \infty)$ for some $t_2 \geq t_1$. If $z(t) \leq 0$ on $[t_2, \infty)$, then

$$y(t) + \sum_{i=1}^{\ell} h_i(t) y(\rho_i(t)) \leq Q(t), \quad t \geq t_2. \tag{7.2.9}$$

The left hand side of (7.2.9) is positive, whereas the right hand side of (7.2.9) is oscillatory at $t = \infty$. This is a contradiction. Hence, we conclude that $z(t) > 0$ on $[t_2, \infty)$. Since $z''(t) \leq 0$, $z(t) > 0$ $(t \geq t_2)$, we find that $z'(t) \geq 0$ $(t \geq t_3)$ for some $t_3 \geq t_2$. We easily obtain

$$y(t) \leq z(t) + Q(t), \quad t \geq t_1$$

and hence

$$\begin{aligned} y(t) &= z(t) - \sum_{i=1}^{\ell} h_i(t) y(\rho_i(t)) + Q(t) \\ &\geq z(t) - \sum_{i=1}^{\ell} h_i(t)\big(z(\rho_i(t)) + Q(\rho_i(t))\big) + Q(t) \\ &\geq \left(1 - \sum_{i=1}^{\ell} h_i(t)\right) z(t) + H(t), \quad t \geq t_4 \end{aligned}$$

in view of the fact that $z(t)$ is nondecreasing, where t_4 is some number satisfying $t_4 \geq t_3$. Since $y(t) > 0$ for $t \geq t_4$, we observe that

$$y(t) \geq \left[\left(1 - \sum_{i=1}^{\ell} h_i(t) \right) z(t) + H(t) \right]_+, \quad t \geq t_4.$$

The hypothesis (ii) implies that

$$\varphi_j\big(y(\sigma_j(t))\big) \geq \varphi_j \left(\left[\left(1 - \sum_{i=1}^{\ell} h_i(\sigma_j(t)) \right) z(\sigma_j(t)) + H(\sigma_j(t)) \right]_+ \right) \tag{7.2.10}$$

for $t > t_5$ ($\geq t_4$). Combining (7.2.8) with (7.2.10), we conclude that $z(t)$ is an eventually positive solution of (7.2.7). This contradicts the hypothesis and completes the proof. □

Lemma 7.2.2. *Assume that the hypothesis* (ii) *of Lemma* 7.2.1 *holds. If*

$$\int_{t_0}^{\infty} p_j(t)\varphi_j \left(\left[\left(1 - \sum_{i=1}^{\ell} h_i(\sigma_j(t)) \right) c + H(\sigma_j(t)) \right]_+ \right) dt = \infty \tag{7.2.11}$$

for any $c > 0$, then (7.2.7) *has no eventually positive solution.*

Proof. Let $y(t)$ be an eventually positive solution of (7.2.7). It follows from (7.2.7) that $y''(t) \leq 0$ ($t \geq t_1$) for some $t_1 \geq t_0$. Then we find that $y'(t) \geq 0$ ($t \geq t_2$) for some $t_2 \geq t_1$, and therefore

$$y(t) \geq y(t_2) > 0 \quad \text{for } t \geq t_2.$$

Then there exists a number $T \geq t_2$ such that

$$y(\sigma_j(t)) \geq y(t_2) \quad \text{for } t \geq T. \tag{7.2.12}$$

Integrating (7.2.7) over $[T, t]$ and taking account of (7.2.12), we obtain

$$\begin{aligned} &\int_T^t p_j(s)\varphi_j \left(\left[\left(1 - \sum_{i=1}^{\ell} h_i(\sigma_j(s)) \right) y(t_2) + H(\sigma_j(s)) \right]_+ \right) ds \\ &\quad \leq - \int_T^t y''(s)\, ds = y'(T) - y'(t) \leq y'(T) \end{aligned}$$

which contradicts the hypothesis (7.2.11). This completes the proof. □

Theorem 7.2.4. *Assume that* (H6.1-4), (H6.3-3), (H7.2-1), (H7.2-2), (i), (ii) *of Lemma* 7.2.1 *are satisfied, and assume, moreover, that there exists a* *function* $\Theta(t) \in C^2((0,\infty);\mathbb{R})$ *such that* $\Theta''(t) = G(t)$ [*resp.* $\Theta''(t) = \tilde{G}(t)$] *and* $\Theta(t)$ *is oscillatory at* $t = \infty$. *If the functional differential inequalities*

$$y''(t) + p_j(t)\varphi_j\left(\left[\left(1 - \sum_{i=1}^{\ell} h_i(\sigma_j(t))\right) y(\sigma_j(t)) \pm \left(\Theta(\sigma_j(t)) - \sum_{i=1}^{\ell} h_i(\sigma_j(t))\Theta\big(\rho_i(\sigma_j(t))\big)\right)\right]_+\right) \leq 0$$

have no eventually positive solutions, then every solution u of the boundary value problem (7.2.1), (B_1) [*resp.* (7.2.1), (B_2)] *is oscillatory in* Ω.

Proof. We note that Theorem 7.2.1 remains true if (7.2.2) is replaced by

$$\frac{d^2}{dt^2}\left(y(t) + \sum_{i=1}^{\ell} h_i(t)y(\rho_i(t))\right) + \sum_{i=1}^{m} p_i(t)\varphi_i\big(y(\sigma_i(t))\big) \leq \pm G(t).$$

The conclusion follows by using Theorems 7.2.1, 7.2.2 and Lemma 7.2.1.□

Theorem 7.2.5. *Assume that* (H6.1-4), (H6.3-3), (H7.2-1), (H7.2-2), (i), (ii) *of Lemma* 7.2.1 *are satisfied. Every solution u of the boundary value problem* (7.2.1), (B_1) [*resp.* (7.2.1), (B_2)] *is oscillatory in* Ω *if there exists a function* $\Theta(t) \in C^2((0,\infty);\mathbb{R})$ *such that* $\Theta''(t) = G(t)$ [*resp.* $\Theta''(t) = \tilde{G}(t)$], $\Theta(t)$ *is oscillatory at* $t = \infty$, *and*

$$\int_{t_0}^{\infty} p_j(t)\varphi_j\left(\left[\left(1 - \sum_{i=1}^{\ell} h_i(\sigma_j(t))\right) c \pm \left(\Theta(\sigma_j(t)) - \sum_{i=1}^{\ell} h_i(\sigma_j(t))\Theta\big(\rho_i(\sigma_j(t))\big)\right)\right]_+\right) dt = \infty$$

for some $t_0 > 0$ *and any* $c > 0$.

Proof. The conclusion follows from Theorem 7.2.4 and Lemma 7.2.2. □

Example 7.2.1. We consider the problem

$$\frac{\partial^2}{\partial t^2}\left(u(x,t) + \frac{1}{3}u(x,t-\pi)\right) - \frac{\partial^2 u}{\partial x^2}(x,t) - \frac{\partial^2 u}{\partial x^2}(x,t-\pi) + u(x,t-2\pi) = -\frac{1}{3}(\sin x)\cos t, \quad (x,t) \in (0,\pi)\times(0,\infty), \tag{7.2.13}$$

$$u(0,t) = u(\pi,t) = 0, \quad t > 0. \tag{7.2.14}$$

Here $n = 1$, $G = (0,\pi)$, $\Omega = (0,\pi)\times(0,\infty)$, $\ell = k = m = \tilde{m} = 1$, $h_1(t) = 1/3$, $\rho_1(t) = t - \pi$, $a(t) = 1$, $b_1(t) = 1$, $\tau_1(t) = t - \pi$, $p_1(t) = 1$, $\sigma_1(t) =$

$t - 2\pi$, $\varphi_1(\xi) = \xi$ and $f(x,t) = -(1/3)(\sin x)\cos t$. We easily see that $\lambda_1 = 1$, $\Phi(x) = \sin x$, $\Psi(t) = 0$ and $F(t) = G(t) = -(\pi/12)\cos t$. Choosing $\Theta(t) = (\pi/12)\cos t$, we find that $\Theta(\sigma_1(t)) = (\pi/12)\cos t$, $\Theta(\rho_1(\sigma_1(t))) = -(\pi/12)\cos t$, and $\Theta(t)$ is oscillatory at $t = \infty$. An easy computation shows that

$$\int_1^\infty \left[\left(1 - \frac{1}{3}\right) c \pm \left(\frac{\pi}{12}\cos t - \frac{1}{3}\left(-\frac{\pi}{12}\cos t\right)\right)\right]_+ dt$$
$$= \int_1^\infty \left[\frac{2}{3}c \pm \frac{\pi}{9}\cos t\right]_+ dt$$
$$\geq \frac{\pi}{9}\int_1^\infty [\pm\cos t]_+ \, dt = \infty$$

for any $c > 0$. Hence, it follows from Theorem 7.2.5 that every solution u of the problem (7.2.13), (7.2.14) is oscillatory in $(0,\pi) \times (0,\infty)$. One such solution is $u(x,t) = -(\sin x)\cos t$.

Example 7.2.2. We consider the problem

$$\frac{\partial^2}{\partial t^2}\left(u(x,t) + \frac{1}{2}u(x,t-\pi)\right) - \frac{\partial^2 u}{\partial x^2}(x,t) - \frac{\partial^2 u}{\partial x^2}(x,t-2\pi)$$
$$+u(x,t-\pi) = -\frac{1}{2}(\cos x)\sin t, \quad (x,t) \in (0,\pi)\times(0,\infty), \quad (7.2.15)$$
$$-\frac{\partial u}{\partial x}(0,t) = \frac{\partial u}{\partial x}(\pi,t) = 0, \quad t > 0. \quad (7.2.16)$$

Here $n = 1$, $G = (0,\pi)$, $\Omega = (0,\pi)\times(0,\infty)$, $\ell = k = m = \tilde{m} = 1$, $h_1(t) = 1/2$, $\rho_1(t) = t - \pi$, $a(t) = 1$, $b_1(t) = 1$, $\tau_1(t) = t - 2\pi$, $p_1(t) = 1$, $\sigma_1(t) = t - \pi$, $\varphi_1(\xi) = \xi$ and $f(x,t) = -(1/2)(\cos x)\sin t$. It is easy to check that

$$\tilde{F}(t) = \tilde{G}(t) = \frac{1}{\pi}\int_0^\pi \left(-\frac{1}{2}\right)(\cos x)\sin t \, dx = 0.$$

We choose $\Theta(t) = 0$ and observe that $\Theta''(t) = 0$ and $\Theta(t)$ is oscillatory at $t = \infty$. Since

$$\int_1^\infty \left[\left(1 - \frac{1}{2}\right)c\right]_+ dt = \infty$$

for any $c > 0$, Theorem 7.2.5 implies that every solution u of the problem (7.2.15), (7.2.16) is oscillatory in $(0,\pi)\times(0,\infty)$. In fact, $u(x,t) = -(\cos x)\sin t$ is such a solution.

Remark 7.2.1. The second statement of Theorem 7.2.3 was established by Mishev and Bainov [207] in the case where $h_i(t) = 0$ $(i = 1, 2, ..., \ell)$,

$a(t) = 1$, $k = 1$, $b_1(t)$ is constant, $\tau_1(t) = t - \tau$ (τ is constant), $m = \tilde{m} = 1$, $\sigma_1(t) = t$ and $\mu = 0$. The second statement of Theorem 7.2.3 was obtained by Mishev and Bainov [208] in the case where $\ell = 1$, $h_1(t)$ is constant, $\rho_1(t) = t - \rho$ (ρ is constant), $a(t) = 1$, $b_i(t) = 0$ $(i = 1, 2, ..., k)$, $m = \tilde{m} = 1$, $\sigma_1(t) = t$ and $\mu = 0$. In case $\ell = 1$, $a(t) = 1$, $k = 1$, $m = 1$, $m_1 = \tilde{m} = 2$ and $\sigma_1(t) = t$, Theorem 7.2.3 reduces to a result of Mishev [202].

Remark 7.2.2. When specialized to the case where $\rho_i(t) = t - \rho_i$ ($\rho_i > 0$; $i = 1, 2, ..., \ell$), $a(t) = 1$, $\tau_i(t) = t - \tau_i(\tau_i > 0$; $i = 1, 2, ..., k)$, $\sigma_i(t) = t - \sigma_i$ ($\sigma_i \geq 0$; $i = 1, 2, ..., m$), $f(x, t) = 0$, $m = \tilde{m}$, $\varphi_i(s) = s$ $(i = 1, 2, ..., m)$, $\psi = \tilde{\psi} = 0$, Theorem 7.2.5 was derived by Mishev [205]. The case where $\ell = 1$, $\rho_1(t) = t - \rho_1$ ($\rho_1 > 0$), $b_i(t) = 0$ $(i = 1, 2, ..., k)$, $f(x, t) = 0$, $m = 1$, $\varphi_1(s) \geq \beta s$ for some $\beta > 0$ was treated by Lalli, Yu and Cui [164].

7.3 Equations with Forcing Terms II

In this section we investigate the case where the coefficients $h_i(t)$ of Eq. (7.2.1) of Section 7.2 are nonpositive, that is, we treat the hyperbolic equation with functional arguments

$$\begin{aligned}
&\frac{\partial^2}{\partial t^2}\left(u(x,t) - \sum_{i=1}^{\ell} h_i(t)u(x,\rho_i(t))\right) \\
&\quad -a(t)\Delta u(x,t) - \sum_{i=1}^{k} b_i(t)\Delta u(x,\tau_i(t)) \\
&\quad +c(x,t,(z_i[u](x,t))_{i=1}^{\tilde{m}}) = f(x,t), \quad (x,t) \in \Omega := G \times (0,\infty), \quad (7.3.1)
\end{aligned}$$

where G is a bounded domain in $\mathbb{R}^n$ with piecewise smooth boundary ∂G.

It is assumed that the hypotheses (H6.1-4) of Section 6.1, (H6.3-3) of Section 6.3, (H7.2-1), (H7.2-2) of Section 7.2 are satisfied.

We consider the boundary conditions (B_i) $(i = 1, 2)$ which are given in Section 6.3.

By a *solution* of Eq. (7.3.1) we mean a function $u(x, t) \in C^2(\overline{G} \times [t_{-1}, \infty); \mathbb{R}) \cap C(\overline{G} \times [\tilde{t}_{-1}, \infty); \mathbb{R})$ which satisfies (7.3.1), where t_{-1} and $\tilde{t}_{-1}$ are those given in Section 7.2.

We note that the functions $F(t)$, $\Psi(t)$, $\tilde{F}(t)$, $\tilde{\Psi}(t)$, $G(t)$, $\tilde{G}(t)$ appearing in this section are the same as those given in Section 6.3, and that λ_1 is the first eigenvalue of the eigenvalue problem (EVP) appearing in Section 6.1 and $\Phi(x)$ is the eigenfunction corresponding to λ_1.

We recall that:

$$G(t) = F(t) - a(t)\Psi(t) - \sum_{i=1}^{k} b_i(t)\Psi(\tau_i(t)),$$

$$\tilde{G}(t) = \tilde{F}(t) + a(t)\tilde{\Psi}(t) + \sum_{i=1}^{k} b_i(t)\tilde{\Psi}(\tau_i(t)),$$

$$U(t) = \left(\int_G \Phi(x)dx\right)^{-1} \int_G u(x,t)\Phi(x)\,dx,$$

$$\tilde{U}(t) = \frac{1}{|G|}\int_G u(x,t)\,dx \quad \left(|G| = \int_G dx\right)$$

(see, (6.3.3), (6.3.10) of Section 6.3, (6.2.3), (6.2.16) of Section 6.2).

We can obtain the analogues of Theorems 7.2.1 and 7.2.2 of Section 7.2.

Theorem 7.3.1. *Assume that the hypotheses* (H6.1-4), (H6.3-3), (H7.2-1) *and* (H7.2-2) *are satisfied. If the functional differential inequalities*

$$\frac{d^2}{dt^2}\left(y(t) - \sum_{i=1}^{\ell} h_i(t)y(\rho_i(t))\right) + \lambda_1 a(t)y(t)$$
$$+\lambda_1 \sum_{i=1}^{k} b_i(t)y(\tau_i(t)) + \sum_{i=1}^{m} p_i(t)\varphi_i\big(y(\sigma_i(t))\big) \le \pm G(t) \qquad (7.3.2)$$

have no eventually positive unbounded solutions, then every solution u of the boundary value problem (7.3.1), (B_1) *with unbounded $U(t)$ is oscillatory in Ω.*

Proof. Suppose to the contrary that there exists a nonoscillatory solution u of the problem (7.3.1), (B_1) with the property that $U(t)$ is unbounded. First we assume that $u > 0$ in $G \times [t_0, \infty)$ for some $t_0 > 0$. Proceeding as in the proof of Theorem 6.3.1 of Section 6.3, we observe that $U(t)$ is an eventually positive unbounded solution of (7.3.2) with $+G(t)$. This contradicts the hypothesis. The case where $u < 0$ in $G \times [t_0, \infty)$ for some $t_0 > 0$ can be treated similarly, and we are also led to a contradiction. The proof is complete. □

Theorem 7.3.2. *Assume that the hypotheses* (H6.1-4), (H6.3-3), (H7.2-1) *and* (H7.2-2) *are satisfied. If the functional differential inequalities*

$$\frac{d^2}{dt^2}\left(y(t) - \sum_{i=1}^{\ell} h_i(t)y(\rho_i(t))\right) + \sum_{i=1}^{m} p_i(t)\varphi_i\big(y(\sigma_i(t))\big) \le \pm\tilde{G}(t) \qquad (7.3.3)$$

have no eventually positive unbounded solutions, then every solution u of the boundary value problem (7.3.1), (B_2) *with unbounded $\tilde{U}(t)$ is oscillatory in Ω.*

Proof. Arguing as in the proofs of Theorem 7.3.1 and Theorem 6.3.2 of Section 6.3, we see that the theorem follows. □

Now we derive sufficient conditions for the functional differential inequality

$$\frac{d^2}{dt^2}\left(y(t) - \sum_{i=1}^{\ell} h_i(t)y(\rho_i(t))\right) + \sum_{i=1}^{m} p_i(t)\varphi_i\big(y(\sigma_i(t))\big) \leq H(t) \tag{7.3.4}$$

to have no eventually positive unbounded solution, where $H(t)$ is a continuous function on $[t_0, \infty)$ for some $t_0 > 0$.

It is assumed that:

(H7.3-1) there exists a positive constant h_0 satisfying

$$\sum_{i=1}^{\ell} h_i(t) \leq h_0 < 1;$$

(H7.3-2) $\rho_i(t) \leq t$ for $t > 0$ $(i = 1, 2, ..., \ell)$;
(H7.3-3) $\tilde{\sigma}(t) := \min_{1\leq i\leq m} \sigma_i(t)$ is a nondecreasing continuous function;
(H7.3-4) $\varphi_i(\xi) > 0$ for $\xi > 0$ and $\varphi_i(\xi)$ are nondecreasing for $\xi > 0$ $(i = 1, 2, ..., m)$.

Theorem 7.3.3. *Assume that the hypotheses* (H6.3-3), (H7.2-1), (H7.2-2), (H7.3-1)–(H7.3-4) *are satisfied, and that the following hypothesis is satisfied:*

(H7.3-5) *there is a C^2-function $\Theta(t)$ such that $\Theta(t)$ is bounded and*

$$\Theta''(t) = H(t).$$

If

$$\int_c^{\infty} \sum_{i=1}^{m} p_i(t)\, dt = \infty \tag{7.3.5}$$

for some $c > 0$, then (7.3.4) *has no eventually positive unbounded solution.*

Proof. Suppose that (7.3.4) has an eventually positive unbounded solution $y(t)$. Letting

$$z(t) = y(t) - \sum_{i=1}^{\ell} h_i(t)y(\rho_i(t)) - \Theta(t)$$

and taking into account (H7.3-5), we find that

$$\begin{aligned} z''(t) &\le -\sum_{i=1}^{m} p_i(t)\varphi_i\big(y(\sigma_i(t))\big) \\ &\le 0. \end{aligned} \tag{7.3.6}$$

Therefore, $z'(t) \ge 0$ or $z'(t) < 0$ eventually. Hence, $z(t)$ is a monotone function, and $z(t) > 0$ or $z(t) \le 0$ eventually. We claim that $\lim_{t\to\infty} z(t) = \infty$. Since $y(t)$ is unbounded from above, there exists a sequence $\{t_n\}_{n=1}^{\infty}$ satisfying $\lim_{n\to\infty} t_n = \infty$, $\lim_{n\to\infty} y(t_n) = \infty$ and $\max_{t_0\le t\le t_n} y(t) = y(t_n)$. The hypotheses (H7.3-1) and (H7.3-2) imply that

$$\begin{aligned} z(t_n) &= y(t_n) - \sum_{i=1}^{\ell} h_i(t_n)y(\rho_i(t_n)) - \Theta(t_n) \\ &\ge y(t_n) - y(t_n)\sum_{i=1}^{\ell} h_i(t_n) - \Theta(t_n) \\ &= \left(1 - \sum_{i=1}^{\ell} h_i(t_n)\right) y(t_n) - \Theta(t_n) \\ &\ge (1 - h_0)y(t_n) - \Theta(t_n) \end{aligned}$$

for sufficiently large n. Since $\Theta(t)$ is bounded and $\lim_{n\to\infty}(1 - h_0)y(t_n) = \infty$, we find that $\lim_{t\to\infty} z(t_n) = \infty$. This combined with the monotonicity property of $z(t)$ implies that $\lim_{t\to\infty} z(t) = \infty$. Hence, $z(t) > 0$ eventually. In this case it is easily seen that $z'(t) \ge 0$ eventually. Since $\Theta(t)$ is bounded and $\lim_{t\to\infty} z(t) = \infty$, for any $\varepsilon > 0$ there is a sufficiently large number T such that $\Theta(t) \ge -\varepsilon z(t)$ $(t \ge T)$. Hence we see that

$$y(t) \ge z(t) + \Theta(t) \ge (1 - \varepsilon)z(t)$$

and therefore

$$y(\sigma_i(t)) \ge (1 - \varepsilon)z(\sigma_i(t)).$$

The inequality (7.3.6) implies that

$$\begin{aligned}
z''(t) &\leq -\sum_{i=1}^{m} p_i(t)\varphi_i\big((1-\varepsilon)z(\sigma_i(t))\big) \\
&\leq -\sum_{i=1}^{m} p_i(t)\varphi_i\big((1-\varepsilon)z(\tilde{\sigma}(t))\big) \\
&\leq -\sum_{i=1}^{m} p_i(t)\varphi_i\big((1-\varepsilon)z(\tilde{\sigma}(T))\big) \\
&\leq -C_0\sum_{i=1}^{m} p_i(t), \quad t \geq T, \qquad (7.3.7)
\end{aligned}$$

where $T > 0$ sufficiently large and $C_0 = \min_{1\leq i\leq m}\{\varphi_i\big((1-\varepsilon)z(\tilde{\sigma}(T))\big)\} > 0$. Integrating (7.3.7) over $[T, t]$, we obtain

$$z'(t) - z'(T) \leq -C_0 \int_T^t \sum_{i=1}^{m} p_i(s)\,ds$$

which yields

$$z'(T) \geq C_0 \int_T^t \sum_{i=1}^{m} p_i(s)\,ds.$$

Letting $t \to \infty$ in the above inequality, we obtain

$$\int_T^{\infty} \sum_{i=1}^{m} p_i(s)\,ds \leq \frac{1}{C_0} z'(T) < \infty$$

which contradicts the hypothesis (7.3.5). The proof is complete. □

Theorem 7.3.4. *Assume that the hypotheses* (H6.1-4), (H6.3-3), (H7.2-1), (H7.2-2), (H7.3-1)–(H7.3-4) *hold, and that there exists a* C^2*-function* $\Theta(t)$ *such that* $\Theta(t)$ *is bounded and*

$$\Theta''(t) = G(t).$$

If the condition (7.3.5) *is satisfied, then every solution* u *of the boundary value problem* (7.3.1), (B_1) *with unbounded* $U(t)$ *is oscillatory in* Ω.

Proof. The conclusion follows by combining Theorem 7.3.1 with Theorem 7.3.3. □

Theorem 7.3.5. *Assume that the hypotheses* (H6.1-4), (H6.3-3), (H7.2-1), (H7.2-2), (H7.3-1)–(H7.3-4) *hold, and that there exists a* C^2*-function* $\Theta(t)$ *such that* $\Theta(t)$ *is bounded and*

$$\Theta''(t) = \tilde{G}(t).$$

If the condition (7.3.5) *is satisfied, then every solution* u *of the boundary value problem* (7.3.1), (B_2) *with unbounded* $\tilde{U}(t)$ *is oscillatory in* Ω.

Proof. A combination of Theorem 7.3.2 and Theorem 7.3.3 yields the conclusion. □

Example 7.3.1. We consider the problem

$$\frac{\partial^2}{\partial t^2}\left(u(x,t)-\frac{1}{2}\,u(x,t-2\pi)\right)-2e^{-t}\frac{\partial^2 u}{\partial x^2}(x,t)-e^{\pi}e^{-t}\frac{\partial^2 u}{\partial x^2}(x,t-\pi)$$
$$+e^{(\pi/2)}(2-e^{-2\pi})u(x,t-\tfrac{\pi}{2})=(\sin x)\sin t,\ (x,t)\in(0,\pi)\times(0,\infty),\quad (7.3.8)$$
$$u(0,t)=u(\pi,t)=0,\quad t>0. \quad (7.3.9)$$

Here $n=1$, $G=(0,\pi)$, $\Omega=(0,\pi)\times(0,\infty)$, $\ell=k=m=\tilde{m}=1$, $h_1(t)=1/2$, $\rho_1(t)=t-2\pi$, $a(t)=2e^{-t}$, $b_1(t)=e^{\pi}e^{-t}$, $\tau_1(t)=t-\pi$, $p_1(t)=e^{(\pi/2)}(2-e^{-2\pi})$, $\sigma_1(t)=t-(\pi/2)$, $\varphi_1(\xi)=\xi$ and $f(x,t)=(\sin x)\sin t$. It is readily seen that $\lambda_1=1$, $\Phi(x)=\sin x$, $\Psi(t)=0$ and $F(t)=G(t)=(\pi/4)\sin t$. We choose $\Theta(t)=-(\pi/4)\cos t$, and find that $\Theta(t)$ is bounded. Since

$$\int_1^{\infty}p_1(t)\,dt=\int_1^{\infty}e^{(\pi/2)}(2-e^{-2\pi})\,dt=\infty,$$

Theorem 7.3.4 implies that every solution u of the problem (7.3.8), (7.3.9) with unbounded $U(t)$ is oscillatory in Ω. One such solution is $u(x,t)=(\sin x)e^t\sin t$.

Example 7.3.2. We consider the problem

$$\frac{\partial^2}{\partial t^2}\left(u(x,t)-\frac{1}{2}\,u(x,t-\pi)\right)-e^{-t}\frac{\partial^2 u}{\partial x^2}(x,t)-\frac{1}{2}e^{\pi}e^{-t}\frac{\partial^2 u}{\partial x^2}(x,t-\pi)$$
$$+e^{(\pi/2)}(2+e^{-\pi})u(x,t-\tfrac{\pi}{2})=(\cos x)\cos t,\ (x,t)\in(0,\pi)\times(0,\infty),\quad (7.3.10)$$
$$-\frac{\partial u}{\partial x}(0,t)=\frac{\partial u}{\partial x}(\pi,t)=0,\quad t>0. \quad (7.3.11)$$

Here $n=1$, $G=(0,\pi)$, $\Omega=(0,\pi)\times(0,\infty)$, $\ell=k=m=\tilde{m}=1$, $h_1(t)=1/2$, $\rho_1(t)=t-\pi$, $a(t)=e^{-t}$, $b_1(t)=(1/2)e^{\pi}e^{-t}$, $\tau_1(t)=t-\pi$, $p_1(t)=e^{(\pi/2)}(2+e^{-\pi})$, $\sigma_1(t)=t-(\pi/2)$, $\varphi_1(\xi)=\xi$ and $f(x,t)=(\cos x)\cos t$. It is easily checked that $\tilde{\Psi}(t)=0$ and $\tilde{F}(t)=\tilde{G}(t)=0$. We can choose $\Theta(t)=0$, and observe that

$$\int_1^{\infty}p_1(t)\,dt=\int_1^{\infty}e^{(\pi/2)}(2+e^{-\pi})\,dt=\infty.$$

It follows from Theorem 7.3.5 that every solution u of the problem (7.3.10), (7.3.11) with unbounded $\tilde{U}(t)$ is oscillatory in Ω. For example, the function $u(x,t)=2(\cos x+1)e^t\cos t$ is such a solution.

7.4 Impulsive Hyperbolic Equations

In 1996 impulsive hyperbolic differential-functional equations were studied by Bainov, Kamont and Minchev [21], and comparison results and a uniqueness criterion were obtained. Oscillation results for impulsive hyperbolic equations were investigated by numerous authors. We refer the reader to Fu and Liu [81], Zhang [326], Luo [184] for impulsive hyperbolic equations without delays, and to Cui, Liu and Deng [58], Liu, Xiao and Liu [180], Fu and Zhang [84] for impulsive hyperbolic equations with delays.

Let $\{t_k\}_{k=0}^{\infty}$ be a sequence of real numbers such that

$$0 = t_0 < t_1 < \cdots < t_k < \cdots$$

and $\lim_{k\to\infty} t_k = \infty$.

The following notation will be used:

$$\begin{aligned}\Omega &= G \times (0,\infty),\\ J_{\text{imp}} &= \{t_k;\ k = 1,2,...\},\\ \Omega_{\text{imp}} &= \{(x,t) \in \Omega;\ t \in J_{\text{imp}}\},\\ \overline{\Omega}_{\text{imp}} &= \{(x,t) \in \overline{\Omega};\ t \in J_{\text{imp}}\},\end{aligned}$$

where G is a bounded domain in $\mathbb{R}^n$ with piecewise smooth boundary ∂G.

We investigate the impulsive hyperbolic equation

$$\frac{\partial^2 u}{\partial t^2}(x,t) - a(t)\Delta u(x,t) + p(x,t)\varphi\big(u(x,t-\sigma)\big) = 0,\ (x,t) \in \Omega \setminus \Omega_{\text{imp}}, \tag{7.4.1}$$

$$u(x,t^+) - u(x,t^-) = h(x,t,u(x,t)), \quad (x,t) \in \overline{\Omega}_{\text{imp}}, \tag{7.4.2}$$

$$\frac{\partial u}{\partial t}(x,t^+) - \frac{\partial u}{\partial t}(x,t^-) = \tilde{h}\big(x,t,\tfrac{\partial u}{\partial t}(x,t)\big), \quad (x,t) \in \overline{\Omega}_{\text{imp}}. \tag{7.4.3}$$

We assume that the hypotheses (H6.5-1)–(H6.5-4) of Section 6.5 hold, and the following hypotheses are satisfied:

(H7.4-1) $h(x,t,\xi) \in C(\overline{\Omega}_{\text{imp}} \times \mathbb{R}, \mathbb{R})$ and

$$h(x,t_k,\xi) = M_k\xi \quad \text{for } x \in \overline{G},\ \xi \in \mathbb{R},\ k = 1,2,...,$$

where $M_k \geq 0$ are constants;

(H7.4-2) $\tilde{h}(x,t,\eta) \in C(\overline{\Omega}_{\text{imp}} \times \mathbb{R}, \mathbb{R})$ and

$$\tilde{h}(x,t_k,\eta) = \tilde{M}_k\eta \quad \text{for } x \in \overline{G},\ \eta \in \mathbb{R},\ k = 1,2,...,$$

where $\tilde{M}_k \geq 0$ are constants.

We consider two kinds of boundary conditions:

(B_0) $\quad u = 0 \quad$ on $\partial G \times ([0,\infty) \setminus J_{\mathrm{imp}})$,

$(\tilde{B}_0)$ $\quad \dfrac{\partial u}{\partial \nu} + \mu u = 0 \quad$ on $\partial G \times ([0,\infty) \setminus J_{\mathrm{imp}})$,

where $\mu \in C_{\mathrm{imp}}(\partial G \times [0,\infty); [0,\infty))$ and ν denotes the unit exterior normal vector to ∂G (*cf.* Section 6.5).

Definition 7.4.1. A real-valued function $u(x,t)$ on $\overline{G} \times [-\sigma,\infty)$ is said to be a *solution* of the problem (7.4.1)–(7.4.3), (B_0) [or (7.4.1)–(7.4.3), $(\tilde{B}_0)$] if the following conditions are satisfied:

(i) $u(x,t) \in C_{\mathrm{imp}(-\sigma)}(\overline{G} \times [-\sigma,\infty); \mathbb{R}) \cap C^1(\overline{\Omega} \setminus \overline{\Omega}_{\mathrm{imp}})$ and there exist the continuous derivatives $\frac{\partial^2 u}{\partial t^2}(x,t)$, $\frac{\partial^2 u}{\partial x_i^2}(x,t)$ $(i = 1, 2, \ldots n)$ for $(x,t) \in \Omega \setminus \Omega_{\mathrm{imp}}$;

(ii) $\dfrac{\partial u}{\partial t}(x,t) = \dfrac{\partial u}{\partial t}(x,t^-) \quad$ for $(x,t) \in \overline{\Omega}_{\mathrm{imp}}$;

(iii) $u(x,t)$ satisfies (7.4.1)–(7.4.3), (B_0) [or (7.4.1)–(7.4.3), $(\tilde{B}_0)$],

where the function space $C_{\mathrm{imp}(-\sigma)}(\overline{G} \times [-\sigma,\infty); \mathbb{R})$ is defined in Definition 6.5.6 of Section 6.5.

Definition 7.4.2. A solution u of (7.4.1) is said to be *nonoscillatory* in Ω if it is either eventually positive or eventually negative (*cf.* Definition 6.5.5 of Section 6.5). Otherwise, it is called *oscillatory* in Ω.

Let λ_1 be the first eigenvalue of the eigenvalue problem (EVP) appearing in Section 6.1 and $\Phi(x)$ be the corresponding eigenfunction such that $\Phi(x) > 0$ in G.

First we consider the boundary condition (B_0).

Lemma 7.4.1. *Assume that the hypotheses* (H6.5-1)–(H6.5-4) *of Section* 6.5, (H7.4-1), (H7.4-2) *hold. If u is an eventually positive solution of the problem* (7.4.1)–(7.4.3), (B_0), *then we obtain*

$$\begin{aligned} &U''(t) + \lambda_1 a(t) U(t) + K P(t) U(t-\sigma) \le 0, \quad t \ne t_k, \\ &U(t_k^+) = (1 + M_k) U(t_k), \\ &U'(t_k^+) = (1 + \tilde{M}_k) U'(t_k) \end{aligned}$$

eventually, where

$$P(t) = \min_{x \in \overline{G}} p(x,t),$$

$$U(t) = \int_G u(x,t) \Phi(x)\, dx.$$

Proof. Suppose that $u(x,t) > 0$ in $G \times [T, \infty)$ for some $T \geq 0$, and let $t \geq T + \sigma$, where σ is defined in the hypothesis (H6.5-1). We multiply (7.4.1) by $\Phi(x)$ and then integrate over G to obtain

$$\frac{d^2}{dt^2}\int_G u(x,t)\Phi(x)\,dx - a(t)\int_G \Delta u(x,t)\Phi(x)\,dx + \int_G p(x,t)\varphi\big(u(x,t-\sigma)\big)\Phi(x)\,dx = 0, \quad t \neq t_k. \tag{7.4.4}$$

From Green's formula it follows that

$$\int_G \Delta u(x,t)\Phi(x)\,dx = -\lambda_1 \int_G u(x,t)\Phi(x)\,dx, \quad t \neq t_k \tag{7.4.5}$$

(see, for example, (6.2.5)). The hypothesis (H6.5-4) implies that

$$\int_G p(x,t)\varphi\big(u(x,t-\sigma)\big)\Phi(x)\,dx \geq P(t)K\int_G u(x,t-\sigma)\Phi(x)\,dx, \quad t \neq t_k. \tag{7.4.6}$$

Combining (7.4.4)–(7.4.6), we have

$$U''(t) + \lambda_1 a(t)U(t) + KP(t)U(t-\sigma) \leq 0, \quad t \neq t_k$$

which is the desired inequality. For $t = t_k$ we obtain

$$U(t_k^+) - U(t_k^-) = M_k \int_G u(x,t_k)\Phi(x)\,dx = M_k U(t_k),$$

$$U'(t_k^+) - U'(t_k^-) = \tilde{M}_k \int_G \frac{\partial u}{\partial t}(x,t_k)\Phi(x)\,dx = \tilde{M}_k U'(t_k)$$

and hence

$$U(t_k^+) = (1 + M_k)U(t_k),$$
$$U'(t_k^+) = (1 + \tilde{M}_k)U'(t_k).$$

The proof is complete. □

Theorem 7.4.1. *Assume that the hypotheses* (H6.5-1)–(H6.5-4), (H7.4-1) *and* (H7.4-2) *hold. If the impulsive differential inequality*

$$y''(t) + \lambda_1 a(t)y(t) + KP(t)y(t-\sigma) \leq 0, \quad t \neq t_k, \tag{7.4.7}$$
$$y(t_k^+) = (1 + M_k)y(t_k), \tag{7.4.8}$$
$$y'(t_k^+) = (1 + \tilde{M}_k)y'(t_k) \tag{7.4.9}$$

has no eventually positive solution, then every solution u of the problem (7.4.1)–(7.4.3), (B_0) *is oscillatory in* Ω.

Proof. Suppose to the contrary that there exists a nonoscillatory solution $u(x,t)$ of the problem (7.4.1)–(7.4.3), (B_0). Without loss of generality we may assume that $u(x,t) > 0$ in $G \times [T, \infty)$ for some $T > 0$. Applying Lemma 7.4.1, we find that $U(t)$ is an eventually positive solution of the impulsive differential inequality (7.4.7)–(7.4.9). This contradicts the hypothesis, and the proof is complete. □

Now we consider the impulsive differential inequality

$$y''(t) + Q(t)y(t-\sigma) \le 0, \quad t \ne t_k, \tag{7.4.10}$$

$$y(t_k^+) = (1 + L_k)y(t_k), \tag{7.4.11}$$

$$y'(t_k^+) = (1 + \tilde{L}_k)y'(t_k), \tag{7.4.12}$$

where $Q(t) \in C_{\text{imp}}([0,\infty); [0,\infty))$, and $L_k, \tilde{L}_k$ are nonnegative constants.

Let $PC([0,\infty); \mathbb{R})$ denote the class of all real-valued piecewise continuous functions on $[0,\infty)$, with discontinuities of the first kind only at $t = t_k$, $k = 1, 2, ...$, and $PC^1([0,\infty); \mathbb{R})$ is defined to be the space $PC([0,\infty); \mathbb{R}) \cap C^1((0,\infty) \setminus J_{\text{imp}})$.

Theorem 7.4.2. [163, Theorem 1.4.1] *Assume that:*

(i) *the sequence* $\{t_k\}$ *satisfies* $0 \le t_0 < t_1 < t_2 < \cdots$ *and* $\lim_{k\to\infty} t_k = \infty$;

(ii) $z(t) \in PC^1([0,\infty); \mathbb{R})$ *and* $z(t)$ *is left-continuous at* t_k, $k = 1, 2, ...$;

(iii) *for* $k = 1, 2, ...,$ $t \ge t_0$,

$$z'(t) \le p(t)z(t) + q(t), \; t \ne t_k,$$
$$z(t_k^+) \le d_k z(t_k) + b_k,$$

where $p(t), q(t) \in C([0,\infty); \mathbb{R})$, $d_k \ge 0$ *and* b_k *are constants.*
Then

$$\begin{aligned} z(t) \le z(t_0) & \prod_{t_0 < t_k < t} d_k \exp\left(\int_{t_0}^t p(s)\,ds\right) \\ & + \sum_{t_0 < t_k < t} \left(\prod_{t_k < t_j < t} d_j \exp\left(\int_{t_k}^t p(s)\,ds\right)\right) b_k \\ & + \int_{t_0}^t \prod_{s < t_k < t} d_k \exp\left(\int_s^t p(r)\,dr\right) q(s)\,ds, \quad t \ge t_0. \end{aligned}$$

Lemma 7.4.2. *Let* $y(t)$ *be an eventually positive solution of the problem* (7.4.10)–(7.4.12). *Assume that there exists a number* $T \ge t_0$ *such that*

$y(t) > 0$ and $y(t-\sigma) > 0$ for $t \geq T$. If

$$\int_{t_0}^{\infty} \prod_{t_0<t_k<t} \frac{1+\tilde{L}_k}{1+L_k}\, dt = \infty, \tag{7.4.13}$$

then we see that $y'(t) \geq 0$ for $t \in (T, t_N] \cup (\bigcup_{k=N}^{\infty}(t_k, t_{k+1}])$, where $N = \min\{k;\, t_k \geq T\}$.

Proof. First we claim that $y'(t_k) \geq 0$ for any $k \geq N$. If it is not true, then there is an integer m such that $m \geq N$ and $y'(t_m) < 0$. From (7.4.12) we see that

$$y'(t_m^+) = (1+\tilde{L}_m)y'(t_m) < 0.$$

We let $y'(t_m^+) = -\delta$, where $\delta > 0$. Since

$$y''(t) = -Q(t)y(t-\sigma) \leq 0 \quad \text{for } t \neq t_k,$$

we find that $y'(t)$ is nonincreasing in $(t_{m+i-1}, t_{m+i}]$ $(i = 1, 2, \ldots)$. Hence we obtain

$$\begin{aligned}
y'(t_{m+1}) &\leq y'(t_m^+) = -\delta < 0,\\
y'(t_{m+2}) &\leq y'(t_{m+1}^+) = (1+\tilde{L}_{m+1})y'(t_{m+1})\\
&\leq -\delta(1+\tilde{L}_{m+1}) < 0,\\
y'(t_{m+3}) &\leq y'(t_{m+2}^+) = (1+\tilde{L}_{m+2})y'(t_{m+2})\\
&\leq -\delta(1+\tilde{L}_{m+1})(1+\tilde{L}_{m+2}) < 0.
\end{aligned}$$

Proceeding in this fashion, we can show that

$$y'(t_{m+j}) \leq -\delta \prod_{i=1}^{j-1}(1+\tilde{L}_{m+i}) < 0$$

for any positive integer $j \geq 2$. Here we consider the impulsive differential inequalities

$$\begin{aligned}
&y''(t) \leq 0, \quad t > t_m,\ t \neq t_k,\ k = m+1, m+2, \ldots,\\
&y'(t_k^+) = (1+\tilde{L}_k)y'(t_k), \quad k = m+1, m+2, \ldots.
\end{aligned}$$

Letting $z(t) = y'(t)$, we obtain

$$\begin{aligned}
&z'(t) \leq 0, \quad t > t_m,\ t \neq t_k,\ k = m+1, m+2, \ldots,\\
&z(t_k^+) = (1+\tilde{L}_k)z(t_k), \quad k = m+1, m+2, \ldots.
\end{aligned}$$

It follows from Theorem 7.4.2 that

$$z(t) \leq z(t_m^+) \prod_{t_m<t_k<t} (1+\tilde{L}_k)$$

or

$$y'(t) \le y'(t_m^+) \prod_{t_m<t_k<t} (1+\tilde{L}_k). \tag{7.4.14}$$

Using Theorem 7.4.2, we see from (7.4.11) and (7.4.14) that

$$\begin{aligned} y(t) &\le y(t_m^+) \prod_{t_m<t_k<t} (1+L_k) \\ &\quad + \int_{t_m^+}^t \prod_{s<t_k<t} (1+L_k) \left(y'(t_m^+) \prod_{t_m<t_k<s} (1+\tilde{L}_k) \right) ds \\ &= \prod_{t_m<t_k<t} (1+L_k) \left[y(t_m^+) - \delta \int_{t_m^+}^t \prod_{t_m<t_k<s} \frac{1+\tilde{L}_k}{1+L_k} ds \right]. \end{aligned} \tag{7.4.15}$$

Since $y(t) > 0$ for $t \ge T$ and the right side of (7.4.15) is negative for sufficiently large t by the hypothesis (7.4.13), we are led to a contradiction. Hence, it was shown that $y'(t_k) \ge 0$ for $k \ge N$. From (7.4.12) we find that

$$y'(t_k^+) = (1+\tilde{L}_k) y'(t_k) \ge 0$$

for any $k \ge N$. Since $y'(t)$ is nonincreasing in $(t_k, t_{k+1}]$, it can be shown that $y'(t) \ge y'(t_{k+1}) \ge 0$ in $(t_k, t_{k+1}]$ $(k \ge N)$. It is clear that $y'(t) \ge y'(t_N) \ge 0$ in $(T, t_N]$. Consequently, we observe that $y'(t) \ge 0$ in $(T, t_N] \cup (\bigcup_{k=N}^{\infty} (t_k, t_{k+1}])$. □

Theorem 7.4.3. *Assume that the hypotheses* (H6.5-1)–(H6.5-4), (H7.4-1) *and* (H7.4-2) *hold. If the conditions*

$$\int_{t_0}^{\infty} \prod_{t_0<t_k<t} \frac{1+\tilde{M}_k}{1+M_k} dt = \infty, \tag{7.4.16}$$

$$\int_{t_0}^{\infty} \left(\prod_{t_0<t_k<t} \frac{1+M_k}{1+\tilde{M}_k} \right) a(t)\, dt = \infty \tag{7.4.17}$$

are satisfied, then every solution u of the problem (7.4.1)–(7.4.3), (B$_0$) *is oscillatory in* Ω.

Proof. Suppose to the contrary that there exists a nonoscillatory solution $u(x,t)$ of the problem (7.4.1)–(7.4.3), (B$_0$). Without loss of generality we may assume that $u(x,t) > 0$ in $G \times [T_0, \infty)$ for some $T_0 > 0$. Then there is a number $\tilde{T} := T_0 + \sigma$ such that $u(x, t-\sigma) > 0$ in $G \times [\tilde{T}, \infty)$. Lemma 7.4.1 implies that $U(t)$ is an eventually positive solution of (7.4.7)–(7.4.9). From Lemma 7.4.2 we see that $U'(t) \ge 0$ on $[T, \infty)$ for some $T \ge \tilde{T}$. Letting

$$W(t) = \frac{U'(t)}{U(t)}, \quad t \ge T,$$

we observe that $W(t) \geq 0$ for $t \geq T$. It is easily seen that

$$\begin{aligned} W'(t) &= \frac{U''(t)}{U(t)} - \left(\frac{U'(t)}{U(t)}\right)^2 \\ &\leq \frac{U''(t)}{U(t)} \\ &\leq -\lambda_1 a(t), \quad t \neq t_k,\ t \geq T \end{aligned} \tag{7.4.18}$$

and

$$W(t_k^+) = \frac{1+\tilde{M}_k}{1+M_k} W(t_k).$$

It follows from Theorem 7.4.2 that

$$\begin{aligned} W(t) &\leq W(T) \prod_{T<t_k<t} \frac{1+\tilde{M}_k}{1+M_k} + \int_T^t \prod_{s<t_k<t} \frac{1+\tilde{M}_k}{1+M_k}(-\lambda_1 a(s))\,ds \\ &= \prod_{T<t_k<t} \frac{1+\tilde{M}_k}{1+M_k}\left[W(T) - \lambda_1 \int_T^t \left(\prod_{T<t_k<s} \frac{1+M_k}{1+\tilde{M}_k}\right) a(s)\,ds\right]. \end{aligned} \tag{7.4.19}$$

Since $W(t) \geq 0$ for $t \geq T$, the inequality (7.4.19) contradicts the hypothesis (7.4.17). The proof is complete. □

Theorem 7.4.4. *Let* $\sigma = 0$. *Assume that the hypotheses* (H6.5-2)–(H6.5-4), (H7.4-1), (H7.4-2) *hold. If the condition* (7.4.16) *holds and the following condition*

$$\int_{t_0}^{\infty} \left(\prod_{t_0<t_k<t} \frac{1+M_k}{1+\tilde{M}_k}\right)(\lambda_1 a(t) + KP(t))\,dt = \infty$$

is satisfied, then every solution u *of the problem* (7.4.1)–(7.4.3), (B_0) *is oscillatory in* Ω.

Proof. The proof is quite similar to that of Theorem 7.4.3, except that we use the inequality

$$W'(t) \leq -\lambda_1 a(t) - KP(t), \quad t \neq t_k,\ t \geq T$$

instead of (7.4.18). The proof is complete. □

Theorem 7.4.5. *Assume that the hypotheses* (H6.5-1)–(H6.5-4), (H7.4-1) *and* (H7.4-2) *hold. If the condition* (7.4.16) *holds and the following condition*

$$\int_{t_0}^{\infty} \frac{P(t)}{\prod_{t_0<t_k<t}(1+\tilde{M}_k)}\,dt = \infty \tag{7.4.20}$$

is satisfied, then every solution u *of the problem* (7.4.1)–(7.4.3), (B_0) *is oscillatory in* Ω.

Proof. Suppose that there is an eventually positive solution $u(x,t)$ of the problem (7.4.1)–(7.4.3), (B_0). Proceeding as in the proof of Theorem 7.4.3, we find that $U'(t) \geq 0$ and $U(t-\sigma) > 0$ on $[T,\infty)$ for some $T > 0$. Hence, we observe that $U(t-\sigma) \geq K_0$ on $[T,\infty)$ for some $K_0 > 0$, in view of the fact that $U(t_k^+) = (1+M_k)U(t_k) \geq U(t_k)$. Hence we obtain

$$U''(t) \leq -KP(t)U(t-\sigma) \leq -KK_0P(t), \quad t \neq t_k,\ t \geq T,$$
$$U'(t_k^+) = (1+\tilde{M}_k)U'(t_k).$$

Letting $Z(t) = U'(t)$, we have

$$Z'(t) \leq -KK_0P(t), \quad t \neq t_k,\ t \geq T,$$
$$Z(t_k^+) = (1+\tilde{M}_k)Z(t_k).$$

From Theorem 7.4.2 we see that

$$\begin{aligned} Z(t) &\leq Z(T) \prod_{T<t_k<t} (1+\tilde{M}_k) + \int_T^t \prod_{s<t_k<t} (1+\tilde{M}_k)(-KK_0P(s))\,ds \\ &= \prod_{T<t_k<t} (1+\tilde{M}_k) \left[Z(T) - KK_0 \int_T^t \frac{P(s)}{\prod_{T<t_k<s}(1+\tilde{M}_k)}\,ds \right]. \end{aligned}$$

Since $Z(t) \geq 0$, the above inequality contradicts the hypothesis (7.4.20). This completes the proof. □

Next we consider the boundary condition ($\tilde{\mathrm{B}}_0$).

Lemma 7.4.3. *Assume that the hypotheses* (H6.5-1)–(H6.5-4), (H7.4-1) *and* (H7.4-2) *hold. If u is an eventually positive solution of the problem* (7.4.1)–(7.4.3), ($\tilde{\mathrm{B}}_0$), *then we obtain*

$$\tilde{U}''(t) + KP(t)\tilde{U}(t-\sigma) \leq 0, \quad t \neq t_k, \tag{7.4.21}$$
$$\tilde{U}(t_k^+) = (1+M_k)\tilde{U}(t_k), \tag{7.4.22}$$
$$\tilde{U}'(t_k^+) = (1+\tilde{M}_k)\tilde{U}'(t_k) \tag{7.4.23}$$

eventually, where

$$\tilde{U}(t) = \int_G u(x,t)\,dx.$$

Proof. Suppose that $u(x,t) > 0$ in $G \times [T,\infty)$ for some $T \geq 0$, and let $t \geq T + \sigma$. We integrate (7.4.1) over G and obtain

$$\begin{aligned} &\frac{d^2}{dt^2} \int_G u(x,t)\,dx - a(t) \int_G \Delta u(x,t)\,dx \\ &\quad + \int_G p(x,t)\varphi\big(u(x,t-\sigma)\big)\,dx = 0, \quad t \neq t_k. \end{aligned}$$

Arguing as in the proof of Lemma 6.5.3 of Section 6.5, we find that (7.4.21) holds. For $t = t_k$ it is easy to check that (7.4.22) and (7.4.23) hold. This completes the proof. □

Theorem 7.4.6. *Assume that the hypotheses* (H6.5-1)–(H6.5-4), (H7.4-1) *and* (H7.4-2) *hold. If the impulsive differential inequality*

$$\begin{aligned} & y''(t) + KP(t)y(t-\sigma) \leq 0, \quad t \neq t_k, \\ & y(t_k^+) = (1 + M_k)y(t_k), \\ & y'(t_k^+) = (1 + \tilde{M}_k)y'(t_k) \end{aligned}$$

has no eventually positive solution, then every solution u of the problem (7.4.1)–(7.4.3), $(\tilde{\mathrm{B}}_0)$ *is oscillatory in* Ω.

Proof. The conclusion follows by using the same arguments as in the proof of Theorem 7.4.1, and will be omitted. □

Theorem 7.4.7. *Let* $\sigma = 0$. *Assume that the hypotheses* (H6.5-2)–(H6.5-4), (H7.4-1), (H7.4-2) *hold. If the condition* (7.4.16) *holds and the following condition*

$$\int_{t_0}^{\infty} \left(\prod_{t_0 < t_k < t} \frac{1 + M_k}{1 + \tilde{M}_k} \right) P(t)\, dt = \infty$$

is satisfied, then every solution u of the problem (7.4.1)–(7.4.3), $(\tilde{\mathrm{B}}_0)$ *is oscillatory in* Ω.

Proof. The proof is quite similar to that of Theorem 7.4.4, and is omitted. □

Theorem 7.4.8. *Assume that the hypotheses* (H6.5-1)–(H6.5-4), (H7.4-1) *and* (H7.4-2) *hold. If the conditions* (7.4.16) *and* (7.4.20) *hold, then every solution u of the problem* (7.4.1)–(7.4.3), $(\tilde{\mathrm{B}}_0)$ *is oscillatory in* Ω.

Proof. The conclusion follows by the same arguments as were used in Theorem 7.4.5, and is omitted. □

Remark 7.4.1. If there exists a subsequence $\{n_k\} \subset \{n\}$ such that $M_{n_k} < -1$ $(k = 1, 2, \ldots)$, then every solution u of (7.4.1), (7.4.2) is oscillatory in Ω.

7.5 Higher Order Equations

In 1976 Onose and Yokoyama [232] investigated oscillations of a class of higher order partial differential equations without functional arguments. We refer the reader to Jin, Dong and Li [119], Kiguradze, Kusano and Yoshida [126], Kiguradze and Stavroulakis [127], Li and Debnath [173], Liu and Fu [183], Onose and Yokoyama [232], Yoshida [313] for boundary value problems for higher order hyperbolic equations, and to Kusano and Yoshida [158], Travis and Yoshida [266] for characteristic initial value problems for higher order hyperbolic equations.

In this section we study the partial functional differential equation of Nth order

$$\begin{aligned}&\frac{\partial^N}{\partial t^N}\left(u(x,t)+\sum_{i=1}^{\ell} h_i(t)u(x,\rho_i(t))\right)\\&-\sum_{j=1}^{K}(-1)^{j-1}a_j(t)\Delta^j u(x,t)-\sum_{i=1}^{k}\sum_{j=1}^{K}(-1)^{j-1}b_{ij}(t)\Delta^j u(x,\tau_{ij}(t))\\&+c(x,t,(z_i[u](x,t))_{i=1}^{\tilde{m}})=f(x,t),\quad (x,t)\in\Omega:=G\times(0,\infty),\qquad(7.5.1)\end{aligned}$$

where G is a bounded domain of $\mathbb{R}^n$ with piecewise smooth boundary ∂G, Δ is the Laplacian in $\mathbb{R}^n$ and Δ^j is the jth iterated Laplacian.

The boundary conditions to be considered are the following:

$$(\mathrm{B}_D)\quad \Delta^j u=\psi_{j+1}\quad\text{on } \partial G\times(0,\infty)\ \ (j=0,1,...,K-1),$$

$$(\mathrm{B}_R)\quad \frac{\partial\Delta^j u}{\partial\nu}+(-1)^j\mu_{j+1}u=\tilde{\psi}_{j+1}\quad\text{on } \partial G\times(0,\infty)\ \ (j=0,1,...,K-1),$$

where $\psi_{j+1},\ \tilde{\psi}_{j+1}\in C(\partial G\times(0,\infty);\mathbb{R})$, $\mu_{j+1}\in C(\partial G\times(0,\infty);[0,\infty))$ and ν denotes the unit exterior normal vector to ∂G.

Let λ_1 be the first eigenvalue of the eigenvalue problem (EVP) appearing in Section 6.1 and $\Phi(x)$ be the corresponding eigenfunction such that $\Phi(x)>0$ in G.

The following notation will be used:

$$F(t)=\left(\int_G\Phi(x)\,dx\right)^{-1}\int_G f(x,t)\Phi(x)\,dx,$$

$$\Psi_j(t)=\left(\int_G\Phi(x)\,dx\right)^{-1}\int_{\partial G}\psi_j\frac{\partial\Phi}{\partial\nu}(x)\,dS\quad(j=1,2,...,K),$$

$$\tilde{F}(t) = \frac{1}{|G|}\int_G f(x,t)\,dx,$$

$$\tilde{\Psi}_j(t) = \frac{1}{|G|}\int_{\partial G} \tilde{\psi}_j\,dS \quad (j = 1,2,...,K),$$

where $|G| = \int_G dx$ and $F(t)$, $\tilde{F}(t)$ are appeared in Section 6.3.

It is assumed that (H6.1-4) of Section 6.1 and (H6.3-3) of Section 6.3 are satisfied, and that the following hypotheses hold:

(H7.5-1) $h_i(t) \in C^N([0,\infty);[0,\infty))$ $(i = 1,2,...,\ell)$, $a_j(t) \in C([0,\infty);[0,\infty))$ $(j = 1,2,...,K)$, $b_{ij}(t) \in C([0,\infty);[0,\infty))$ $(i = 1,2,...,k;\ j = 1,2,...,K)$ and $f(x,t) \in C(\overline{\Omega};\mathbb{R})$;

(H7.5-2) $\rho_i(t) \in C^N([0,\infty);\mathbb{R})$, $\lim_{t\to\infty} \rho_i(t) = \infty$ $(i = 1,2,...,\ell)$, $\tau_{ij}(t) \in C([0,\infty);\mathbb{R})$, $\lim_{t\to\infty} \tau_{ij}(t) = \infty$ $(i = 1,2,...,k;\ j = 1,2,...,K)$.

By a *solution* of Eq. (7.5.1) we mean a function $u(x,t) \in C^M(\overline{G} \times [t_{-1},\infty);\mathbb{R}) \cap C(\overline{G} \times [\tilde{t}_{-1},\infty);\mathbb{R})$ which satisfies (7.5.1), where

$$t_{-1} = \min\left\{0,\ \min_{1\le i\le \ell}\left\{\inf_{t\ge 0} \rho_i(t)\right\},\ \min_{\substack{1\le i\le k\\ 1\le j\le K}}\left\{\inf_{t\ge 0} \tau_{ij}(t)\right\}\right\},$$

$$\tilde{t}_{-1} = \min\left\{0,\ \min_{1\le i\le m}\left\{\inf_{t\ge 0} \sigma_i(t)\right\},\ \min_{\substack{m_1+1\le i\le \tilde{m}\\ 1\le j\le N_i}}\left\{\inf_{t\ge 0} \sigma_{ij}(t)\right\}\right\}$$

and $M = \max\{N, 2K\}$.

Lemma 7.5.1. *If $u(x,t) \in C^{2K}(\overline{G} \times [0,\infty);\mathbb{R})$, then the following identities hold for $j = 1,2,...,K$:*

$$K_\Phi \int_G \left(\Delta^j u(x,t)\right)\Phi(x)\,dx = -\sum_{p=0}^{j-1} (-\lambda_1)^p\, \Psi_{j-p}(t) + (-\lambda_1)^j\, U(t), \quad (7.5.2)$$

where $U(t)$ is given by (6.2.3),

$$K_\Phi = \left(\int_G \Phi(x)\,dx\right)^{-1},$$

$$\Psi_{j-p}(t) = \left(\int_G \Phi(x)\,dx\right)^{-1} \int_{\partial G} \psi_{j-p}\frac{\partial \Phi(x)}{\partial \nu}\,dS.$$

Proof. It follows from Green's formula that

$$\begin{aligned}&\int_G \left(\Delta^j u(x,t)\right)\Phi(x)\,dx\\&=\int_{\partial G}\left(\frac{\partial}{\partial\nu}\left(\Delta^{j-1}u(x,t)\right)\Phi(x)-\left(\Delta^{j-1}u(x,t)\right)\frac{\partial\Phi(x)}{\partial\nu}\right)dS\\&\qquad+\int_G\left(\Delta^{j-1}u(x,t)\right)\Delta\Phi(x)\,dx\\&=-\int_{\partial G}\psi_j\frac{\partial\Phi(x)}{\partial\nu}\,dS-\lambda_1\int_G\left(\Delta^{j-1}u(x,t)\right)\Phi(x)\,dx.\end{aligned}\tag{7.5.3}$$

Analogously we obtain

$$\begin{aligned}\int_G\left(\Delta^{j-1}u(x,t)\right)\Phi(x)\,dx=-\int_{\partial G}&\psi_{j-1}\frac{\partial\Phi(x)}{\partial\nu}\,dS\\&-\lambda_1\int_G\left(\Delta^{j-2}u(x,t)\right)\Phi(x)\,dx.\end{aligned}\tag{7.5.4}$$

Combining (7.5.3) with (7.5.4) yields

$$\begin{aligned}\int_G\left(\Delta^j u(x,t)\right)\Phi(x)\,dx=-\int_{\partial G}\psi_j\frac{\partial\Phi(x)}{\partial\nu}\,dS-\lambda_1\left(-\int_{\partial G}\psi_{j-1}\frac{\partial\Phi(x)}{\partial\nu}\,dS\right)\\+(-\lambda_1)^2\int_G\left(\Delta^{j-2}u(x,t)\right)\Phi(x)\,dx.\end{aligned}$$

Repeating this procedure, we have

$$\begin{aligned}\int_G\left(\Delta^j u(x,t)\right)\Phi(x)\,dx=\sum_{p=0}^{j-1}(-\lambda_1)^p\left(-\int_{\partial G}\psi_{j-p}\frac{\partial\Phi(x)}{\partial\nu}\,dS\right)\\+(-\lambda_1)^j\int_G u(x,t)\Phi(x)\,dx.\end{aligned}\tag{7.5.5}$$

Multiplying (7.5.5) by K_Φ yields the desired identity (7.5.2). □

Theorem 7.5.1. *Assume that the hypotheses* (H6.1-4), (H6.3-3), (H7.5-1) *and* (H7.5-2) *are satisfied. If the functional differential inequalities*

$$\begin{aligned}&\frac{d^N}{dt^N}\left(y(t)+\sum_{i=1}^{\ell}h_i(t)y(\rho_i(t))\right)+\left(\sum_{j=1}^{K}\lambda_1^j a_j(t)\right)y(t)\\&\quad+\sum_{i=1}^{k}\sum_{j=1}^{K}\lambda_1^j b_{ij}(t)y(\tau_{ij}(t))+\sum_{i=1}^{m}p_i(t)\varphi_i\big(y(\sigma_i(t))\big)\le\pm H(t)\end{aligned}\tag{7.5.6}$$

have no eventually positive solutions, then every solution u of the boundary value problem (7.5.1), (B_D) *is oscillatory in* Ω, *where*

$$H(t) = F(t) + \sum_{j=1}^{K}\sum_{p=0}^{j-1}(-1)^{j+p}\lambda_1^p a_j(t)\Psi_{j-p}(t)$$
$$+\sum_{i=1}^{k}\sum_{j=1}^{K}\sum_{p=0}^{j-1}(-1)^{j+p}\lambda_1^p b_{ij}(t)\Psi_{j-p}(\tau_{ij}(t)).$$

Proof. Suppose to the contrary that there is a solution $u(x,t)$ of the problem (7.5.1), (B_D) which has no zero in $G \times [t_0, \infty)$ for some $t_0 > 0$. First we suppose that $u(x,t) > 0$ in $G \times [t_0, \infty)$. Proceeding as in the proof of Theorem 6.3.1, we obtain

$$\frac{\partial^N}{\partial t^N}\left(u(x,t) + \sum_{i=1}^{\ell} h_i(t)u(x,\rho_i(t))\right)$$
$$-\sum_{j=1}^{K}(-1)^{j-1}a_j(t)\Delta^j u(x,t) - \sum_{i=1}^{k}\sum_{j=1}^{K}(-1)^{j-1}b_{ij}(t)\Delta^j u(x,\tau_{ij}(t))$$
$$+\sum_{i=1}^{m} p_i(t)\varphi_i\big(u(x,\sigma_i(t))\big) \le f(x,t) \quad \text{in } G \times [t_1, \infty) \tag{7.5.7}$$

for some $t_1 \ge t_0$. Multiplying (7.5.7) by $\left(\int_G \Phi(x)\,dx\right)^{-1}\Phi(x)$ and then integrating over G, we observe that

$$\frac{d^N}{dt^N}\left(U(t) + \sum_{i=1}^{\ell} h_i(t)U(\rho_i(t))\right)$$
$$-\left[\sum_{j=1}^{K}(-1)^{j-1}a_j(t)K_\Phi\int_G \Delta^j u(x,t)\Phi(x)\,dx\right.$$
$$\left.+\sum_{i=1}^{k}\sum_{j=1}^{K}(-1)^{j-1}b_{ij}(t)K_\Phi\int_G \Delta^j u(x,\tau_{ij}(t))\Phi(x)\,dx\right]$$
$$+\sum_{i=1}^{m} p_i(t)K_\Phi\int_G \varphi_i\big(u(x,\sigma_i(t))\big)\Phi(x)\,dx \le F(t), \quad t \ge t_1. \tag{7.5.8}$$

Using Lemma 7.5.1, we obtain

$$\sum_{j=1}^{K}(-1)^{j-1}a_j(t)K_\Phi\int_G \Delta^j u(x,t)\Phi(x)\,dx$$
$$= \sum_{j=1}^{K}\sum_{p=0}^{j-1}(-1)^{j+p}\lambda_1^p a_j(t)\Psi_{j-p}(t) - \sum_{j=1}^{K}\lambda_1^j a_j(t)U(t). \tag{7.5.9}$$

Similarly we see that

$$\sum_{i=1}^{k}\sum_{j=1}^{K}(-1)^{j-1}b_{ij}(t)K_\Phi\int_G \Delta^j u(x,\tau_{ij}(t))\Phi(x)\,dx$$

$$=\sum_{i=1}^{k}\sum_{j=1}^{K}\sum_{p=0}^{j-1}(-1)^{j+p}\lambda_1^p b_{ij}(t)\Psi_{j-p}(\tau_{ij}(t))-\sum_{i=1}^{k}\sum_{j=1}^{K}\lambda_1^j b_{ij}(t)U(\tau_{ij}(t)). \quad (7.5.10)$$

An application of Jensen's inequality shows that

$$K_\Phi\int_G \varphi_i\big(u(x,\sigma_i(t))\big)\Phi(x)\,dx \geq \varphi_i\big(U(\sigma_i(t))\big). \quad (7.5.11)$$

Combining (7.5.8)–(7.5.11), we observe that $U(t)$ is an eventually positive solution of (7.5.6) with $+H(t)$. This contradicts the hypothesis. The case where $u<0$ in $G\times[t_0,\infty)$ can be handled analogously, and we are also led to a contradiction. This completes the proof. □

Theorem 7.5.2. *Assume that the hypotheses* (H6.1-4), (H6.3-3), (H7.5-1) *and* (H7.5-2) *are satisfied. If the functional differential inequalities*

$$\frac{d^N}{dt^N}\left(y(t)+\sum_{i=1}^{\ell}h_i(t)y(\rho_i(t))\right)+\sum_{i=1}^{m}p_i(t)\varphi_i\big(y(\sigma_i(t))\big)\leq \pm\tilde{H}(t) \quad (7.5.12)$$

have no eventually positive solutions, then every solution u of the boundary value problem (7.5.1), (B_R) *is oscillatory in Ω, where*

$$\tilde{H}(t)=\tilde{F}(t)+\sum_{j=1}^{K}(-1)^{j-1}a_j(t)\tilde{\Psi}_j(t)+\sum_{i=1}^{k}\sum_{j=1}^{K}(-1)^{j-1}b_{ij}(t)\tilde{\Psi}_j(\tau_{ij}(t)).$$

Proof. Suppose that there exists a nonoscillatory solution u of the problem (7.5.1), (B_R). First we assume $u>0$ in $G\times[t_0,\infty)$ for some $t_0>0$. Proceeding as in the proof of Theorem 7.5.1, we find that (7.5.7) holds. Integrating (7.5.7) over G and then dividing by $|G|$, we obtain

$$\frac{d^N}{dt^N}\left(\tilde{U}(t)+\sum_{i=1}^{\ell}h_i(t)\tilde{U}(\rho_i(t))\right)$$
$$-\left[\sum_{j=1}^{M}(-1)^{j-1}a_j(t)\frac{1}{|G|}\int_G \Delta^j u(x,t)\,dx\right.$$
$$\left.+\sum_{i=1}^{k}\sum_{j=1}^{M}(-1)^{j-1}b_{ij}(t)\frac{1}{|G|}\int_G \Delta^j u(x,\tau_{ij}(t))\,dx\right]$$
$$+\sum_{i=1}^{m}p_i(t)\frac{1}{|G|}\int_G \varphi_i\big(u(x,\sigma_i(t))\big)\,dx\leq \tilde{F}(t),\quad t\geq t_1 \quad (7.5.13)$$

for some $t_1 \geq t_0$, where $\tilde{U}(t)$ is given by (6.2.16). In view of the boundary condition (B_R) we obtain

$$\begin{aligned}\int_G (-1)^{j-1} \Delta^j u(x,t)\, dx &= \int_{\partial G} (-1)^{j-1} \frac{\partial \Delta^{j-1} u}{\partial \nu}\, dS \\ &= \int_{\partial G} (-1)^{j-1}(-(-1)^{j-1}\mu_j u + \tilde{\psi}_j)\, dS \\ &= \int_{\partial G} (-\mu_j u + (-1)^{j-1}\tilde{\psi}_j)\, dS \\ &\leq |G|(-1)^{j-1}\tilde{\Psi}_j(t) \quad (j = 1, 2, ..., K)\end{aligned}$$

and hence

$$\sum_{j=1}^{K}(-1)^{j-1}a_j(t)\frac{1}{|G|}\int_G \Delta^j u(x,t)\, dx \leq \sum_{j=1}^{K}(-1)^{j-1}a_j(t)\tilde{\Psi}_j(t). \tag{7.5.14}$$

Analogously we have

$$\begin{aligned}&\sum_{i=1}^{k}\sum_{j=1}^{K}(-1)^{j-1}b_{ij}(t)\frac{1}{|G|}\int_G \Delta^j u(x,\tau_{ij}(t))\, dx \\ &\quad \leq \sum_{i=1}^{k}\sum_{j=1}^{K}(-1)^{j-1}b_{ij}(t)\tilde{\Psi}_j(\tau_{ij}(t)).\end{aligned} \tag{7.5.15}$$

Application of Jensen's inequality yields

$$\frac{1}{|G|}\int_G \varphi_i\big(u(x,\sigma_i(t))\big)\, dx \geq \varphi_i\big(\tilde{U}(\sigma_i(t))\big). \tag{7.5.16}$$

Combining (7.5.13)–(7.5.16), we conclude that $\tilde{U}(t)$ is an eventually positive solution of (7.5.12) with $+\tilde{H}(t)$. The case where $u < 0$ in $G \times [t_0, \infty)$ can be treated similarly, and we are also led to a contradiction. This completes the proof. □

Now we study the functional differential inequality

$$\frac{d^N}{dt^N}\left(y(t) + \sum_{i=1}^{\ell} h_i(t) y(\rho_i(t))\right) + A(t)y(t) + \sum_{i=1}^{m} p_i(t)\varphi_i\big(y(\sigma_i(t))\big) \leq q(t), \tag{7.5.17}$$

where $A(t) \in C([t_0, \infty); [0, \infty))$ and $q(t) \in C([t_0, \infty); \mathbb{R})$ for some $t_0 > 0$.

It is assumed that:

(H7.5-3) N is an even integer;

(H7.5-4) $\sum_{i=1}^{\ell} h_i(t) \le 1$ and $\rho_i(t) \le t$ $(i = 1, 2, ..., \ell)$.

We recall that:

$$[M(t)]_{\pm} = \max\{\pm M(t),\ 0\}$$

for any continuous functions $M(t)$.

Theorem 7.5.3. *Assume that the hypotheses* (H6.3-3), (H7.5-1), (H7.5-2) *are satisfied. The differential inequality* (7.5.17) *has no eventually positive solution if*

$$\liminf_{t\to\infty} \int_T^t \left(1 - \frac{s}{t}\right)^{N-1} q(s)\, ds = -\infty$$

for all large T.

Proof. Suppose that there is a positive solution $y(t)$ of (7.5.17) on $[T_0, \infty)$ for some $T_0 > 0$. From the hypothesis we see that

$$\frac{d^N}{dt^N}\left(y(t) + \sum_{i=1}^{\ell} h_i(t) y(\rho_i(t))\right) \le q(t), \quad t \ge T \tag{7.5.18}$$

for some $T \ge \max\{t_0, T_0\}$. Integrating (7.5.18) over $[T, t]$ yields

$$Y^{(N-1)}(t) - Y^{(N-1)}(T) \le \int_T^t q(s)\, ds, \quad t \ge T, \tag{7.5.19}$$

where $Y(t) = y(t) + \sum_{i=1}^{\ell} h_i(t) y(\rho_i(t))$. We integrate (7.5.19) over $[T, t]$ to obtain

$$Y^{(N-2)}(t) - Y^{(N-2)}(T) - Y^{(N-1)}(T)(t - T) \le \int_T^t \left(\int_T^{s_2} q(s_1)\, ds_1\right) ds_2$$

for $t \ge T$. Repetition of this procedure yields finally the inequality

$$Y(t) - \sum_{i=0}^{N-1} \frac{(t-T)^i}{i\,!} Y^{(i)}(T) \le \int_T^t \int_T^{s_N} \cdots \int_T^{s_2} q(s_1)\, ds_1 \cdots ds_{N-1} ds_N \tag{7.5.20}$$

for $t \ge T$. It is easily seen that

$$\int_T^t \int_T^{s_N} \cdots \int_T^{s_2} q(s_1)\, ds_1 \cdots ds_{N-1} ds_N = \frac{1}{(N-1)!} \int_T^t (t-s)^{N-1} q(s)\, ds. \tag{7.5.21}$$

Substituting (7.5.21) into (7.5.20) and then dividing (7.5.20) by t^{N-1}, we obtain

$$\frac{Y(t)}{t^{N-1}} - \sum_{i=0}^{N-1} \frac{Y^{(i)}(T)}{i\,!} \left(1 - \frac{T}{t}\right)^i \frac{1}{t^{N-i-1}} \le \frac{1}{(N-1)!} \int_T^t \left(1 - \frac{s}{t}\right)^{N-1} q(s)\,ds \tag{7.5.22}$$

for $t \ge T$. Since $Y(t) > 0$ on $[T_1, \infty)$ for some $T_1 \ge T$, the left hand side of (7.5.22) is bounded from below on $[T_1, \infty)$. However, the hypothesis implies that the right hand side of (7.5.22) is not bounded from below. The contradiction establishes the theorem. □

Theorem 7.5.4. *Let the hypotheses* (H6.3-3), (H7.5-1)–(H7.5-4) *be satisfied. Assume that there exists a function* $Q(t) \in C^N([t_0, \infty); \mathbb{R})$ *such that* $Q^{(N)}(t) = q(t)$, $Q(t)$ *is oscillatory at* $t = \infty$, *and assume, moreover, that there exists an integer* $j \in \{1, 2, ..., m\}$ *such that* $\varphi_j(\xi)$ *is nondecreasing for* $\xi > 0$. *If*

$$\int_{t_0}^{\infty} p_j(s)\varphi_j \left(\left[\left(1 - \sum_{i=1}^{\ell} h_i(\sigma_j(s)) \right) c + \hat{Q}(\sigma_j(s)) \right]_+ \right) ds = \infty \tag{7.5.23}$$

for some $t_0 > 0$ *and any* $c > 0$, *then* (7.5.17) *has no eventually positive solution, where*

$$\hat{Q}(t) = Q(t) - \sum_{i=1}^{\ell} h_i(t) Q(\rho_i(t)).$$

Proof. Suppose that $y(t)$ is an eventually positive solution of (7.5.17). Then we see that $y(t) > 0$, $y(\sigma_i(t)) > 0$ $(i = 1, 2, ..., m)$, $y(\rho_i(t)) > 0$ $(i = 1, 2, ..., \ell)$ on $[t_1, \infty)$ for some $t_1 \ge t_0$. Since

$$\frac{d^N}{dt^N} \left(y(t) + \sum_{i=1}^{\ell} h_i(t) y(\rho_i(t)) - Q(t) \right) \le 0, \quad t \ge t_1,$$

we observe that $z^{(i)}(t)$ $(i = 0, 1, ..., N-1)$ are monotone functions and either $z(t) > 0$ or $z(t) \le 0$ eventually, where

$$z(t) = y(t) + \sum_{i=1}^{\ell} h_i(t) y(\rho_i(t)) - Q(t).$$

If $z(t) \le 0$ eventually, then we have

$$0 < y(t) + \sum_{i=1}^{\ell} h_i(t) y(\rho_i(t)) \le Q(t).$$

Since $Q(t)$ is oscillatory at $t = \infty$, the above inequality cannot be valid. Hence, $z(t) > 0$ must hold eventually. From a result of Kiguradze [125] it follows that there exists an odd integer k $(0 \le k \le N)$ and a number $t_2 \ge t_1$ such that

$$z^{(i)}(t) > 0 \ (0 \le i \le k), \quad t \ge t_2,$$
$$(-1)^{i-k} z^{(i)}(t) > 0 \ (k \le i \le N), \quad t \ge t_2.$$

Hence we find that

$$z'(t) > 0, \quad z^{(N-1)}(t) > 0, \quad t \ge t_2.$$

Using the inequality

$$y(t) \le z(t) + Q(t), \quad t \ge t_2,$$

we obtain

$$\begin{aligned} y(t) &= z(t) - \sum_{i=1}^{\ell} h_i(t) y(\rho_i(t)) + Q(t) \\ &\ge z(t) - \sum_{i=1}^{\ell} h_i(t) \left[z(\rho_i(t)) + Q(\rho_i(t)) \right] + Q(t) \\ &\ge \left(1 - \sum_{i=1}^{\ell} h_i(t) \right) z(t) + Q(t) - \sum_{i=1}^{\ell} h_i(t) Q(\rho_i(t)) \\ &\ge \left(1 - \sum_{i=1}^{\ell} h_i(t) \right) z(t_3) + \hat{Q}(t), \quad t \ge t_3 \end{aligned}$$

for some $t_3 \ge t_2$, in view of the fact that $z(t)$ is an increasing function. Since $y(t)$ is positive, we observe that

$$y(t) \ge \left[\left(1 - \sum_{i=1}^{\ell} h_i(t) \right) z(t_3) + \hat{Q}(t) \right]_+, \quad t \ge t_3$$

and therefore

$$y(\sigma_j(t)) \ge \left[\left(1 - \sum_{i=1}^{\ell} h_i(\sigma_j(t)) \right) z(t_3) + \hat{Q}(\sigma_j(t)) \right]_+, \quad t \ge t_4 \qquad (7.5.24)$$

for some $t_4 \ge t_3$. Combining (7.5.17) with (7.5.24), we obtain

$$z^{(N)}(t) + p_j(t) \varphi_j \left(\left[\left(1 - \sum_{i=1}^{\ell} h_i(\sigma_j(t)) \right) z(t_3) + \hat{Q}(\sigma_j(t)) \right]_+ \right) \le 0 \qquad (7.5.25)$$

for $t \geq t_4$. Integrating (7.5.25) over $[t_4, t]$ yields

$$\begin{aligned} & z^{(N-1)}(t) - z^{(N-1)}(t_4) + \\ & \quad \int_{t_4}^{t} p_j(s)\varphi_j\left(\left[\left(1 - \sum_{i=1}^{\ell} h_i(\sigma_j(s))\right) z(t_3) + \hat{Q}(\sigma_j(s))\right]_+\right) ds \leq 0 \end{aligned}$$

and hence

$$\begin{aligned} & \int_{t_4}^{t} p_j(s)\varphi_j\left(\left[\left(1 - \sum_{i=1}^{\ell} h_i(\sigma_j(s))\right) z(t_3) + \hat{Q}(\sigma_j(s))\right]_+\right) ds \\ & \leq -z^{(N-1)}(t) + z^{(N-1)}(t_4) \\ & \leq z^{(N-1)}(t_4). \end{aligned}$$

This contradicts the condition (7.5.23) and the proof is complete. □

Next we consider the case where $q(t) = 0,\ h_i(t) = 0\ (i = 1, 2, ..., \ell)$, that is,

$$\frac{d^N y}{dt^N}(t) + A(t)y(t) + \sum_{i=1}^{m} p_i(t)\varphi_i\big(y(\sigma_i(t))\big) \leq 0. \tag{7.5.26}$$

Theorem 7.5.5. *Let the hypotheses* (H6.3-3), (H7.5-1)–(H7.5-3) *be satisfied. The differential inequality* (7.5.26) *has no eventually positive solution if*

$$\begin{aligned} & A(t) > 0 \quad \text{on } [t_0, \infty) \text{ for some } t_0 > 0, \\ & \int_{t_0}^{\infty} t^{N-1-\varepsilon} A(t)\, dt = \infty \quad \text{for some } \varepsilon > 0. \end{aligned}$$

Proof. By the hypothesis the differential equation

$$\frac{d^N y}{dt^N}(t) + A(t)y(t) = 0$$

has no eventually positive solution (see Mikusiński [198]). Using the results of Kartsatos [123] and Onose [229], we observe that the differential inequality

$$\frac{d^N y}{dt^N}(t) + A(t)y(t) \leq 0$$

has no eventually positive solution. Hence, (7.5.26) has no eventually positive solution. □

We are now in a position to obtain oscillation results for the boundary value problems (7.5.1), (B_D) [or (7.5.1), (B_R)] by combining the above theorems.

Theorem 7.5.6. *Assume that the hypotheses* (H6.1-4), (H6.3-3), (H7.5-1) *and* (H7.5-2) *are satisfied. If*

$$\liminf_{t\to\infty} \int_T^t \left(1-\frac{s}{t}\right)^{N-1} H(s)\,ds = -\infty,$$

$$\limsup_{t\to\infty} \int_T^t \left(1-\frac{s}{t}\right)^{N-1} H(s)\,ds = \infty$$

for all large T, then every solution u of the boundary value problem (7.5.1), (B_D) *is oscillatory in Ω.*

Proof. Theorem 7.5.3 implies that the functional differential inequalities

$$\frac{d^N}{dt^N}\left(y(t)+\sum_{i=1}^{\ell} h_i(t)y(\rho_i(t))\right)+\sum_{i=1}^{m} p_i(t)\varphi_i\big(y(\sigma_i(t))\big) \le \pm H(t)$$

have no eventually positive solutions. Hence, the inequality (7.5.6) also have no eventually positive solutions in view of the fact that $b_{ij}(t) \ge 0$ ($i = 1,2,...,k$; $j = 1,2,...,K$). The conclusion follows from Theorem 7.5.1. □

Theorem 7.5.7. *Assume that the hypotheses* (H6.1-4), (H6.3-3), (H7.5-1) *and* (H7.5-2) *are satisfied. If*

$$\liminf_{t\to\infty} \int_T^t \left(1-\frac{s}{t}\right)^{N-1} \tilde{H}(s)\,ds = -\infty,$$

$$\limsup_{t\to\infty} \int_T^t \left(1-\frac{s}{t}\right)^{N-1} \tilde{H}(s)\,ds = \infty$$

for all large T, then every solution u of the boundary value problem (7.5.1), (B_R) *is oscillatory in Ω.*

Proof. The conclusion follows from Theorem 7.5.2 and Theorem 7.5.3. □

Theorem 7.5.8. *Let the hypotheses* (H6.1-4), (H6.3-3), (H7.5-1)–(H7.5-4) *be satisfied. Assume that there exists a function $\Theta(t) \in C^N((0,\infty);\mathbb{R})$ such that $\Theta^{(N)}(t) = H(t)$ [resp. $\Theta^{(N)}(t) = \tilde{H}(t)$], $\Theta(t)$ is oscillatory at $t=\infty$, and assume, moreover, that there exists an integer $j \in \{1,2,...,m\}$ such that $\varphi_j(\xi)$ is nondecreasing for $\xi > 0$. If*

$$\int_{t_0}^{\infty} p_j(s)\varphi_j\left(\left[\left(1-\sum_{i=1}^{\ell} h_i(\sigma_j(s))\right)c+\hat{\Theta}(\sigma_j(s))\right]_+\right)ds = \infty,$$

$$\int_{t_0}^{\infty} p_j(s)\varphi_j\left(-\left[\left(1-\sum_{i=1}^{\ell} h_i(\sigma_j(s))\right)c-\hat{\Theta}(\sigma_j(s))\right]_+\right)ds = -\infty$$

for some $t_0 > 0$ *and any* $c > 0$, *then every solution* u *of the boundary value problem* (7.5.1), (B_D) [*resp.* (7.5.1), (B_R)] *is oscillatory in* Ω, *where*

$$\hat{\Theta}(t) = \Theta(t) - \sum_{i=1}^{\ell} h_i(t)\Theta(\rho_i(t)).$$

Proof. The conclusion follows by combining Theorems 7.5.1 and 7.5.2 with Theorem 7.5.4. □

Theorem 7.5.9. *Let the hypotheses* (H6.1-4), (H6.3-3), (H7.5-1)–(H7.5-4) *be satisfied. Assume that there exists a function* $\Theta(t) \in C^N((0,\infty);\mathbb{R})$ *such that* $\Theta^{(N)}(t) = H(t)$, $\Theta(t)$ *is oscillatory at* $t = \infty$. *If*

$$\int_{t_0}^{\infty} \left(\sum_{j=1}^{K} \lambda_1^j a_j(t)\right) \left[\left(1 - \sum_{i=1}^{\ell} h_i(\sigma_j(s))\right) c \pm \hat{\Theta}(\sigma_j(s))\right]_+ ds = \infty$$

for some $t_0 > 0$ *and any* $c > 0$, *then every solution* u *of the boundary value problem* (7.5.1), (B_D) *is oscillatory in* Ω.

Proof. It follows from Theorem 7.5.4 that

$$\frac{d^N}{dt^N}\left(y(t) + \sum_{i=1}^{\ell} h_i(t) y(\rho_i(t))\right) + \left(\sum_{j=1}^{K} \lambda_1^j a_j(t)\right) y(t) \le \pm H(t)$$

have no eventually positive solutions, and therefore the inequalities (7.5.6) have no eventually positive solutions. The conclusion follows from Theorem 7.5.1. □

Theorem 7.5.10. *Let* $H(t) = 0$, $h_i(t) = 0$ $(i = 1, 2, ..., \ell)$. *Assume that the hypotheses* (H6.1-4), (H6.3-3), (H7.5-1)–(H7.5-4) *are satisfied. Every solution* u *of the problem* (7.5.1), (B_D) *is oscillatory in* Ω *if*

$$\sum_{j=1}^{K} \lambda_1^j a_j(t) > 0 \quad \text{on } [t_0, \infty) \text{ for some } t_0 > 0,$$

$$\int_{t_0}^{\infty} t^{N-1-\varepsilon} \left(\sum_{j=1}^{K} \lambda_1^j a_j(t)\right) dt = \infty \quad \text{for some } \varepsilon > 0.$$

Proof. Theorem 7.5.5 implies that

$$\frac{d^N y}{dt^N}(t) + \left(\sum_{j=1}^{K} \lambda_1^j a_j(t)\right) y(t) + \sum_{i=1}^{m} p_i(t)\varphi_i\big(y(\sigma_i(t))\big) \le 0$$

has no eventually positive solution, and therefore (7.5.6) with $H(t) = 0$, $h_i(t) = 0$ $(i = 1, 2, ..., \ell)$ has no eventually positive solution. The conclusion follows from Theorem 7.5.1. □

Example 7.5.1. We consider the problem

$$\frac{\partial^3}{\partial t^3}\Big(u(x,t)+u(x,t-\tfrac{\pi}{2})\Big)-\frac{\partial^2 u}{\partial x^2}(x,t)-\frac{\partial^2 u}{\partial x^2}(x,t-\tfrac{3}{2}\pi)$$
$$+t^3u(x,t-\tfrac{\pi}{2})=(\sin x)\,t^3\sin t,\quad (x,t)\in(0,\pi)\times(0,\infty),\qquad (7.5.27)$$
$$u(0,t)=u(\pi,t)=0,\quad t>0.\qquad (7.5.28)$$

Here $n=1$, $G=(0,\pi)$, $N=3$, $\ell=K=k=m=\tilde{m}=1$, $h_1(t)=a_1(t)=b_{11}(t)=1$, $\rho_1(t)=t-(\pi/2)$, $\tau_{11}(t)=t-(3/2)\pi$, $\sigma_1(t)=t-(\pi/2)$, $f(x,t)=(\sin x)\,t^3\sin t$ and $\psi_1=0$. It is easily seen that $\lambda_1=1$, $\Phi(x)=\sin x$, $\Psi_1(t)=0$ and $F(t)=H(t)=(\pi/4)\,t^3\sin t$. A simple calculation shows that

$$\int_T^t\Big(1-\frac{s}{t}\Big)^2 H(s)\,ds=\frac{\pi}{2}\,t\cos t+B(t,T),$$

where $B(t,T)$ is bounded as t tends to ∞. Since $A(t)=\lambda_1a_1(t)=1>0$, $B_{11}(t)=\lambda_1b_{11}(t)=1>0$ and

$$\liminf_{t\to\infty}\int_T^t\Big(1-\frac{s}{t}\Big)^2 H(s)\,ds=-\infty,$$
$$\limsup_{t\to\infty}\int_T^t\Big(1-\frac{s}{t}\Big)^2 H(s)\,ds=\infty$$

for all large T, it follows from Theorem 7.5.6 that every solution u of the problem (7.5.27), (7.5.28) is oscillatory in $(0,\pi)\times(0,\infty)$. One such solution is $u=(\sin x)\cos t$.

Example 7.5.2. We consider the problem

$$\frac{\partial^4}{\partial t^4}\left(u(x,t)+\frac{1}{2}\,u(x,t-\pi)\right)-\frac{\partial^2 u}{\partial x^2}(x,t)+\frac{1}{2}\,\frac{\partial^4 u}{\partial x^4}(x,t)$$
$$+u(x,t-2\pi)=3(\sin x)\sin t,\quad (x,t)\in(0,\pi)\times(0,\infty),\qquad (7.5.29)$$
$$u(0,t)=u(\pi,t)=\frac{\partial^2 u}{\partial x^2}(0,t)=\frac{\partial^2 u}{\partial x^2}(\pi,t)=0,\quad t>0.\qquad (7.5.30)$$

Here $n=1$, $G=(0,\pi)$, $\Omega=(0,\pi)\times(0,\infty)$, $N=4$, $\ell=1$, $h_1(t)=1/2$, $\rho_1(t)=t-\pi$, $K=2$, $a_1(t)=1$, $a_2(t)=1/2$, $b_{ij}(t)=0$ ($i=1,2,...,k$; $j=1,2$), $m=1$, $p_1(t)=1$, $\varphi_1(\xi)=\xi$, $\sigma_1(t)=t-2\pi$, $f(x,t)=3(\sin x)\sin t$ and $\psi_1=\psi_2=0$. It is easy to see that $\lambda_1=1$, $\Phi(x)=\sin x$, $\Psi_1(t)=\Psi_2(t)=0$ and $H(t)=F(t)=(3/4)\pi\sin t$. Choosing $\Theta(t)=(3/4)\pi\sin t$, we observe that $\Theta^{(4)}(t)=H(t)$ and $\Theta(t)$ is oscillatory at $t=\infty$. Furthermore, we find that $\hat{\Theta}(t)=\hat{\Theta}(t-2\pi)=(9/8)\pi\sin t$. A simple calculation yields

$$\int_1^\infty\left[\frac{1}{2}c\pm\frac{9}{8}\pi\sin s\right]_+ds=\infty$$

for any $c > 0$. Hence, it follows from Theorem 7.5.8 that every solution u of the problem (7.5.29), (7.5.30) is oscillatory in $(0, \pi) \times (0, \infty)$. Indeed, $u = (\sin x) \sin t$ is such a solution.

Example 7.5.3. We consider the problem

$$\frac{\partial^6}{\partial t^6}\left(u(x,t) + \frac{1}{3}\,u(x, t - \tfrac{\pi}{2})\right) - \frac{\partial^2 u}{\partial x^2}(x,t) + \frac{\partial^4 u}{\partial x^4}(x,t)$$

$$+\frac{1}{3}\,u(x, t - \tfrac{\pi}{2}) = (\cos x)\sin t, \quad (x,t) \in (0,\pi) \times (0,\infty), \qquad (7.5.31)$$

$$-\frac{\partial u}{\partial x}(0,t) = \frac{\partial u}{\partial x}(\pi,t) = -\frac{\partial^3 u}{\partial x^3}(0,t) = \frac{\partial^3 u}{\partial x^3}(\pi,t) = 0, \quad t > 0. \quad (7.5.32)$$

Here $n = 1$, $G = (0,\pi)$, $\Omega = (0,\pi) \times (0,\infty)$, $N = 6$, $\ell = 1$, $h_1(t) = 1/3$, $\rho_1(t) = t - (\pi/2)$, $K = 2$, $a_1(t) = a_2(t) = 1$, $b_{ij}(t) = 0$ $(i = 1, 2, ..., k;\ j = 1, 2)$, $m = 1$, $p_1(t) = 1/3$, $\varphi_1(\xi) = \xi$, $\sigma_1(t) = t - (\pi/2)$, $f(x,t) = (\cos x)\sin t$, $\mu_1 = \mu_2 = 0$ and $\tilde{\psi}_1 = \tilde{\psi}_2 = 0$. It is easily checked that $\tilde{\Psi}_1(t) = \tilde{\Psi}_2(t) = 0$ and $\tilde{H}(t) = \tilde{F}(t) = 0$. We can choose $\Theta(t) = 0$, and observe that $\Theta^{(6)}(t) = \tilde{H}(t)$, $\Theta(t)$ is oscillatory at $t = \infty$ and $\hat{\Theta}(t) = 0$. An easy computation shows that

$$\int_1^\infty \frac{1}{3}\left[\left(1 - \frac{1}{3}\right)c\right]_+ ds = \infty$$

for any $c > 0$. Therefore, Theorem 7.5.8 implies that every solution u of the problem (7.5.31), (7.5.32) is oscillatory in $(0,\pi) \times (0,\infty)$. One such solution is $u = (\cos x)\sin t$.

Example 7.5.4. We consider the problem

$$\frac{\partial^4 u}{\partial t^4}(x,t) + \frac{\partial^4 u}{\partial x^4}(x,t) + \frac{\partial^4 u}{\partial x^4}(x, t - \pi)$$

$$+u(x, t - \pi) = 0, \ (x,t) \in (0,\pi) \times (0,\infty), \qquad (7.5.33)$$

$$u(0,t) = \frac{\partial^2 u}{\partial x^2}(0,t) = u(\pi,t) = \frac{\partial^2 u}{\partial x^2}(\pi,t) = 0, \quad t > 0. \quad (7.5.34)$$

Here $n = 1$, $G = (0,\pi)$, $N = 4$, $K = 2$, $k = m = \tilde{m} = 1$, $a_1(t) = 0$, $a_2(t) = -1$, $b_{11}(t) = 0$, $b_{12}(t) = -1$, $\tau_{12}(t) = \sigma_1(t) = t - \pi$, $f(x,t) = 0$, $\psi_1 = \psi_2 = 0$ and $\lambda_1 = 1$. Since $A(t) = -\lambda_1^2 a_2(t) = 1 > 0$, $B_{11}(t) = 0$, $B_{12}(t) = -\lambda_1^2 b_{12}(t) = 1 > 0$, $H(t) = 0$ and

$$\int_1^\infty t^{4-1-\varepsilon} A(t)\,dt = \infty$$

for any ε with $0 < \varepsilon \leq 4$, Theorem 7.5.10 implies that every solution u of the problem (7.5.33), (7.5.34) is oscillatory in $(0,\pi) \times (0,\infty)$. Indeed, $u = (\sin x)\sin t$ is such a solution.

Remark 7.5.1. If $N = K = 2$, $n = 1$, $a_2(t) = -\alpha$ (α is a positive constant), then Eq. (7.5.1) reduces to the beam equation with deviating arguments

$$\frac{\partial^2}{\partial t^2}\left(u(x,t) + \sum_{i=1}^{\ell} h_i(t)u(x,\rho_i(t))\right) + \alpha\frac{\partial^4 u}{\partial x^4}(x,t) - \sum_{i=1}^{k} b_{i2}(t)\frac{\partial^4 u}{\partial x^4}(x,\tau_{i2}(t))$$
$$-a_1(t)\frac{\partial^2 u}{\partial x^2}(x,t) - \sum_{i=1}^{k} b_{i1}(t)\frac{\partial^2 u}{\partial x^2}(x,\tau_{i1}(t))$$
$$+c(x,t,(z_i[u](x,t))_{i=1}^{\tilde{m}}) = f(x,t), \quad (x,t) \in (0,L) \times (0,\infty),$$

where L is a positive number. In case $\psi_1 = \psi_2 = 0$ on $\partial G \times (0,\infty)$, the boundary condition (B_D) reduces to the hinged ends

$$u(0,t) = \frac{\partial^2 u}{\partial x^2}(0,t) = u(L,t) = \frac{\partial^2 u}{\partial x^2}(L,t) = 0, \quad t > 0.$$

We note that oscillations of beam equations without deviating arguments were investigated in Section 4.1, and various end conditions including hinged ends were considered.

Remark 7.5.2. We cannot apply Theorem 7.5.6 to the case where $f(x,t) = 0$ and $\psi_j = 0$ ($j = 1, 2, ..., K$). However, Theorem 7.5.10 is applicable to this case.

Remark 7.5.3. In the case where $\ell = j = K = m = \tilde{m} = 1$, $\sigma_1(t) = t - \sigma$ ($\sigma > 0$), $\rho_1(t) = t - \rho$ ($\rho > 0$), $\psi_1 = \tilde{\psi}_1 = 0$ and $f(x,t) = 0$, Eq. (7.5.1) was studied by Liu and Fu [183].

7.6 Hyperbolic Systems I

In Section 6.6, H-oscillation of vector parabolic equations with functional arguments were studied, where H is a unit vector in $\mathbb{R}^N$. Analogous results for vector hyperbolic equations are derived in this section. We refer the reader to Li, Han and Meng [174] and Minchev and Yoshida [200].

We consider the vector hyperbolic equation with functional arguments

$$\frac{\partial^2}{\partial t^2}\left(U(x,t) + \sum_{i=1}^{\ell} h_i(t)U(x,\rho_i(t))\right) - a(t)\Delta U(x,t) - \sum_{i=1}^{k} b_i(t)\Delta U(x,\tau_i(t))$$
$$+\sum_{i=1}^{m} c_i(x,t,U(x,\sigma_i(t)))U(x,\sigma_i(t)) = F(x,t) \qquad (7.6.1)$$

for $(x,t) \in \Omega := G \times (0,\infty)$, where G is a bounded domain in $\mathbb{R}^n$ with piecewise smooth boundary ∂G.

It is assumed that the hypothesis (H6.6-3) of Section 6.6 is satisfied, and the following hypotheses hold:

(H7.6-1) $h_i(t) \in C^2([0,\infty);\mathbb{R})$ $(i = 1,2,...,\ell)$,
$b_i(t) \in C([0,\infty);[0,\infty))$ $(i = 1,2,...,k)$,
$a(t) \in C([0,\infty);[0,\infty))$ and $F(x,t) \in C(\overline{\Omega};\mathbb{R}^N)$;

(H7.6-2) $\rho_i(t) \in C^2([0,\infty);\mathbb{R})$, $\lim_{t\to\infty} \rho_i(t) = \infty$ $(i = 1,2,...,\ell)$,
$\tau_i(t) \in C([0,\infty);\mathbb{R})$, $\lim_{t\to\infty} \tau_i(t) = \infty$ $(i = 1,2,...,k)$,
$\sigma_i(t) \in C([0,\infty);\mathbb{R})$, $\lim_{t\to\infty} \sigma_i(t) = \infty$ $(i = 1,2,...,m)$.

Definition 7.6.1. By a *solution* of Eq. (7.6.1) we mean a function $U(x,t) \in C^2(\overline{G}\times[t_{-1},\infty);\mathbb{R}^N)\cap C(\overline{G}\times[\hat{t}_{-1},\infty);\mathbb{R}^N)$ which satisfies (7.6.1), where

$$t_{-1} = \min\left\{0,\ \min_{1\le i\le \ell}\left\{\inf_{t\ge 0}\rho_i(t)\right\},\ \min_{1\le i\le k}\left\{\inf_{t\ge 0}\tau_i(t)\right\}\right\},$$
$$\hat{t}_{-1} = \min\left\{0,\ \min_{1\le i\le m}\left\{\inf_{t\ge 0}\sigma_i(t)\right\}\right\}.$$

Definition 7.6.2. Let H be a unit vector in $\mathbb{R}^N$. A solution $U(x,t)$ of (7.6.1) is said to be *H-oscillatory* in Ω if the scalar product $\langle U(x,t), H\rangle$ has a zero in $G \times (t,\infty)$ for any $t > 0$.

We consider two kinds of boundary conditions:

$$(\mathrm{B}_1)\qquad U = \Psi \quad \text{on } \partial G \times (0,\infty),$$
$$(\mathrm{B}_2)\qquad \frac{\partial U}{\partial \nu} + \mu U = \tilde{\Psi} \quad \text{on } \partial G \times (0,\infty),$$

where $\Psi,\ \tilde{\Psi} \in C(\partial G \times (0,\infty);\mathbb{R}^N)$, $\mu \in C(\partial G \times (0,\infty);[0,\infty))$ and ν denotes the unit exterior normal vector to ∂G.

Two examples which satisfy the hypothesis (H6.6-3) are shown in Section 6.6 (*cf.* [224]).

The following notation will be used:

$$u_H(x,t) = \langle U(x,t), H\rangle,$$
$$f_H(x,t) = \langle F(x,t), H\rangle,$$

where H is a unit vector in $\mathbb{R}^N$ and $\langle U, V\rangle$ denotes the scalar product of $U, V \in \mathbb{R}^N$.

Theorem 7.6.1. *Assume that the hypotheses* (H6.6-3), (H7.6-1), (H7.6-2) *hold. Let $U(x,t)$ be a solution of* (7.6.1). *If $u_H(x,t)$ is eventually positive, then $u_H(x,t)$ satisfies the scalar hyperbolic differential inequality*

$$\frac{\partial^2}{\partial t^2}\left(u_H(x,t)+\sum_{i=1}^{\ell} h_i(t)u_H(x,\rho_i(t))\right)-a(t)\Delta u_H(x,t)$$
$$-\sum_{i=1}^{k} b_i(t)\Delta u_H(x,\tau_i(t))+\sum_{i=1}^{m} p_i(t)\varphi_i\big(u_H(x,\sigma_i(t))\big)\le f_H(x,t), \quad (7.6.2)$$

where $\varphi_i(\xi)=\xi\psi_i(|\xi|)$. If $u_H(x,t)$ is eventually negative, then $v_H(x,t):=-u_H(x,t)$ satisfies the scalar hyperbolic differential inequality

$$\frac{\partial^2}{\partial t^2}\left(v_H(x,t)+\sum_{i=1}^{\ell} h_i(t)v_H(x,\rho_i(t))\right)-a(t)\Delta v_H(x,t)$$
$$-\sum_{i=1}^{k} b_i(t)\Delta v_H(x,\tau_i(t))+\sum_{i=1}^{m} p_i(t)\varphi_i\big(v_H(x,\sigma_i(t))\big)\le -f_H(x,t). \quad (7.6.3)$$

Proof. Let $u_H(x,t)$ be eventually positive. The scalar product of (7.6.1) and the unit vector H yields

$$\frac{\partial^2}{\partial t^2}\left(\langle U(x,t),H\rangle+\sum_{i=1}^{\ell} h_i(t)\langle U(x,\rho_i(t)),H\rangle\right)$$
$$-a(t)\langle\Delta U(x,t),H\rangle-\sum_{i=1}^{k} b_i(t)\langle\Delta U(x,\tau_i(t)),H\rangle$$
$$+\sum_{i=1}^{m} c_i(x,t,U(x,\sigma_i(t)))\langle U(x,\sigma_i(t)),H\rangle=\langle F(x,t),H\rangle. \quad (7.6.4)$$

Proceeding as in the proof of Theorem 6.6.1, we observe that

$$c_i(x,t,U(x,\sigma_i(t)))\langle U(x,\sigma_i(t)),H\rangle \;\ge\; p_i(t)\varphi_i\big(u_H(x,\sigma_i(t))\big) \quad (7.6.5)$$

(see (6.6.7) of Section 6.6). Combining (7.6.4) with (7.6.5), we conclude that $u_H(x,t)$ satisfies the inequality (7.6.2). If $u_H(x,t)$ is eventually negative, we see that

$$c_i(x,t,U(x,\sigma_i(t)))\langle U(x,\sigma_i(t)),H\rangle \;\le\; -p_i(t)\varphi_i\big(-u_H(x,\sigma_i(t))\big) \quad (7.6.6)$$

(see (6.6.8) of Section 6.6). We combine (7.6.4) and (7.6.6) to conclude that $v_H(x,t)$ satisfies the inequality (7.6.3). □

Associated with the boundary conditions (B_i) ($i=1,2$), we consider the following scalar boundary conditions:

$(\tilde{\mathrm{B}}_1)$ $\quad u = \psi_H \quad \text{on} \quad \partial G \times (0, \infty),$

$(\tilde{\mathrm{B}}_2)$ $\quad \dfrac{\partial u}{\partial \nu} + \mu u = \tilde{\psi}_H \quad \text{on} \quad \partial G \times (0, \infty),$

where

$$\psi_H = \langle \Psi, H \rangle,$$
$$\tilde{\psi}_H = \langle \tilde{\Psi}, H \rangle.$$

Theorem 7.6.2. *Assume that the hypotheses* (H6.6-3), (H7.6-1), (H7.6-2) *hold. If the scalar hyperbolic differential inequalities*

$$\frac{\partial^2}{\partial t^2}\left(u(x,t) + \sum_{i=1}^{\ell} h_i(t) u(x, \rho_i(t)) \right) - a(t)\Delta u(x,t)$$
$$- \sum_{i=1}^{k} b_i(t) \Delta u(x, \tau_i(t)) + \sum_{i=1}^{m} p_i(t) \varphi_i\big(u(x, \sigma_i(t))\big) \le \pm f_H(x,t) \quad (7.6.7)$$

have no eventually positive solutions satisfying the boundary conditions $(\tilde{\mathrm{B}}_i)$ $(i = 1, 2)$, *then every solution* $U(x,t)$ *of the boundary value problems* (7.6.1), (B_i) $(i = 1, 2)$ *is* H*-oscillatory in* Ω*, respectively.*

Proof. Suppose to the contrary that there is a solution $U(x,t)$ of the problem (7.6.1), (B_i) which is not H-oscillatory in Ω. If $u_H(x,t)$ is eventually positive, then $u_H(x,t)$ satisfies (7.6.7) with $+f_H(x,t)$ by Theorem 7.6.1. It is easy to see that $u_H(x,t)$ satisfies the boundary conditions $(\tilde{\mathrm{B}}_i)$. This contradicts the hypothesis. If $u_H(x,t)$ is eventually negative, then $v_H(x,t) = -u_H(x,t)$ is an eventually positive solution of (7.6.7) with $-f_H(x,t)$ satisfying the boundary conditions $(\tilde{\mathrm{B}}_i)$. This also contradicts the hypothesis. □

The following notation will be used:

$$F_H(t) = \left(\int_G \Phi(x)\, dx \right)^{-1} \int_G f_H(x,t) \Phi(x)\, dx,$$
$$\Psi_H(t) = \left(\int_G \Phi(x)\, dx \right)^{-1} \int_{\partial G} \psi_H \frac{\partial \Phi}{\partial \nu}(x)\, dS,$$
$$\tilde{F}_H(t) = \frac{1}{|G|} \int_G f_H(x,t)\, dx,$$
$$\tilde{\Psi}_H(t) = \frac{1}{|G|} \int_{\partial G} \tilde{\psi}_H\, dS,$$

where $\Phi(x)$ (> 0 in G) is the eigenfunction corresponding to the first eigenvalue λ_1 of the eigenvalue problem (EVP) of Section 6.1 and $|G| = \int_G dx$.

Theorem 7.6.3. *Assume that the hypotheses* (H6.6-3), (H7.6-1), (H7.6-2) *hold, and the following hypothesis* (H7.6-3) *holds:*

(H7.6-3) $\varphi_i(\xi) = \xi\psi_i(|\xi|)$ $(i = 1, 2, ..., m)$ *are convex in* $(0, \infty)$.

If the functional differential inequalities

$$\frac{d^2}{dt^2}\left(y(t) + \sum_{i=1}^{\ell} h_i(t)y(\rho_i(t))\right) + \lambda_1 a(t)y(t)$$
$$+\lambda_1 \sum_{i=1}^{k} b_i(t)y(\tau_i(t)) + \sum_{i=1}^{m} p_i(t)\varphi_i\big(y(\sigma_i(t))\big) \le \pm G_H(t) \tag{7.6.8}$$

have no eventually positive solutions, then every solution $U(x,t)$ *of the boundary value problem* (7.6.1), (B_1) *is* H*-oscillatory in* Ω, *where*

$$G_H(t) = F_H(t) - a(t)\Psi_H(t) - \sum_{i=1}^{k} b_i(t)\Psi_H(\tau_i(t)).$$

Proof. Suppose to the contrary that there is a solution $U(x,t)$ of the problem (7.6.1), (B_1) which is not H-oscillatory in Ω. First we consider the case where $u_H(x,t) > 0$ in $G \times [t_0, \infty)$ for some $t_0 > 0$. We observe that $\varphi_i(\xi) \in C(\mathbb{R};\mathbb{R})$ $(i = 1, 2, ..., m)$, $\varphi_i(-\xi) = -\varphi_i(\xi)$, $\varphi_i(\xi) > 0$ for $\xi > 0$, and $\varphi_i(\xi)$ are nondecreasing in $(0, \infty)$. We easily see that

$$U_H(t) := \left(\int_G \Phi(x)\,dx\right)^{-1} \int_G u_H(x,t)\Phi(x)\,dx$$

is an eventually positive solution of the inequality (7.6.8) with $+G_H(t)$ (*cf.* [256, Theorem 3.1], [319, Theorem 1]). Hence, we are led to a contradiction. The case where $u_H(x,t) < 0$ in $G \times [t_0, \infty)$ can be treated similarly, and we are also led to a contradiction. This completes the proof. □

Theorem 7.6.4. *Assume that the hypotheses* (H6.6-3), (H7.6-1)–(H7.6-3) *hold. If the functional differential inequalities*

$$\frac{d^2}{dt^2}\left(y(t) + \sum_{i=1}^{\ell} h_i(t)y(\rho_i(t))\right) + \sum_{i=1}^{m} p_i(t)\varphi_i\big(y(\sigma_i(t))\big) \le \pm\tilde{G}_H(t) \tag{7.6.9}$$

have no eventually positive solutions, then every solution $U(x,t)$ *of the boundary value problem* (7.6.1), (B_2) *is* H*-oscillatory in* Ω, *where*

$$\tilde{G}_H(t) = \tilde{F}_H(t) + a(t)\tilde{\Psi}_H(t) + \sum_{i=1}^{k} b_i(t)\tilde{\Psi}_H(\tau_i(t)).$$

Proof. The proof is quite similar to that of Theorem 7.6.3, and hence will be omitted. □

Theorem 7.6.5. *Assume that the hypotheses* (H6.6-3), (H7.6-1)–(H7.6-3) *hold, and assume, moreover, that:*

(H7.6-4) $h_i(t) \geq 0$ $(i = 1, 2, ..., \ell)$, $\sum_{i=1}^{\ell} h_i(t) \leq 1$;

(H7.6-5) $\rho_i(t) \leq t$ $(i = 1, 2, ..., \ell)$;

(H7.6-6) *there exists a function* $\Theta(t) \in C^2([t_0, \infty); \mathbb{R})$ *such that* $\Theta(t)$ *is oscillatory at* $t = \infty$ *and* $\Theta''(t) = G_H(t)$ [*resp.* $\Theta''(t) = \tilde{G}_H(t)$], *where* $t_0 > 0$ *is some number*;

(H7.6-7) $$\int_{t_0}^{\infty} p_j(s)\varphi_j\left(\left[\left(1 - \sum_{i=1}^{\ell} h_i(\sigma_j(s))\right) c \pm \hat{\Theta}(\sigma_j(s))\right]_+\right) ds = \infty$$
for some $j \in \{1, 2, ..., m\}$ *and every* $c > 0$, *where*

$$[M(s)]_+ = \max\{M(s), 0\},$$

$$\hat{\Theta}(t) = \Theta(t) - \sum_{i=1}^{\ell} h_i(t)\Theta(\rho_i(t)).$$

Then every solution $U(x,t)$ *of the boundary value problem* (7.6.1), (B_1) [*resp.* (7.6.1), (B_2)] *is* H*-oscillatory in* Ω.

Proof. The conclusion follows by combining a result of Tanaka [256, Theorem 2.1] with Theorems 7.6.3 and 7.6.4. □

Example 7.6.1. We consider the boundary value problem

$$\begin{aligned}
&\frac{\partial^2}{\partial t^2}\left(U(x,t) + \frac{1}{2}U(x,t-\pi)\right) - \frac{\partial^2 U}{\partial x^2}(x,t) \\
&\quad - \frac{\partial^2 U}{\partial x^2}(x,t-\pi) + 2U(x,t-2\pi) \\
&\quad = \begin{pmatrix} \frac{3}{2}(\cos x)\sin t \\ \left(1 + \frac{1}{2}e^{-\pi} + 2e^{-2\pi}\right)e^t \end{pmatrix}, \quad (x,t) \in (0,\pi) \times (0,\infty),
\end{aligned} \tag{7.6.10}$$

$$-\frac{\partial U}{\partial x}(0,t) = \frac{\partial U}{\partial x}(\pi,t) = \begin{pmatrix} 0 \\ 0 \end{pmatrix}, \quad t > 0. \tag{7.6.11}$$

Here $n = 1$, $G = (0,\pi)$, $\Omega = (0,\pi) \times (0,\infty)$, $\ell = k = m = 1$, $N = 2$, $h_1(t) = 1/2$, $\rho_1(t) = t - \pi$, $a(t) = b_1(t) = 1$, $\tau_1(t) = t - \pi$, $\sigma_1(t) = t - 2\pi$,

$c_1(x,t,\Xi) = 1$, $p_1(t) = 2$, $\psi_1(\xi) = 1$, $\varphi_1(\xi) = \xi$, $\mu = 0$, $\tilde{\Psi} = \begin{pmatrix} 0 \\ 0 \end{pmatrix}$ and

$$F(x,t) = \begin{pmatrix} \frac{3}{2}(\cos x)\sin t \\ \left(1 + \frac{1}{2}e^{-\pi} + 2e^{-2\pi}\right)e^t \end{pmatrix}.$$

Letting $H = e_1 = \begin{pmatrix} 1 \\ 0 \end{pmatrix}$, we see that $\tilde{F}_{e_1}(t) = \tilde{\Psi}_{e_1}(t) = 0$, and hence $\tilde{G}_{e_1}(t) = 0$. We can choose $\Theta(t) = 0$, and observe that $\hat{\Theta}(t) = 0$ and

$$\int_1^\infty 2 \cdot \frac{1}{2} c\, ds = \infty$$

for every $c > 0$. Hence, Theorem 7.6.5 implies that every solution $U(x,t)$ of the problem (7.6.10), (7.6.11) is e_1-oscillatory in $(0,\pi) \times (0,\infty)$. In fact

$$U(x,t) = \begin{pmatrix} (\cos x)\sin t \\ e^t \end{pmatrix}$$

is such a solution. We note that the above solution $U(x,t)$ is not e_2-oscillatory in $(0,\pi) \times (0,\infty)$, where $e_2 = \begin{pmatrix} 0 \\ 1 \end{pmatrix}$.

7.7 Hyperbolic Systems II

Parhi and Kirane [237] investigated the oscillatory properties of solutions of coupled hyperbolic equations in 1994. Oscillation of hyperbolic systems with deviating arguments was studied by Li [178] in 1997, and then oscillation results have been developed by several authors, see, for example, Li [167], Li and Meng [176], Agarwal, Meng and Li [7] and the references cited therein.

We study the system of hyperbolic equations with functional arguments

$$\begin{aligned} &\frac{\partial^2}{\partial t^2}\Big(U(x,t) + \sum_{i=1}^{\ell} H_i(t)U(x,\rho_i(t))\Big) - A(t)\Delta U(x,t) \\ &\quad - \sum_{i=1}^{K} B_i(t)\Delta U(x,\tau_i(t)) + \sum_{i=1}^{m} P_i(x,t)\varphi_i\big(U(x,\sigma_i(t))\big) \\ &= F(x,t), \quad (x,t) \in \Omega := G \times (0,\infty), \qquad (7.7.1) \end{aligned}$$

where G is a bounded domain in $\mathbb{R}^n$ with piecewise smooth boundary ∂G, Δ is the Laplacian in $\mathbb{R}^n$, and

$$H_i(t) = \big(h_{ijk}(t)\big)_{j,k=1}^{M}, \ A(t) = \big(a_{jk}(t)\big)_{j,k=1}^{M}, \ B_i(t) = \big(b_{ijk}(t)\big)_{j,k=1}^{M},$$

$$P_i(x,t) = \big(p_{ijk}(x,t)\big)_{j,k=1}^{M},$$

$$U(x,t) = \big(u_1(x,t),...,u_M(x,t)\big)^T,$$

$$U(x,\rho_i(t)) = \big(u_1(x,\rho_i(t)),...,u_M(x,\rho_i(t))\big)^T,$$

$$U(x,\tau_i(t)) = \big(u_1(x,\tau_{i1}(t)),...,u_M(x,\tau_{iM}(t))\big)^T,$$

$$\varphi_i(U(x,\sigma_i(t))) = \big(\varphi_{i1}(u_1(x,\sigma_i(t))),...,\varphi_{iM}(u_M(x,\sigma_i(t)))\big)^T,$$

$$F(x,t) = \big(f_1(x,t),...,f_M(x,t)\big)^T$$

(the superscript T denotes the transpose).

It is easy to check that (7.7.1) is equivalent to the following system:

$$\frac{\partial^2}{\partial t^2}\Big(u_j(x,t) + \sum_{i=1}^{\ell}\sum_{k=1}^{M} h_{ijk}(t)u_k(x,\rho_i(t))\Big)$$

$$-\sum_{k=1}^{M} a_{jk}(t)\Delta u_k(x,t) - \sum_{i=1}^{K}\sum_{k=1}^{M} b_{ijk}(t)\Delta u_k(x,\tau_{ik}(t))$$

$$+\sum_{i=1}^{m}\sum_{k=1}^{M} p_{ijk}(x,t)\varphi_{ik}\big(u_k(x,\sigma_i(t))\big) = f_j(x,t) \ \ (j=1,2,...,M) \quad (7.7.2)$$

for $(x,t) \in \Omega$.

We consider the following boundary condition:

(BC) $\alpha(\hat{x})\dfrac{\partial u_j}{\partial \nu} + \mu(\hat{x})\,u_j = \alpha(\hat{x})\tilde{\psi}_j + \mu(\hat{x})\psi_j$ on $\partial G \times (0,\infty)$ $(j = 1,2,...,M)$,

where $\psi_j, \tilde{\psi}_j \in C(\partial G \times (0,\infty);\mathbb{R})$, $\alpha(\hat{x}), \mu(\hat{x}) \in C(\partial G;[0,\infty))$, $\alpha(\hat{x})^2 + \mu(\hat{x})^2 \neq 0$, and ν denotes the unit exterior normal vector to ∂G. We note that the boundary condition (BC) includes Dirichlet, Neumann and Robin's boundary conditions (*cf.* Section 6.7).

We assume that the hypotheses (H6.7-3)–(H6.7-7), (H6.7-9) hold, and the following two hypotheses are satisfied:

(H7.7-1) $h_{ijk}(t) \in C^2([0,\infty);[0,\infty))$ $(i = 1,2,...,\ell;\ j,k = 1,2,...,M)$,
$a_{jk}(t) \in C([0,\infty);[0,\infty))$ $(j,k = 1,2,...,M)$,
$b_{ijk}(t) \in C([0,\infty);[0,\infty))$ $(i = 1,2,...,K;\ j,k = 1,2,...,M)$,
$p_{ijk}(x,t) \in C(\overline{G}\times[0,\infty);[0,\infty))$ $(i = 1,2,...,m;\ j,k = 1,2,...,M)$,
$f_j(x,t) \in C(\overline{G} \times [0,\infty);\mathbb{R})$ $(j = 1,2,...,M)$;

(H7.7-2) $\rho_i(t) \in C^2([0,\infty);\mathbb{R}),\ \lim_{t\to\infty} \rho_i(t) = \infty\ (i = 1,2,...,\ell)$,
$\tau_{ik}(t) \in C([0,\infty);\mathbb{R}),\ \lim_{t\to\infty} \tau_{ik}(t) = \infty\ (i = 1,2,...,K;\ k = 1,2,...,M)$,
$\sigma_i(t) \in C([0,\infty);\mathbb{R}),\ \lim_{t\to\infty} \sigma_i(t) = \infty\ (i = 1,2,...,m)$.

Definition 7.7.1. By a *solution* of system (7.7.2) we mean a vector function $\left(u_1(x,t),...,u_M(x,t)\right)$ such that $u_j(x,t) \in C^2(\overline{G}\times[t_{-1},\infty);\mathbb{R}) \cap C(\overline{G}\times[\tilde{t}_{-1},\infty);\mathbb{R})$, and $u_j(x,t)$ $(j = 1,2,...,M)$ satisfy (7.7.2) in Ω, where

$$t_{-1} = \min\left\{0,\ \min_{\substack{1\le i\le K\\1\le k\le M}}\left\{\inf_{t\ge 0}\tau_{ik}(t)\right\},\ \min_{1\le i\le \ell}\left\{\inf_{t\ge 0}\rho_i(t)\right\}\right\},$$

$$\tilde{t}_{-1} = \min\left\{0,\ \min_{1\le i\le m}\left\{\inf_{t\ge 0}\sigma_i(t)\right\}\right\}.$$

Definition 7.7.2. A solution $\left(u_1(x,t),...,u_M(x,t)\right)$ of system (7.7.2) is called *weakly oscillatory* in Ω if at least one of its components is oscillatory in Ω (*cf.* Ladde, Lakshmikantham and Zhang [162, Definition 6.2.1]).

Let λ_1 (≥ 0) be the smallest eigenvalue of the eigenvalue problem

$$-\Delta w = \lambda w \quad \text{in } G,$$
$$\alpha(\hat{x})\frac{\partial w}{\partial \nu} + \mu(\hat{x})\,w = 0 \quad \text{on } \partial G$$

and $\Phi(x)$ be the corresponding eigenfunction with the property that $\Phi(x) > 0$ in G (*cf.* Section 6.7).

We use the notation:

$$\Gamma = \left\{\left((-1)^{\alpha_1},(-1)^{\alpha_2},...,(-1)^{\alpha_M}\right);\ \alpha_j = 0,1\ (j = 1,2,...,M)\right\},$$

$$\left[\Theta(t)\right]_{\pm} = \max\{\pm\Theta(t),0\},$$

$$K_\Phi = \left(\int_G \Phi(x)\,dx\right)^{-1},$$

$$F_j(t) = K_\Phi\int_G f_j(x,t)\Phi(x)\,dx \quad (j = 1,2,...,M),$$

$$\Psi_j(t) = K_\Phi\left(-\int_{\Gamma_1}\psi_j\frac{\partial\Phi(x)}{\partial\nu}\,dS + \int_{\Gamma_2}\left(\tilde{\psi}_j + \frac{\mu(\hat{x})}{\alpha(\hat{x})}\psi_j\right)\Phi(x)\,dS\right),$$

where

$$\Gamma_1 = \{\,\hat{x}\in\partial G;\ \alpha(\hat{x}) = 0\},$$
$$\Gamma_2 = \{\,\hat{x}\in\partial G;\ \alpha(\hat{x}) \ne 0\}.$$

Theorem 7.7.1. *Assume that the hypotheses* (H6.7-3)–(H6.7-7), (H6.7-9), (H7.7-1) *and* (H7.7-2) *hold. If the following conditions are satisfied:*

(H7.7-3) $\rho_i(t) \le t \ (i = 1, 2, ..., \ell)$;

(H7.7-4) $\sum_{i=1}^{\ell}\sum_{j=1}^{M} \tilde{h}_{ij}(t) \le 1$ *on* $[t_0, \infty)$ *for some* $t_0 > 0$;

(H7.7-5) *there exist functions* $\Theta_\gamma(t) \in C^2([t_0, \infty); \mathbb{R})$ $(\gamma \in \Gamma)$ *such that* $\Theta_\gamma(t)$ *are oscillatory at* $t = \infty$ *and* $\Theta''_\gamma(t) = \gamma \cdot \left(G_j(t)\right)_{j=1}^{M}$ *(the dot* $\cdot$ *denotes the scalar product),*

and if for some $j_0 \in \{1, 2, ..., m\}$ *and for any* $c > 0$

$$\int_{t_0}^{\infty} \hat{p}_{j_0}(t)\hat{\varphi}_{j_0}\left(\left[\left(1 - \sum_{i=1}^{\ell}\sum_{j=1}^{M} \tilde{h}_{ij}(\sigma_{j_0}(t))\right) c + \hat{\Theta}_\gamma(\sigma_{j_0}(t))\right]_+\right) dt = \infty, \tag{7.7.3}$$

then every solution $(u_1(x,t), ..., u_M(x,t))$ *of the boundary value problem* (7.7.2), (BC) *is weakly oscillatory in* Ω, *where*

$$\tilde{h}_{ij}(t) = \max_{1 \le k \le M} h_{ijk}(t),$$

$$G_j(t) = \frac{1}{M}\left(F_j(t) + \sum_{k=1}^{M} a_{jk}(t)\Psi_k(t) + \sum_{i=1}^{K}\sum_{k=1}^{M} b_{ijk}(t)\Psi_k(\tau_{ik}(t))\right),$$

$$\hat{p}_i(t) = \min_{x \in \overline{G}}\left\{\min_{1 \le k \le M}\left(p_{ikk}(x,t) - \sum_{\substack{j=1 \\ j \ne k}}^{M} p_{ijk}(x,t)\right)\right\},$$

$$\hat{\Theta}_\gamma(t) = \Theta_\gamma(t) - \sum_{i=1}^{\ell}\sum_{j=1}^{M} \tilde{h}_{ij}(t)\Theta_\gamma(\rho_i(t)).$$

Proof. Assume, for the sake of contradiction, that there is a solution $(u_1(x,t), ..., u_M(x,t))$ of the problem (7.7.2), (BC) which is not weakly oscillatory in Ω. Then each component $u_j(x,t)$ is nonoscillatory in Ω. It is easy to check that there exists a number $t_1 \ge t_0$ such that $|u_j(x,t)| > 0$ for $x \in G, t \ge t_1$ $(j = 1, 2, ..., M)$. Letting

$$w_j(x,t) = \delta_j u_j(x,t),$$

where $\delta_j = \text{sgn } u_j(x,t)$, we find that $w_j(x,t) = |u_j(x,t)| > 0$ in $G \times [t_1, \infty)$. There is a number $t_2 \ge t_1$ for which $w_j(x,t) > 0$, $w_j(x, \rho_i(t)) > 0$,

$w_j(x, \tau_{ij}(t)) > 0$, $w_j(x, \sigma_i(t)) > 0$ in $G \times [t_2, \infty)$. Arguing as in the proof of Theorem 6.7.1, we conclude that the following identity holds:

$$\frac{d^2}{dt^2}\left(V(t) + \frac{1}{M}\sum_{i=1}^{\ell}\sum_{j,k=1}^{M}\delta_j\delta_k h_{ijk}(t)W_k(\rho_i(t))\right) + \sum_{i=1}^{m}\hat{p}_i(t)\hat{\varphi}_i\big(V(\sigma_i(t))\big)$$
$$\leq \sum_{j=1}^{M}\delta_j G_j(t), \quad t \geq t_2, \tag{7.7.4}$$

where

$$V(t) = \frac{\sum\limits_{k=1}^{M} W_k(t)}{M}.$$

There exists a $\gamma \in \Gamma$ such that $\sum_{j=1}^{M}\delta_j G_j(t) = \gamma \cdot \big(G_j(t)\big)_{j=1}^{M}$. Letting

$$Y(t) = V(t) + \frac{1}{M}\sum_{i=1}^{\ell}\sum_{j,k=1}^{M}\delta_j\delta_k h_{ijk}(t)W_k(\rho_i(t)) - \Theta_\gamma(t),$$

we obtain

$$Y''(t) \leq -\sum_{i=1}^{m}\hat{p}_i(t)\hat{\varphi}_i\big(V(\sigma_i(t))\big) \leq 0, \quad t \geq t_2. \tag{7.7.5}$$

Hence we see that $Y(t) > 0$ or $Y(t) \leq 0$ on $[t_3, \infty)$ for some $t_3 \geq t_2$. In case $Y(t) \leq 0$ on $[t_3, \infty)$, then we observe that

$$V(t) + \frac{1}{M}\sum_{i=1}^{\ell}\sum_{j,k=1}^{M}\delta_j\delta_k h_{ijk}(t)W_k(\rho_i(t)) \leq \Theta_\gamma(t), \quad t \geq t_3$$

and hence

$$V(t) + \frac{1}{M}\sum_{i=1}^{\ell}\sum_{k=1}^{M}\left(h_{ikk}(t) - \sum_{\substack{j=1\\ j\neq k}}^{M} h_{ijk}(t)\right)W_k(\rho_i(t)) \leq \Theta_\gamma(t), \quad t \geq t_3. \tag{7.7.6}$$

The left hand side of (7.7.6) is positive in light of the hypothesis (H6.7-4), whereas the right hand side of (7.7.6) is oscillatory at $t = \infty$. This yields a contradiction. Therefore, we find that $Y(t) > 0$ on $[t_3, \infty)$. Since $Y''(t) \leq 0$, $Y(t) > 0$ on $[t_3, \infty)$, we see that $Y'(t) \geq 0$ on $[t_4, \infty)$ for some

$t_4 \geq t_3$, which implies that $Y(t) \geq Y(t_4)$ for $t \geq t_4$. In view of the fact that $V(t) \leq Y(t) + \Theta_\gamma(t)$ and $Y(t)$ is nondecreasing, we arrive at

$$\begin{aligned} V(t) &= Y(t) - \frac{1}{M} \sum_{i=1}^{\ell} \sum_{j,k=1}^{M} \delta_j \delta_k h_{ijk}(t) W_k(\rho_i(t)) + \Theta_\gamma(t) \\ &\geq Y(t) - \sum_{i=1}^{\ell} \sum_{j=1}^{M} \tilde{h}_{ij}(t) V(\rho_i(t)) + \Theta_\gamma(t) \\ &\geq Y(t) - \sum_{i=1}^{\ell} \sum_{j=1}^{M} \tilde{h}_{ij}(t) \big(Y(\rho_i(t)) + \Theta_\gamma(\rho_i(t))\big) + \Theta_\gamma(t) \\ &\geq \left(1 - \sum_{i=1}^{\ell} \sum_{j=1}^{M} \tilde{h}_{ij}(t)\right) Y(t) + \hat{\Theta}_\gamma(t), \quad t \geq t_4. \end{aligned} \tag{7.7.7}$$

Since $Y(t) \geq Y(t_4)$ and $V(t) > 0$ for $t \geq t_4$, from (7.7.7) it follows that

$$V(t) \geq \left[\left(1 - \sum_{i=1}^{\ell} \sum_{j=1}^{M} \tilde{h}_{ij}(t)\right) Y(t_4) + \hat{\Theta}_\gamma(t)\right]_+, \quad t \geq t_4$$

and consequently

$$V(\sigma_{j_0}(t)) \geq \left[\left(1 - \sum_{i=1}^{\ell} \sum_{j=1}^{M} \tilde{h}_{ij}(\sigma_{j_0}(t))\right) Y(t_4) + \hat{\Theta}_\gamma(\sigma_{j_0}(t))\right]_+, \quad t \geq T \tag{7.7.8}$$

for some $T \geq t_4$. Since $\hat{\varphi}_{j_0}(t)$ is nondecreasing, from (7.7.5) and (7.7.8) we see that

$$\begin{aligned} &\hat{p}_{j_0}(t) \hat{\varphi}_{j_0} \left(\left[\left(1 - \sum_{i=1}^{\ell} \sum_{j=1}^{M} \tilde{h}_{ij}(\sigma_{j_0}(t))\right) Y(t_4) + \hat{\Theta}_\gamma(\sigma_{j_0}(t))\right]_+ \right) \\ &\leq -Y''(t), \quad t \geq T. \end{aligned} \tag{7.7.9}$$

An integration of (7.7.9) over $[T, t]$ yields

$$\begin{aligned} &\int_T^t \hat{p}_{j_0}(s) \hat{\varphi}_{j_0} \left(\left[\left(1 - \sum_{i=1}^{\ell} \sum_{j=1}^{M} \tilde{h}_{ij}(\sigma_{j_0}(s))\right) Y(t_4) + \hat{\Theta}_\gamma(\sigma_{j_0}(s))\right]_+ \right) ds \\ &\leq -Y'(t) + Y'(T) \leq Y'(T), \quad t \geq T. \end{aligned}$$

This contradicts the hypothesis (7.7.3) and the proof is complete. □

Theorem 7.7.2. *Assume that the hypotheses* (H6.7-3)–(H6.7-7), (H6.7-9), (H7.7-1) *and* (H7.7-2) *hold. Every solution* $(u_1(x,t),...,u_M(x,t))$ *of the boundary value problem* (7.7.2), (BC) *is weakly oscillatory in* Ω *if for any* $\gamma \in \Gamma$

$$\liminf_{t\to\infty} \int_T^t \left(1-\frac{s}{t}\right)\left(\gamma\cdot\left(G_j(s)\right)_{j=1}^M\right) ds = -\infty \tag{7.7.10}$$

for all large T.

Proof. Suppose that there is a solution $(u_1(x,t),...,u_M(x,t))$ of the problem (7.7.2), (BC) which is not weakly oscillatory in Ω. Proceeding as in the proof of Theorem 7.7.1, we find that (7.7.4) holds, and therefore

$$\frac{d^2}{dt^2}\left(V(t)+\frac{1}{M}\sum_{i=1}^{\ell}\sum_{j,k=1}^{M}\delta_j\delta_k h_{ijk}(t)W_k(\rho_i(t))\right) \le \sum_{j=1}^{M}\delta_j G_j(t), \ \ t \ge t_2$$

for some $t_2 > 0$. Setting

$$\tilde{V}(t) = V(t)+\frac{1}{M}\sum_{i=1}^{\ell}\sum_{j,k=1}^{M}\delta_j\delta_k h_{ijk}(t)W_k(\rho_i(t)),$$

we observe that $\tilde{V}(t) > 0$ $(t > t_3)$ for some $t_3 \ge t_2$ (see the proof of Theorem 6.7.2). Using the same arguments as were used in Theorem 7.2.3, we are led to a contradiction. □

Remark 7.7.1. As in Remark 6.7.1, we see that $\Gamma = \left\{\pm\gamma;\ \gamma\in\tilde{\Gamma}\right\}$ for some $\tilde{\Gamma}\subset\Gamma$ with $\#\tilde{\Gamma} = 2^{M-1}$. Therefore, Theorem 7.7.1 holds true if the hypothesis (H7.7-5) and the condition (7.7.3) are replaced by

($\tilde{\text{H}}$7.7-5) there are functions $\Theta_\gamma(t)\in C^2([t_0,\infty);\mathbb{R})$ $(\gamma\in\tilde{\Gamma})$ such that $\Theta_\gamma(t)$ are oscillatory at $t=\infty$ and $\Theta''_\gamma(t) = \gamma\cdot\left(G_j(t)\right)_{j=1}^M$ (the dot $\cdot$ denotes the scalar product)

and

$$\int_{t_0}^{\infty}\hat{p}_{j_0}(t)\hat{\varphi}_{j_0}\left(\left[\left(1-\sum_{i=1}^{\ell}\sum_{j=1}^{M}\tilde{h}_{ij}(\sigma_{j_0}(t))\right)c\pm\hat{\Theta}_\gamma(\sigma_{j_0}(t))\right]_+\right)dt = \infty,$$

respectively.

Remark 7.7.2. Theorem 7.7.2 remains valid if (7.7.10) is replaced by the following hypothesis:

$$\liminf_{t\to\infty} \int_T^t \left(1-\frac{s}{t}\right)\left(\gamma\cdot (G_j(s))_{j=1}^M\right) ds = -\infty,$$

$$\limsup_{t\to\infty} \int_T^t \left(1-\frac{s}{t}\right)\left(\gamma\cdot (G_j(s))_{j=1}^M\right) ds = \infty$$

for any $\gamma \in \tilde{\Gamma}$ and all large T.

Example 7.7.1. We consider the hyperbolic system

$$\begin{aligned}
&\frac{\partial^2}{\partial t^2}\Big(u_1(x,t)+\frac{1}{2}u_1(x,t-2\pi)+\frac{1}{4}u_2(x,t-2\pi)\Big)\\
&\quad -3\frac{\partial^2 u_1}{\partial x^2}(x,t)-\frac{1}{8}e^{2\pi}\frac{\partial^2 u_2}{\partial x^2}(x,t)-\frac{\partial^2 u_1}{\partial x^2}(x,t-\pi)-\frac{1}{8}\frac{\partial^2 u_2}{\partial x^2}(x,t-2\pi)\\
&\quad +\frac{1}{2}u_1(x,t-\pi)+\frac{1}{2}e^{\pi}u_2(x,t-\pi)=e^{2\pi}(\sin x)e^{-t}, \qquad (7.7.11)\\
&\frac{\partial^2}{\partial t^2}\Big(u_2(x,t)+\frac{1}{4}u_1(x,t-2\pi)+\frac{1}{4}u_2(x,t-2\pi)\Big)\\
&\quad -2\frac{\partial^2 u_1}{\partial x^2}(x,t)-e^{2\pi}\frac{\partial^2 u_2}{\partial x^2}(x,t)-\frac{1}{2}\frac{\partial^2 u_1}{\partial x^2}(x,t-\pi)-\frac{1}{4}\frac{\partial^2 u_2}{\partial x^2}(x,t-2\pi)\\
&\quad +\frac{1}{4}u_1(x,t-\pi)+\frac{3}{2}e^{\pi}u_2(x,t-\pi)\\
&\quad =(\sin x)\sin t+(1+3e^{2\pi})(\sin x)e^{-t}, \qquad (7.7.12)
\end{aligned}$$

for $(x,t)\in(0,\pi)\times(0,\infty)$ and the boundary condition

$$u_j(0,t)=u_j(\pi,t)=0, \quad t>0 \ (j=1,2). \qquad (7.7.13)$$

Here $G=(0,\pi)$, $n=1$, $M=2$, $\ell=K=m=1$, $h_{111}(t)=1/2$, $h_{112}(t)=h_{121}(t)=h_{122}(t)=1/4$, $\rho_1(t)=t-2\pi$, $a_{11}(t)=3$, $a_{12}(t)=(1/8)e^{2\pi}$, $a_{21}(t)=2$, $a_{22}(t)=e^{2\pi}$, $b_{111}(t)=1$, $b_{112}(t)=1/8$, $b_{121}(t)=1/2$, $b_{122}(t)=1/4$, $\tau_{11}(t)=t-\pi$, $\tau_{12}(t)=t-2\pi$, $p_{111}(x,t)=1/2$, $p_{112}(t)=(1/2)e^{\pi}$, $p_{121}(t)=1/4$, $p_{122}(t)=(3/2)e^{\pi}$, $\sigma_1(t)=t-\pi$, $\varphi_{11}(\xi)=\varphi_{12}(\xi)=\xi$, $f_1(x,t)=e^{2\pi}(\sin x)e^{-t}$, $f_2(x,t)=(\sin x)\sin t+(1+3e^{2\pi})(\sin x)e^{-t}$, $\psi_j=0$ $(j=1,2)$, $\alpha(\hat{x})=0$ and $\Gamma_2=\emptyset$. It is easy to check that

$$\tilde{h}_{11}(t)=\frac{1}{2},\ \tilde{h}_{12}(t)=\frac{1}{4},\ \hat{p}_1(t)=\frac{1}{4}.$$

We easily see that $\lambda_1=1$, $\Phi(x)=\sin x$, $\Psi_j(t)=0$ $(j=1,2)$, $G_1(t)=(1/2)F_1(t)=(\pi/8)e^{2\pi}e^{-t}$ and $G_2(t)=(1/2)F_2(t)=(\pi/8)\sin t+(\pi/8)(1+3e^{2\pi})e^{-t}$. Choosing

$$\Theta_{(1,1)}(t)=\frac{\pi}{8}\left(-\sin t+\left(1+4e^{2\pi}\right)e^{-t}\right),$$

$$\Theta_{(1,-1)}(t)=\frac{\pi}{8}\left(\sin t-\left(1+2e^{2\pi}\right)e^{-t}\right),$$

we see that

$$\hat{\Theta}_{(1,1)}(t) = \frac{\pi}{8}\left(-\frac{1}{4}\sin t + (1+4e^{2\pi})\left(1-\frac{3}{4}e^{2\pi}\right)e^{-t}\right),$$

$$\hat{\Theta}_{(1,-1)}(t) = \frac{\pi}{8}\left(\frac{1}{4}\sin t - (1+2e^{2\pi})\left(1-\frac{3}{4}e^{2\pi}\right)e^{-t}\right)$$

and hence

$$\int_{t_0}^{\infty}\frac{1}{4}\left[\left(1-\frac{3}{4}\right)c \pm \frac{\pi}{8}\left(-\frac{1}{4}\sin(t-\pi)+(1+4e^{2\pi})\left(1-\frac{3}{4}e^{2\pi}\right)e^{-(t-\pi)}\right)\right]_+ dt$$
$$\geq \frac{\pi}{32}\int_{t_0}^{\infty}\left[\frac{1}{4}\sin t + (1+4e^{2\pi})\left(1-\frac{3}{4}e^{2\pi}\right)e^{\pi}e^{-t}\right]_{\pm} dt$$
$$= \infty$$

and

$$\int_{t_0}^{\infty}\frac{1}{4}\left[\left(1-\frac{3}{4}\right)c \pm \frac{\pi}{8}\left(\frac{1}{4}\sin(t-\pi)-(1+2e^{2\pi})\left(1-\frac{3}{4}e^{2\pi}\right)e^{-(t-\pi)}\right)\right]_+ dt$$
$$\geq \frac{\pi}{32}\int_{t_0}^{\infty}\left[-\frac{1}{4}\sin t - (1+2e^{2\pi})\left(1-\frac{3}{4}e^{2\pi}\right)e^{\pi}e^{-t}\right]_{\pm} dt$$
$$= \infty$$

for any $c > 0$. From Theorem 7.7.1 and Remark 7.7.1 it follows that every solution $(u_1(x,t), u_2(x,t))$ of the problem (7.7.11)–(7.7.13) is weakly oscillatory in $(0,\pi)\times(0,\infty)$. One such solution is

$$U(x,t) = \begin{pmatrix}(\sin x)\sin t\\ (\sin x)e^{-t}\end{pmatrix}.$$

Example 7.7.2. We consider the hyperbolic system

$$\frac{\partial^2}{\partial t^2}\left(u_1(x,t)+\frac{1}{2}u_1(x,t-\pi)+\frac{1}{2}u_2(x,t-\pi)\right)$$
$$-\frac{\partial^2 u_1}{\partial x^2}(x,t)-\frac{\partial^2 u_2}{\partial x^2}(x,t)-e^{(3/2)\pi}\frac{\partial^2 u_1}{\partial x^2}(x,t+\tfrac{\pi}{2})-\frac{\partial^2 u_2}{\partial x^2}(x,t+\pi)$$
$$+\frac{3}{2}u_1(x,t-\pi)+\frac{1}{2}u_2(x,t-\pi) = (2+3e^{\pi})(\cos x)e^{-t}, \tag{7.7.14}$$

$$\frac{\partial^2}{\partial t^2}\left(u_2(x,t)+\frac{1}{3}u_1(x,t-\pi)+\frac{1}{2}u_2(x,t-\pi)\right)$$
$$-\frac{\partial^2 u_1}{\partial x^2}(x,t)-\frac{\partial^2 u_2}{\partial x^2}(x,t)-\frac{2}{3}e^{(3/2)\pi}\frac{\partial^2 u_1}{\partial x^2}(x,t+\tfrac{\pi}{2})-\frac{3}{2}\frac{\partial^2 u_2}{\partial x^2}(x,t+\pi)$$
$$+u_1(x,t-\pi)+u_2(x,t-\pi)$$
$$= -2(\cos x)\sin t + (2e^{\pi}+1)(\cos x)e^{-t}, \tag{7.7.15}$$

for $(x,t) \in (0,\pi/2) \times (0,\infty)$ and the boundary condition

$$-\frac{\partial u_1}{\partial x}(0,t) = 0, \quad \frac{\partial u_1}{\partial x}(\tfrac{\pi}{2},t) = -e^{-t}, \quad t > 0, \tag{7.7.16}$$

$$-\frac{\partial u_2}{\partial x}(0,t) = 0, \quad \frac{\partial u_2}{\partial x}(\tfrac{\pi}{2},t) = -\sin t, \quad t > 0. \tag{7.7.17}$$

Here $G = (0,\pi/2)$, $n = 1$, $M = 2$, $\ell = K = m = 1$, $h_{111}(t) = h_{112}(t) = 1/2$, $h_{121}(t) = 1/3$, $h_{122}(t) = 1/2$, $\rho_1(t) = t - \pi$, $a_{11}(t) = a_{12}(t) = a_{21}(t) = a_{22}(t) = 1$, $b_{111}(t) = e^{(3/2)\pi}$, $b_{112}(t) = 1$, $b_{121}(t) = (2/3)e^{(3/2)\pi}$, $b_{122}(t) = 3/2$, $\tau_{11}(t) = t + (\pi/2)$, $\tau_{12}(t) = t + \pi$, $p_{111}(x,t) = 3/2$, $p_{112}(x,t) = 1/2$, $p_{121}(x,t) = p_{122}(x,t) = 1$, $\sigma_1(t) = t - \pi$, $\varphi_{11}(\xi) = \varphi_{12}(\xi) = \xi$, $f_1(x,t) = (2+3e^{\pi})(\cos x)e^{-t}$, $f_2(x,t) = -2(\cos x)\sin t + (2e^{\pi}+1)(\cos x)e^{-t}$, $\alpha(\hat{x}) = 1$, $\mu(\hat{x}) = 0$ and $\Gamma_1 = \emptyset$. We observe that

$$\tilde{h}_{11}(t) = \tilde{h}_{12}(t) = \frac{1}{2}, \ \hat{p}_1(t) = \frac{1}{2}.$$

It is easily seen that $\lambda_1 = 0$, $\Phi(x) = 1$ and that

$$G_1(t) = \frac{1}{2\pi}\,(2 + 4e^{\pi})\,e^{-t},$$

$$G_2(t) = \frac{1}{2\pi}\left(-3\sin t + \frac{8}{3}e^{\pi}e^{-t}\right).$$

Letting

$$\Theta_{(1,1)}(t) = \frac{1}{2\pi}\left(3\sin t + \left(\frac{20}{3}e^{\pi} + 2\right)e^{-t}\right),$$

$$\Theta_{(1,-1)}(t) = \frac{1}{2\pi}\left(-3\sin t + \left(\frac{4}{3}e^{\pi} + 2\right)e^{-t}\right),$$

we obtain

$$\hat{\Theta}_{(1,1)}(t) = \frac{1}{2\pi}\left(6\sin t + \left(\frac{20}{3}e^{\pi} + 2\right)(1 - e^{\pi})e^{-t}\right),$$

$$\hat{\Theta}_{(1,-1)}(t) = \frac{1}{2\pi}\left(-6\sin t + \left(\frac{4}{3}e^{\pi} + 2\right)(1 - e^{\pi})e^{-t}\right).$$

Since

$$\begin{aligned}
&\int_{t_0}^{\infty} \frac{1}{2}\left[\pm\hat{\Theta}_{(1,1)}(t-\pi)\right]_{+} dt \\
&= \frac{1}{4\pi}\int_{t_0}^{\infty}\left[-6\sin t + \left(\frac{20}{3}e^{\pi} + 2\right)(e^{\pi} - e^{2\pi})e^{-t}\right]_{\pm} dt \\
&= \infty
\end{aligned}$$

and

$$\int_{t_0}^{\infty} \frac{1}{2}\left[\pm\hat{\Theta}_{(1,-1)}(t-\pi)\right]_+ dt$$
$$= \frac{1}{4\pi}\int_{t_0}^{\infty}\left[6\sin t + \left(\frac{4}{3}e^{\pi}+2\right)(e^{\pi}-e^{2\pi})e^{-t}\right]_{\pm} dt$$
$$= \infty,$$

we see from Theorem 7.7.1 and Remark 7.7.1 that every solution $(u_1(x,t), u_2(x,t))$ of (7.7.14)–(7.7.17) is weakly oscillatory in $(0,\pi/2)\times(0,\infty)$. For instance,

$$U(x,t) = \begin{pmatrix} (\cos x)e^{-t} \\ (\cos x)\sin t \end{pmatrix}$$

is such a solution.

7.8 Notes

The oscillation results in Section 7.1 are slight generalizations of those given by Yoshida [306]. Hyperbolic equations with continuous deviating arguments were treated by Deng, Ge and Wang [62].

Theorem 7.2.5 is found in Tanaka [256]. Forced oscillation of hyperbolic equations with continuous deviating arguments were studied by Liu and Fu [182], Wang [278, 279], and hyperbolic equations without forcing terms were investigated by Deng [59], Deng and Ge [60], Li and Cui [171], Tao and Yoshida [260], Wang and Yu [281].

Hyperbolic equations of the form (7.3.1) were studied by Tao and Yoshida [259] and Yoshida [319]. The oscillation results in Section 7.3 are based on Tao and Yoshida [259] which deals with more general hyperbolic equations including Stieltjes integrals.

The results in Section 7.4 are based on Cui, Liu and Deng [58], Fu and Zhang [84], Liu, Xiao and Liu [180] and Luo [184]. Theorem 7.4.2 is due to Lakshmikantham, Bainov and Simeonov [163]. Lemma 7.4.2 is extracted from Luo [184, Lemma 3.4]. Theorems 7.4.1 and 7.4.6 are corollaries of the results of Fu and Zhang [84] which treats nonlinear impulsive hyperbolic equations. The arguments used in Theorems 7.4.3–7.4.5, 7.4.7 and 7.4.8 are those adopted by Liu, Xiao and Liu [180] and Luo [184].

The results in Section 7.5 are taken from Kiguradze, Kusano and Yoshida [126] and Yoshida [313].

The results in Section 7.6 are taken from Minchev and Yoshida [199]. Li, Han and Meng [174] investigated H-oscillation of vector hyperbolic equations without neutral term, but the hyperbolic equations considered by them include the Stieltjes integral.

The results in Section 7.7 are from Shoukaku and Yoshida [246].

Oscillations of characteristic initial value problems for hyperbolic equations with delays were investigated by Georgiou and Kreith [87], Mishev [201]. For characteristic initial value problems for delay hyperbolic equations of neutral type, we refer to Yan and Yoshida [292].

Chapter 8

Picone Identities and Applications

8.1 Half-Linear Elliptic Equations I

Since the pioneering work of Picone [240, 241], efforts have been made by numerous authors to extend the Picone identity in various directions, and the generalized Picone-type identities play an important role in the study of qualitative theory of differential equations. We refer the reader to Jaroš and Kusano [110,111] for Picone identities of half-linear ordinary differential equations of the form

$$\left(p(r)|y'|^{\alpha-1}y'\right)' + q(r)|y|^{\alpha-1}y = 0$$

and to Allegretto [14, 15], Allegretto and Huang [16, 17], Bognár and Došlý [32], Došlý [68], [69], Došlý and Řehák [71], Dunninger [73], Kusano, Jaroš and Yoshida [151] and Yoshida [320] for Picone identities of p-Laplacian or half-linear partial differential operators.

We establish a Picone identity for the half-linear elliptic operators p_α and P_α defined by

$$p_\alpha[u] = \nabla \cdot \left(a(x)|\nabla u|^{\alpha-1}\nabla u\right) + c(x)|u|^{\alpha-1}u, \tag{8.1.1}$$

$$P_\alpha[v] = \nabla \cdot \left(A(x)|\nabla v|^{\alpha-1}\nabla v\right) + C(x)|v|^{\alpha-1}v, \tag{8.1.2}$$

and derive a Sturmian comparison theorem and oscillation theorems for half-linear elliptic equations, where a dot $\cdot$ denotes the scalar product, $\alpha > 0$ is a constant and $|\cdot|$ in (8.1.1) and (8.1.2) denotes the Euclidean length.

Let G be a bounded domain in $\mathbb{R}^n$ with piecewise smooth boundary ∂G, and assume that $a(x) \in C(\overline{G}; (0, \infty))$, $A(x) \in C(\overline{G}; (0, \infty))$, $c(x) \in C(\overline{G}; \mathbb{R})$ and $C(x) \in C(\overline{G}; \mathbb{R})$.

The domain $\mathcal{D}_{p_\alpha}(G)$ of p_α is defined to be the set of all functions u of class $C^1(\overline{G}; \mathbb{R})$ with the property that $a(x)|\nabla u|^{\alpha-1}\nabla u \in C^1(G; \mathbb{R}^n) \cap C(\overline{G}; \mathbb{R}^n)$. The domain $\mathcal{D}_{P_\alpha}(G)$ of P_α is defined similarly.

Theorem 8.1.1 (Picone identity). *If $u \in \mathcal{D}_{p_\alpha}(G)$, $v \in \mathcal{D}_{P_\alpha}(G)$ and $v \neq 0$ in G (that is, v has no zero in G), then the following Picone identity holds:*

$$\begin{aligned}
&\nabla \cdot \left(\frac{u}{\varphi(v)} \left[\varphi(v) a(x) |\nabla u|^{\alpha-1} \nabla u - \varphi(u) A(x) |\nabla v|^{\alpha-1} \nabla v \right] \right) \\
&= (a(x) - A(x)) |\nabla u|^{\alpha+1} + \big(C(x) - c(x)\big) |u|^{\alpha+1} \\
&\quad + A(x) \left[|\nabla u|^{\alpha+1} + \alpha \left| \frac{u}{v} \nabla v \right|^{\alpha+1} - (\alpha+1) \left| \frac{u}{v} \nabla v \right|^{\alpha-1} (\nabla u) \cdot \left(\frac{u}{v} \nabla v \right) \right] \\
&\quad + \frac{u}{\varphi(v)} \big(\varphi(v) p_\alpha[u] - \varphi(u) P_\alpha[v] \big), \qquad (8.1.3)
\end{aligned}$$

where $\varphi(s) = |s|^{\alpha-1}s$ $(s \in \mathbb{R})$.

Proof. We easily see that

$$\begin{aligned}
u p_\alpha[u] &= u \nabla \cdot \big(a(x) |\nabla u|^{\alpha-1} \nabla u\big) + c(x) |u|^{\alpha+1} \\
&= \nabla \cdot \big(u\, a(x) |\nabla u|^{\alpha-1} \nabla u\big) - (\nabla u) \cdot \big(a(x) |\nabla u|^{\alpha-1} \nabla u\big) + c(x) |u|^{\alpha+1}
\end{aligned}$$

from which we obtain

$$\begin{aligned}
&\nabla \cdot \left(\frac{u}{\varphi(v)} \left[\varphi(v) a(x) |\nabla u|^{\alpha-1} \nabla u \right] \right) \\
&\qquad = a(x) |\nabla u|^{\alpha+1} - c(x) |u|^{\alpha+1} + \frac{u}{\varphi(v)} \big(\varphi(v) p_\alpha[u] \big) \qquad (8.1.4)
\end{aligned}$$

in view of the fact that $\varphi(v) \neq 0$ in G. A direct computation shows that

$$\begin{aligned}
&\nabla \cdot \left(u \varphi(u) \frac{A(x) |\nabla v|^{\alpha-1} \nabla v}{\varphi(v)} \right) \\
&= \sum_{i=1}^{n} \left[\frac{\partial}{\partial x_i} (u \varphi(u)) \frac{A(x) |\nabla v|^{\alpha-1} \frac{\partial v}{\partial x_i}}{\varphi(v)} \right. \\
&\qquad + u \varphi(u) \left(- \frac{\varphi'(v)}{\varphi(v)^2} \frac{\partial v}{\partial x_i} \right) A(x) |\nabla v|^{\alpha-1} \frac{\partial v}{\partial x_i} \\
&\qquad \left. + u \varphi(u) \frac{\frac{\partial}{\partial x_i} \left(A(x) |\nabla v|^{\alpha-1} \frac{\partial v}{\partial x_i} \right)}{\varphi(v)} \right], \qquad (8.1.5)
\end{aligned}$$

where $\varphi(s) \in C^1(\mathbb{R} \setminus \{0\}; \mathbb{R})$ and $s\varphi'(s) = \alpha \varphi(s) \in C(\mathbb{R}; \mathbb{R})$. It is easy to check that the following identities hold:

$$\sum_{i=1}^{n} \frac{\partial}{\partial x_i}(u\varphi(u)) \frac{A(x)|\nabla v|^{\alpha-1}\frac{\partial v}{\partial x_i}}{\varphi(v)}$$

$$= (\alpha+1)\varphi(u)(\nabla u) \cdot \frac{A(x)|\nabla v|^{\alpha-1}\nabla v}{\varphi(v)}$$

$$= (\alpha+1)A(x)\left|\frac{u}{v}\nabla v\right|^{\alpha-1}(\nabla u)\cdot\left(\frac{u}{v}\nabla v\right), \tag{8.1.6}$$

$$\sum_{i=1}^{n} u\varphi(u)\left(-\frac{\varphi'(v)}{\varphi(v)^2}\frac{\partial v}{\partial x_i}\right)A(x)|\nabla v|^{\alpha-1}\frac{\partial v}{\partial x_i}$$

$$= -\alpha A(x)\frac{u}{v}\varphi\left(\frac{u}{v}\right)(\nabla v)\cdot|\nabla v|^{\alpha-1}\nabla v$$

$$= -\alpha A(x)\left|\frac{u}{v}\nabla v\right|^{\alpha+1}, \tag{8.1.7}$$

$$\sum_{i=1}^{n} u\varphi(u)\frac{\frac{\partial}{\partial x_i}\left(A(x)|\nabla v|^{\alpha-1}\frac{\partial v}{\partial x_i}\right)}{\varphi(v)}$$

$$= -C(x)|u|^{\alpha+1} + \frac{u}{\varphi(v)}\big(\varphi(u)P_\alpha[v]\big). \tag{8.1.8}$$

Combining (8.1.5)–(8.1.8) yields

$$-\nabla\cdot\left(u\varphi(u)\frac{A(x)|\nabla v|^{\alpha-1}\nabla v}{\varphi(v)}\right)$$

$$= -A(x)|\nabla u|^{\alpha+1} + C(x)|u|^{\alpha+1}$$

$$+A(x)\left[|\nabla u|^{\alpha+1} + \alpha\left|\frac{u}{v}\nabla v\right|^{\alpha+1} - (\alpha+1)\left|\frac{u}{v}\nabla v\right|^{\alpha-1}(\nabla u)\cdot\left(\frac{u}{v}\nabla v\right)\right]$$

$$-\frac{u}{\varphi(v)}\big(\varphi(u)P_\alpha[v]\big). \tag{8.1.9}$$

We combine (8.1.4) with (8.1.9) to obtain the desired identity (8.1.3). □

Lemma 8.1.1. *The inequality*

$$|\xi|^{\alpha+1} + \alpha|\eta|^{\alpha+1} - (\alpha+1)|\eta|^{\alpha-1}\xi\cdot\eta \geq 0$$

is valid for any $\xi \in \mathbb{R}^n$ *and* $\eta \in \mathbb{R}^n$*, where the equality holds if and only if* $\xi = \eta$.

For the proof of Lemma 8.1.1, see, for example, Hardy, Littlewood and Pólya [97, Theorem 41] and Kusano, Jaroš and Yoshida [151, Lemma 2.1].

Theorem 8.1.2 (Sturmian comparison theorem). *If there exists a nontrivial solution $u \in \mathcal{D}_{p_\alpha}(G)$ of $p_\alpha[u] = 0$ such that $u = 0$ on ∂G and*

$$V[u] := \int_G \left[(a(x) - A(x)) |\nabla u|^{\alpha+1} + (C(x) - c(x)) |u|^{\alpha+1} \right] dx \geq 0, \quad (8.1.10)$$

then every solution $v \in \mathcal{D}_{P_\alpha}(G)$ of $P_\alpha[v] = 0$ must vanish at some point of $\overline{G}$.

Proof. Suppose that there exists a solution $v \in \mathcal{D}_{P_\alpha}(G)$ of $P_\alpha[v] = 0$ satisfying $v \neq 0$ on $\overline{G}$. We integrate (8.1.3) over G and then use the divergence theorem to obtain

$$\begin{aligned} 0 = V[u] + \int_G A(x) \Bigg[& |\nabla u|^{\alpha+1} + \alpha \left| \frac{u}{v} \nabla v \right|^{\alpha+1} \\ & -(\alpha+1) \left| \frac{u}{v} \nabla v \right|^{\alpha-1} (\nabla u) \cdot \left(\frac{u}{v} \nabla v \right) \Bigg] dx \\ \geq 0 & \end{aligned}$$

and therefore

$$\int_G A(x) \left[|\nabla u|^{\alpha+1} + \alpha \left| \frac{u}{v} \nabla v \right|^{\alpha+1} - (\alpha+1) \left| \frac{u}{v} \nabla v \right|^{\alpha-1} (\nabla u) \cdot \left(\frac{u}{v} \nabla v \right) \right] dx = 0.$$

From Lemma 8.1.1 we see that

$$\nabla u \equiv \frac{u}{v} \nabla v \quad \text{or} \quad v \nabla \left(\frac{u}{v} \right) \equiv 0 \quad \text{in } G.$$

Therefore, there exists a constant k_0 such that $u/v = k_0$ in G and hence on $\overline{G}$ by continuity. Since $u = 0$ on ∂G, we find that $k_0 = 0$, which contradicts the hypothesis that u is nontrivial. The proof is complete. □

Theorem 8.1.3. *Let $\partial G \in C^1$. If there exists a nontrivial function $u \in C^1(\overline{G}; \mathbb{R})$ such that $u = 0$ on ∂G and*

$$M[u] := \int_G \left[A(x) |\nabla u|^{\alpha+1} - C(x) |u|^{\alpha+1} \right] dx \leq 0, \quad (8.1.11)$$

then every solution $v \in \mathcal{D}_{P_\alpha}(G)$ of $P_\alpha[v] = 0$ must vanish at some point of G unless v is a constant multiple of u.

Proof. Suppose that there exists a solution $v \in \mathcal{D}_{P_\alpha}(G)$ of $P_\alpha[v] = 0$ satisfying $v \neq 0$ in G. Since $\partial G \in C^1$, $u \in C^1(\overline{G}; \mathbb{R})$ and $u = 0$ on ∂G, it is seen that u belongs to the Sobolev space $W_0^{1,\alpha+1}(G)$ which is the closure in the norm

$$\|w\| := \left(\int_G \left[|w|^{\alpha+1} + |\nabla w|^{\alpha+1} \right] dx \right)^{\frac{1}{\alpha+1}} \quad (8.1.12)$$

of the class $C_0^\infty(G)$ of infinitely differentiable functions with compact supports in G (*cf.* Adams and Fournier [1, Theorem 5.37], Evans [77, Theorem 2 of Section 5.5] for $\alpha > 0$, and Agmon [8, Lemma 9.10] for $\alpha = 1$). Let $\{u_k\}$ be a sequence of functions in $C_0^\infty(G)$ converging to u in the norm (8.1.12). Integrating (8.1.9) with $u = u_k$ over G and then applying the divergence theorem, we observe that

$$\begin{aligned} M[u_k] = \int_G A(x) \bigg[&|\nabla u_k|^{\alpha+1} + \alpha \left| \frac{u_k}{v} \nabla v \right|^{\alpha+1} \\ &- (\alpha+1) \left| \frac{u_k}{v} \nabla v \right|^{\alpha-1} (\nabla u_k) \cdot \left(\frac{u_k}{v} \nabla v \right) \bigg] dx \\ \geq 0. \end{aligned} \tag{8.1.13}$$

We first claim that $\lim_{k\to\infty} M[u_k] = M[u] = 0$. Since $A(x)$ and $C(x)$ are bounded on $\overline{G}$, there exists a constant $K_1 > 0$ such that

$$\begin{aligned} \big|M[u_k] - M[u]\big| \leq K_1 \int_G &\big| |\nabla u_k|^{\alpha+1} - |\nabla u|^{\alpha+1} \big| \, dx \\ &+ K_1 \int_G \big| |u_k|^{\alpha+1} - |u|^{\alpha+1} \big| \, dx. \end{aligned} \tag{8.1.14}$$

It follows from the mean value theorem that

$$\big| |\nabla u_k|^{\alpha+1} - |\nabla u|^{\alpha+1} \big| \leq (\alpha+1) \left(|\nabla u_k| + |\nabla u| \right)^\alpha |\nabla (u_k - u)|. \tag{8.1.15}$$

Using (8.1.15) and applying Hölder's inequality, we find that

$$\begin{aligned} &\int_G \big| |\nabla u_k|^{\alpha+1} - |\nabla u|^{\alpha+1} \big| \, dx \\ &\leq (\alpha+1) \left(\int_G \left(|\nabla u_k| + |\nabla u| \right)^{\alpha+1} dx \right)^{\frac{\alpha}{\alpha+1}} \left(\int_G |\nabla (u_k - u)|^{\alpha+1} dx \right)^{\frac{1}{\alpha+1}} \\ &\leq (\alpha+1) \big(\|u_k\| + \|u\| \big)^\alpha \|u_k - u\|. \end{aligned} \tag{8.1.16}$$

Similarly we obtain

$$\int_G \big| |u_k|^{\alpha+1} - |u|^{\alpha+1} \big| \, dx \leq (\alpha+1) \big(\|u_k\| + \|u\| \big)^\alpha \|u_k - u\|. \tag{8.1.17}$$

Combining (8.1.14), (8.1.16) and (8.1.17), we have

$$\big|M[u_k] - M[u]\big| \leq K_2 \big(\|u_k\| + \|u\| \big)^\alpha \|u_k - u\|$$

for some positive constant K_2 depending only on K_1 and α, from which it follows that $\lim_{k\to\infty} M[u_k] = M[u]$. We see from (8.1.13) that $M[u] \geq 0$, which together with (8.1.11) implies $M[u] = 0$.

Let B be an arbitrary ball with $\overline{B} \subset G$ and define

$$Q_B[w] = \int_B A(x) \left[|\nabla w|^{\alpha+1} + \alpha \left| \frac{w}{v} \nabla v \right|^{\alpha+1} - (\alpha+1) \left| \frac{w}{v} \nabla v \right|^{\alpha-1} (\nabla w) \cdot \left(\frac{w}{v} \nabla v \right) \right] dx \quad (8.1.18)$$

for $w \in C^1(G; \mathbb{R})$. It is easily verified that

$$0 \le Q_B[u_k] \le Q_G[u_k] = M[u_k], \quad (8.1.19)$$

where $Q_G[u_k]$ denotes the right hand side of (8.1.18) with $w = u_k$ and with B replaced by G. It is easily seen that $\lim_{k\to\infty} Q_B[u_k] = Q_B[u]$, and letting $k \to \infty$ in (8.1.19), we find that $Q_B[u] = 0$. Since $A(x) > 0$ in B, it follows that

$$|\nabla u|^{\alpha+1} + \alpha \left| \frac{u}{v} \nabla v \right|^{\alpha+1} - (\alpha+1) \left| \frac{u}{v} \nabla v \right|^{\alpha-1} (\nabla u) \cdot \left(\frac{u}{v} \nabla v \right) \equiv 0 \quad \text{in } B$$

from which Lemma 8.1.1 implies that

$$\nabla u \equiv \frac{u}{v} \nabla v \quad \text{or} \quad v \nabla \left(\frac{u}{v} \right) \equiv 0 \quad \text{in } B.$$

Hence, we observe that $u/v = k_0$ in B for some constant k_0. Since B is an arbitrary ball with $\overline{B} \subset G$, we conclude that $u/v = k_0$ in G, where $k_0 \neq 0$ in view of the hypothesis that u is nontrivial, and therefore v is a constant multiple of u in G. The proof is complete. □

Corollary 8.1.1 (Sturmian comparison theorem). *Let $\partial G \in C^1$. If there exists a nontrivial solution $u \in \mathcal{D}_{p_\alpha}(G)$ of $p_\alpha[u] = 0$ for which $u = 0$ on ∂G and* (8.1.10) *holds, then every solution $v \in \mathcal{D}_{P_\alpha}(G)$ of $P_\alpha[v] = 0$ must vanish at some point of G unless v is a constant multiple of u.*

Proof. The hypothesis $V[u] \ge 0$ means that

$$\begin{aligned} M[u] &\le \int_G \left[a(x) |\nabla u|^{\alpha+1} - c(x) |u|^{\alpha+1} \right] dx \\ &= \int_G \left[\nabla \cdot \left(u\, a(x) |\nabla u|^{\alpha-1} \nabla u \right) - u p_\alpha[u] \right] dx \\ &= 0. \end{aligned}$$

The conclusion follows from Theorem 8.1.3. □

Remark 8.1.1. In case $M[u] < 0$ in Theorem 8.1.3 [or $V[u] > 0$ in Corollary 8.1.1], then every solution $v \in \mathcal{D}_{P_\alpha}(G)$ of $P_\alpha[v] = 0$ must vanish at some point of G.

Now we investigate oscillation properties of solutions of

$$P_\alpha[v] = \nabla \cdot \left(A(x)|\nabla v|^{\alpha-1}\nabla v\right) + C(x)|v|^{\alpha-1}v = 0 \tag{8.1.20}$$

in Ω, where $\alpha > 0$ and Ω is an exterior domain in $\mathbb{R}^n$, that is, $\Omega \supset \{x \in \mathbb{R}^n;\ |x| \geq r_0\}$ for some $r_0 > 0$. It is assumed that $A(x) \in C(\Omega; (0,\infty))$ and $C(x) \in C(\Omega; \mathbb{R})$.

The domain $\mathcal{D}_{P_\alpha}(\Omega)$ of P_α is defined to be the set of all functions v of class $C^1(\Omega;\mathbb{R})$ with the property that $A(x)|\nabla v|^{\alpha-1}\nabla v \in C^1(\Omega;\mathbb{R}^n)$.

A solution $v \in \mathcal{D}_{P_\alpha}(\Omega)$ of (8.1.20) is said to be *oscillatory* in Ω if it has a zero in Ω_r for any $r > 0$, where

$$\Omega_r = \Omega \cap \{x \in \mathbb{R}^n;\ |x| > r\}$$

(*cf.* Definition 1.2.1 of Section 1.2).

Let $\bar{A}(r)$ and $\bar{C}(r)$ denote the spherical means of $A(x)$ and $C(x)$ over the sphere $S_r = \{x \in \mathbb{R}^n;\ |x| = r\}$, respectively, that is,

$$\bar{A}(r) = \frac{1}{\omega_n r^{n-1}} \int_{S_r} A(x)\, dS = \frac{1}{\omega_n} \int_{S_1} A(r,\theta)\, d\omega,$$

$$\bar{C}(r) = \frac{1}{\omega_n r^{n-1}} \int_{S_r} C(x)\, dS = \frac{1}{\omega_n} \int_{S_1} C(r,\theta)\, d\omega,$$

where ω_n is the surface area of the unit sphere S_1, (r,θ) is the hyperspherical coordinates in $\mathbb{R}^n$ and ω is the measure on S_1.

We note that if $y(r)$ is a solution of the half-linear ordinary differential equation

$$\left(r^{n-1}\bar{A}(r)|y'|^{\alpha-1}y'\right)' + r^{n-1}\bar{C}(r)|y|^{\alpha-1}y = 0, \tag{8.1.21}$$

then $u(x) = y(|x|)$ is a radially symmetric solution of

$$\nabla \cdot \left(\bar{A}(|x|)|\nabla u|^{\alpha-1}\nabla u\right) + \bar{C}(|x|)|u|^{\alpha-1}u = 0. \tag{8.1.22}$$

Theorem 8.1.4. *If the half-linear ordinary differential equation* (8.1.21) *is oscillatory at* $r = \infty$, *then every solution* $v \in \mathcal{D}_{P_\alpha}(\Omega)$ *of the half-linear elliptic equation* (8.1.20) *is oscillatory in* Ω.

Proof. Let $y(r)$ be an oscillatory solution of (8.1.21) and let $\{r_k\}_{k=1}^{\infty}$ be the sequence of its zeros such that $r_0 \leq r_1 < r_2 < \cdots, \lim_{k\to\infty} r_k = \infty$. Letting

$$G_k = \{x \in \mathbb{R}^n;\ r_k < |x| < r_{k+1}\} \quad (k = 1, 2, \ldots)$$

and $u(x) = y(|x|)$, we find that

$$\begin{aligned} V_{G_k}[u] &= \int_{G_k} \left[\left(\bar{A}(|x|) - A(x) \right) |\nabla u|^{\alpha+1} + \left(C(x) - \bar{C}(|x|) \right) |u|^{\alpha+1} \right] dx \\ &= \int_{r_k}^{r_{k+1}} \int_{S_1} \Big[\left(\bar{A}(r) - A(r,\theta) \right) |y'(r)|^{\alpha+1} \\ &\qquad + \left(C(r,\theta) - \bar{C}(r) \right) |y(r)|^{\alpha+1} \Big] r^{n-1} dr d\omega \\ &= \omega_n \int_{r_k}^{r_{k+1}} \Bigg[\left(\bar{A}(r) - \frac{1}{\omega_n} \int_{S_1} A(r,\theta)\, d\omega \right) |y'(r)|^{\alpha+1} \\ &\qquad + \left(\frac{1}{\omega_n} \int_{S_1} C(r,\theta)\, d\omega - \bar{C}(r) \right) |y(r)|^{\alpha+1} \Bigg] r^{n-1} dr \\ &= 0. \end{aligned}$$

Since there is a nontrivial solution $u(x)$ of (8.1.22) in G_k such that $u(x) = 0$ on ∂G_k and $V_{G_k}[u] = 0$, Theorem 8.1.2 implies that every solution v of (8.1.20) has a zero on $\overline{G_k}$ $(k = 1, 2, \ldots)$ and so is oscillatory in Ω. This completes the proof. □

Oscillation results for the half-linear ordinary differential equation

$$\left(p(r)|y'|^{\alpha-1}y' \right)' + q(r)|y|^{\alpha-1}y = 0, \quad r \geq r_0 \tag{8.1.23}$$

are listed in the following lemma (*cf.* Kusano and Naito [153], Kusano, Naito and Ogata [154]).

Lemma 8.1.2. (i) *Suppose that*

$$\int_{r_0}^{\infty} \left(\frac{1}{p(r)} \right)^{1/\alpha} dr = \infty$$

holds. Equation (8.1.23) *is oscillatory at* $r = \infty$ *if* $q(r)$ *is not integrable on* $[r_0, \infty)$, *that is,*

$$\int_{r_0}^{\infty} q(r)\, dr = \infty$$

or if $q(r)$ *is integrable on* $[r_0, \infty)$ *and*

$$\liminf_{r \to \infty} \left(P(r) \right)^{\alpha} \int_r^{\infty} q(s)\, ds > \frac{\alpha^{\alpha}}{(\alpha+1)^{\alpha+1}},$$

where

$$P(r) = \int_{r_0}^{r} \left(\frac{1}{p(s)} \right)^{1/\alpha} ds.$$

(ii) *Suppose that*

$$\int_{r_0}^{\infty} \left(\frac{1}{p(r)} \right)^{1/\alpha} dr < \infty$$

holds. Equation (8.1.23) *is oscillatory at* $r = \infty$ *if* $(\pi(r))^{\alpha+1} q(r)$ *is not integrable on* $[r_0, \infty)$, *that is,*

$$\int_{r_0}^{\infty} (\pi(r))^{\alpha+1} q(r)\, dr = \infty$$

or if $(\pi(r))^{\alpha+1} q(t)$ *is integrable on* $[r_0, \infty)$ *and*

$$\liminf_{r\to\infty} \frac{1}{\pi(r)} \int_r^{\infty} (\pi(s))^{\alpha+1} q(s)\, ds > \left(\frac{\alpha}{\alpha+1} \right)^{\alpha+1},$$

where

$$\pi(r) = \int_r^{\infty} \left(\frac{1}{p(s)} \right)^{1/\alpha} ds.$$

Combining Theorem 8.1.4 with Lemma 8.1.2 applied to (8.1.21) yields the following oscillation results for (8.1.20).

Theorem 8.1.5. *Let* $\bar{A}(r),\ \bar{C}(r) \in C([r_0, \infty); (0, \infty))$. *Suppose that* $\bar{A}(r)$ *satisfies*

$$\int_{r_0}^{\infty} \left(\frac{1}{r^{n-1} \bar{A}(r)} \right)^{1/\alpha} dr = \infty$$

and $\bar{C}(r)$ *satisfies either*

$$\int_{r_0}^{\infty} r^{n-1} \bar{C}(r)\, dr = \infty$$

or

$$\liminf_{r\to\infty} (\tilde{P}(r))^{\alpha} \int_r^{\infty} s^{n-1} \bar{C}(s)\, ds > \frac{\alpha^{\alpha}}{(\alpha+1)^{\alpha+1}},$$

where $\tilde{P}(r)$ *is defined by*

$$\tilde{P}(r) = \int_{r_0}^{r} \left(\frac{1}{s^{n-1} \bar{A}(s)} \right)^{1/\alpha} ds.$$

Then every solution $v \in \mathcal{D}_{P_\alpha}(\Omega)$ *of* (8.1.20) *is oscillatory in* Ω.

Theorem 8.1.6. *Let $\bar{A}(r)$, $\bar{C}(r) \in C([r_0,\infty);(0,\infty))$. Suppose that $\bar{A}(r)$ satisfies*

$$\int_{r_0}^{\infty} \left(\frac{1}{r^{n-1}\bar{A}(r)} \right)^{1/\alpha} dr < \infty$$

and $\bar{C}(r)$ satisfies either

$$\int_{r_0}^{\infty} (\tilde{\pi}(r))^{\alpha+1} r^{n-1}\bar{C}(r)\, dr = \infty$$

or

$$\liminf_{r\to\infty} \frac{1}{\tilde{\pi}(r)} \int_r^{\infty} (\tilde{\pi}(s))^{\alpha+1} s^{n-1}\bar{C}(s)\, ds > \left(\frac{\alpha}{\alpha+1} \right)^{\alpha+1},$$

where $\tilde{\pi}(r)$ is defined by

$$\tilde{\pi}(r) = \int_r^{\infty} \left(\frac{1}{s^{n-1}\bar{A}(s)} \right)^{1/\alpha} ds.$$

Then every solution $v \in \mathcal{D}_{P_\alpha}(\Omega)$ of (8.1.20) is oscillatory in Ω.

Example 8.1.1. We consider the half-linear elliptic equation

$$\nabla \cdot (A_0|\nabla v|^{\alpha-1}\nabla v) + C_0|v|^{\alpha-1}v = 0 \tag{8.1.24}$$

in $E = \{x \in \mathbb{R}^n;\ |x| \geq r_0\}$ $(r_0 > 0)$, where $\alpha > 0$, $A_0 > 0$ and $C_0 > 0$ are constants. The differential equation for radially symmetric solutions for (8.1.24) is the following

$$(r^{n-1}A_0|y'|^{\alpha-1}y')' + r^{n-1}C_0|y|^{\alpha-1}y = 0, \quad r \geq r_0 \tag{8.1.25}$$

which is a special case of (8.1.21) with $\bar{A}(r) = A_0$ and $\bar{C}(r) = C_0$.

(i) Suppose that $n \leq \alpha + 1$. Then we obtain

$$\int_{r_0}^{\infty} \left(\frac{1}{r^{n-1}\bar{A}(r)} \right)^{1/\alpha} dr = \int_{r_0}^{\infty} A_0^{-1/\alpha} r^{-(n-1)/\alpha}\, dr = \infty.$$

Since

$$\int_{r_0}^{\infty} r^{n-1}\bar{C}(r)\, dr = C_0 \int_{r_0}^{\infty} r^{n-1} dr = \infty,$$

Eq. (8.1.25) is oscillatory at $r = \infty$ by Lemma 8.1.2, and hence every solution $v \in \mathcal{D}_{P_\alpha}(E)$ of (8.1.24) is oscillatory in E by Theorem 8.1.4.

(ii) Suppose that $n > \alpha + 1$. Then we obtain

$$\int_{r_0}^{\infty} \left(\frac{1}{r^{n-1}\bar{A}(r)} \right)^{1/\alpha} dr = \int_{r_0}^{\infty} A_0^{-1/\alpha} r^{-(n-1)/\alpha}\, dr < \infty.$$

It is easy to see that

$$\tilde{\pi}(r) = \int_r^\infty \left(\frac{1}{s^{n-1}\bar{A}(s)} \right)^{1/\alpha} ds = \int_r^\infty \left(\frac{1}{s^{n-1}A_0} \right)^{1/\alpha} ds$$
$$= A_0^{-1/\alpha} \frac{\alpha}{n-\alpha-1} r^{-(n-\alpha-1)/\alpha}$$

and therefore

$$\left(\tilde{\pi}(r)\right)^{\alpha+1} r^{n-1}\bar{C}(r) = A_0^{-(\alpha+1)/\alpha} C_0 \left(\frac{\alpha}{n-\alpha-1} \right)^{\alpha+1} r^{n-1-(\alpha+1)(n-\alpha-1)/\alpha}.$$

If $n \le (\alpha+1)^2$, then

$$\int_{r_0}^\infty \left(\tilde{\pi}(r)\right)^{\alpha+1} r^{n-1}\bar{C}(r)\, dr = \infty,$$

and if $n > (\alpha+1)^2$, then

$$\frac{1}{\tilde{\pi}(r)} \int_r^\infty \left(\tilde{\pi}(s)\right)^{\alpha+1} s^{n-1}\bar{C}(s)\, ds = \frac{C_0}{A_0} \frac{1}{n-(\alpha+1)^2} \left(\frac{\alpha}{n-\alpha-1} \right)^\alpha r^{\alpha+1}$$

which tends to ∞ as $r \to \infty$. From Lemma 8.1.2 it follows that (8.1.25) is oscillatory at $r = \infty$, and hence every solution $v \in \mathcal{D}_{P_\alpha}(E)$ of (8.1.24) is oscillatory in E by Theorem 8.1.4.

We conclude that every solution $v \in \mathcal{D}_{P_\alpha}(E)$ of (8.1.24) is oscillatory in E for any natural number $n \in \mathbb{N}$, any positive constants $A_0 > 0$, $C_0 > 0$.

Remark 8.1.2. In (8.1.20) we assume that there exist the constants $A_0 > 0$ and $C_0 > 0$ such that $0 < A(x) \le A_0$ and $C(x) \ge C_0$ in Ω. From the above example we see that (8.1.25) is oscillatory at $r = \infty$, and hence there is an oscillatory solution y of (8.1.25). It is easily checked that $u(x) = y(|x|)$ is a radially symmetric solution of the equation

$$\nabla \cdot \left(A_0 |\nabla u|^{\alpha-1} \nabla u\right) + C_0 |u|^{\alpha-1} u = 0 \qquad (8.1.26)$$

and therefore Eq. (8.1.26) has nodal domains of the form

$$G_k = \{x \in \mathbb{R}^n;\ r_k < |x| < r_{k+1}\} \quad (k = 1, 2, ...).$$

Comparing (8.1.20) with (8.1.26) and using Theorem 8.1.2, we observe that every solution $v \in \mathcal{D}_{P_\alpha}(\Omega)$ of (8.1.20) is oscillatory in Ω.

8.2 Half-Linear Elliptic Equations II

In 2002 Došlý [68] established a Picone identity for the half-linear elliptic equation

$$\sum_{i=1}^{n} \frac{\partial}{\partial x_i}\left(\left|\frac{\partial u}{\partial x_i}\right|^{\alpha-1}\frac{\partial u}{\partial x_i}\right) + c(x)|u|^{\alpha-1}u = 0$$

and derived oscillation results. More general elliptic equations of the form

$$\sum_{i=1}^{n} \frac{\partial}{\partial x_i}\left(a_i(x)\left|\frac{\partial u}{\partial x_i}\right|^{\alpha-1}\frac{\partial u}{\partial x_i}\right) + c(x)|u|^{\alpha-1}u = 0$$

were studied by Bognár and Došlý [32].

Let p_α and P_α be the half-linear elliptic operators defined by

$$p_\alpha[u] = \sum_{i=1}^{n} \frac{\partial}{\partial x_i}\left(a_i(x)\left|\frac{\partial u}{\partial x_i}\right|^{\alpha-1}\frac{\partial u}{\partial x_i}\right) + c(x)|u|^{\alpha-1}u, \tag{8.2.1}$$

$$P_\alpha[v] = \sum_{i=1}^{n} \frac{\partial}{\partial x_i}\left(A_i(x)\left|\frac{\partial v}{\partial x_i}\right|^{\alpha-1}\frac{\partial v}{\partial x_i}\right) + C(x)|v|^{\alpha-1}v, \tag{8.2.2}$$

where $\alpha > 0$ is a constant. We establish a Picone identity for p_α and P_α, and obtain a Sturmian comparison theorem and oscillation theorems.

Let G be a bounded domain in $\mathbb{R}^n$ with piecewise smooth boundary ∂G, and assume that $a_i(x) \in C(\overline{G};(0,\infty))$ $(i = 1,2,...,n)$, $c(x) \in C(\overline{G};\mathbb{R})$, $A_i(x) \in C(\overline{G};(0,\infty))$ $(i = 1,2,...,n)$ and $C(x) \in C(\overline{G};\mathbb{R})$.

The domain $\mathcal{D}_{p_\alpha}(G)$ of p_α is defined to be the set of all functions u of class $C^1(\overline{G};\mathbb{R})$ with the property that $a_i(x)\left|\frac{\partial u}{\partial x_i}\right|^{\alpha-1}\frac{\partial u}{\partial x_i} \in C^1(G;\mathbb{R}) \cap C(\overline{G};\mathbb{R})$ $(i = 1,2,...,n)$. The domain $\mathcal{D}_{P_\alpha}(G)$ of P_α is defined similarly.

Theorem 8.2.1 (Picone identity). *If $u \in \mathcal{D}_{p_\alpha}(G)$, $v \in \mathcal{D}_{P_\alpha}(G)$ and $v \neq 0$ in G, then we obtain the following Picone identity:*

$$\sum_{i=1}^{n} \frac{\partial}{\partial x_i}\left(\frac{u}{\varphi(v)}\left[\varphi(v)a_i(x)\left|\frac{\partial u}{\partial x_i}\right|^{\alpha-1}\frac{\partial u}{\partial x_i} - \varphi(u)A_i(x)\left|\frac{\partial v}{\partial x_i}\right|^{\alpha-1}\frac{\partial v}{\partial x_i}\right]\right)$$
$$= \sum_{i=1}^{n}(a_i(x) - A_i(x))\left|\frac{\partial u}{\partial x_i}\right|^{\alpha+1} + (C(x) - c(x))|u|^{\alpha+1}$$
$$+ \sum_{i=1}^{n} A_i(x)\left[\left|\frac{\partial u}{\partial x_i}\right|^{\alpha+1} + \alpha\left|\frac{u}{v}\frac{\partial v}{\partial x_i}\right|^{\alpha+1} - (\alpha+1)\frac{\partial u}{\partial x_i}\left|\frac{u}{v}\frac{\partial v}{\partial x_i}\right|^{\alpha-1}\frac{u}{v}\frac{\partial v}{\partial x_i}\right]$$
$$+ \frac{u}{\varphi(v)}\left(\varphi(v)p_\alpha[u] - \varphi(u)P_\alpha[v]\right), \tag{8.2.3}$$

where $\varphi(s) = |s|^{\alpha-1}s$ $(s \in \mathbb{R})$.

Proof. It is easy to see that

$$\begin{aligned} up_\alpha[u] &= u\sum_{i=1}^{n}\frac{\partial}{\partial x_i}\left(a_i(x)\left|\frac{\partial u}{\partial x_i}\right|^{\alpha-1}\frac{\partial u}{\partial x_i}\right)+c(x)|u|^{\alpha+1} \\ &= \sum_{i=1}^{n}\frac{\partial}{\partial x_i}\left(u\,a_i(x)\left|\frac{\partial u}{\partial x_i}\right|^{\alpha-1}\frac{\partial u}{\partial x_i}\right)-\sum_{i=1}^{n}a_i(x)\left|\frac{\partial u}{\partial x_i}\right|^{\alpha+1}+c(x)|u|^{\alpha+1} \end{aligned}$$

from which we have

$$\begin{aligned} &\sum_{i=1}^{n}\frac{\partial}{\partial x_i}\left(\frac{u}{\varphi(v)}\left[\varphi(v)a_i(x)\left|\frac{\partial u}{\partial x_i}\right|^{\alpha-1}\frac{\partial u}{\partial x_i}\right]\right) \\ &\quad= \sum_{i=1}^{n}a_i(x)\left|\frac{\partial u}{\partial x_i}\right|^{\alpha+1}-c(x)|u|^{\alpha+1}+\frac{u}{\varphi(v)}(\varphi(v)p_\alpha[u]) \qquad (8.2.4) \end{aligned}$$

in light of the fact that $\varphi(v) \neq 0$. We observe that the following identity holds:

$$\begin{aligned} &-\sum_{i=1}^{n}\frac{\partial}{\partial x_i}\left(u\varphi(u)\frac{A_i(x)\left|\frac{\partial v}{\partial x_i}\right|^{\alpha-1}\frac{\partial v}{\partial x_i}}{\varphi(v)}\right) \\ &= -\sum_{i=1}^{n}A_i(x)\left|\frac{\partial u}{\partial x_i}\right|^{\alpha+1}+C(x)|u|^{\alpha+1} \\ &\quad+\sum_{i=1}^{n}A_i(x)\left[\left|\frac{\partial u}{\partial x_i}\right|^{\alpha+1}+\alpha\left|\frac{u}{v}\frac{\partial v}{\partial x_i}\right|^{\alpha+1}-(\alpha+1)\frac{\partial u}{\partial x_i}\left|\frac{u}{v}\frac{\partial v}{\partial x_i}\right|^{\alpha-1}\frac{u}{v}\frac{\partial v}{\partial x_i}\right] \\ &\quad-\frac{u\varphi(u)}{\varphi(v)}P_\alpha[v] \qquad (8.2.5) \end{aligned}$$

(see Bognár and Došlý [32, the proof of Theorem 1]). Combining (8.2.4) with (8.2.5) yields the desired Picone identity (8.2.3). □

Theorem 8.2.2 (Sturmian comparison theorem). *If there exists a nontrivial solution $u \in \mathcal{D}_{p_\alpha}(G)$ of $p_\alpha[u]=0$ such that $u=0$ on ∂G and*

$$V[u] := \int_G\left[\sum_{i=1}^{n}(a_i(x)-A_i(x))\left|\frac{\partial u}{\partial x_i}\right|^{\alpha+1}+(C(x)-c(x))|u|^{\alpha+1}\right]dx \geq 0, \qquad (8.2.6)$$

then every solution $v \in \mathcal{D}_{P_\alpha}(G)$ of $P_\alpha[v]=0$ must vanish at some point of $\overline{G}$.

Proof. Suppose that there exists a solution $v \in \mathcal{D}_{P_\alpha}(G)$ of $P_\alpha[v] = 0$ satisfying $v \neq 0$ on $\overline{G}$. Integrating (8.2.3) over G and then applying the divergence theorem, we have

$$
\begin{aligned}
0 = V[u] + \int_G \sum_{i=1}^n A_i(x) \Bigg[\left| \frac{\partial u}{\partial x_i} \right|^{\alpha+1} &+ \alpha \left| \frac{u}{v} \frac{\partial v}{\partial x_i} \right|^{\alpha+1} \\
&- (\alpha+1) \frac{\partial u}{\partial x_i} \left| \frac{u}{v} \frac{\partial v}{\partial x_i} \right|^{\alpha-1} \frac{u}{v} \frac{\partial v}{\partial x_i} \Bigg] dx \\
\geq 0
\end{aligned}
$$

and therefore

$$\int_G \sum_{i=1}^n A_i(x) \left[\left| \frac{\partial u}{\partial x_i} \right|^{\alpha+1} + \alpha \left| \frac{u}{v} \frac{\partial v}{\partial x_i} \right|^{\alpha+1} - (\alpha+1) \frac{\partial u}{\partial x_i} \left| \frac{u}{v} \frac{\partial v}{\partial x_i} \right|^{\alpha-1} \frac{u}{v} \frac{\partial v}{\partial x_i} \right] dx = 0.$$

It follows from Lemma 8.1.1 that

$$\frac{\partial u}{\partial x_i} \equiv \frac{u}{v} \frac{\partial v}{\partial x_i} \quad \text{or} \quad v \frac{\partial}{\partial x_i} \left(\frac{u}{v} \right) \equiv 0 \quad \text{in } G \ (i = 1, 2, ..., n).$$

Consequently, there exists a constant k_0 such that $u/v = k_0$ in G and hence on $\overline{G}$ by continuity. Since u is nontrivial, we see that $k_0 \neq 0$, which contradicts the fact that $u = 0$ on ∂G. □

Theorem 8.2.3. *Let $\partial G \in C^1$. If there exists a nontrivial function $u \in C^1(\overline{G}; \mathbb{R})$ such that $u = 0$ on ∂G and*

$$M[u] := \int_G \left[\sum_{i=1}^n A_i(x) \left| \frac{\partial u}{\partial x_i} \right|^{\alpha+1} - C(x)|u|^{\alpha+1} \right] dx \leq 0, \tag{8.2.7}$$

then every solution $v \in \mathcal{D}_{P_\alpha}(G)$ of $P_\alpha[v] = 0$ must vanish at some point of G unless v is a constant multiple of u.

Proof. Suppose to the contrary that there is a solution $v \in \mathcal{D}_{P_\alpha}(G)$ of $P_\alpha[v] = 0$ such that $v \neq 0$ in G. Since $\partial G \in C^1$, $u \in C^1(\overline{G}; \mathbb{R})$ and $u = 0$ on ∂G, we find that u belongs to the Sobolev space $W_0^{1,\alpha+1}(G)$ which is the closure in the norm

$$\|w\| := \left(\int_G \left[|w|^{\alpha+1} + \sum_{i=1}^n \left| \frac{\partial w}{\partial x_i} \right|^{\alpha+1} \right] dx \right)^{\frac{1}{\alpha+1}} \tag{8.2.8}$$

of the class $C_0^\infty(G)$ of infinitely differentiable functions with compact supports in G (*cf.* Adams and Fournier [1, Theorem 5.37], Evans [77, Theorem 2 of Section 5.5] for $\alpha > 0$, and Agmon [8, Lemma 9.10] for $\alpha = 1$). Let $\{u_k\}$ be a sequence of functions in $C_0^\infty(G)$ converging to u in the norm

(8.2.8). We integrate (8.2.5) with $u = u_k$ over G and apply the divergence theorem to obtain

$$M[u_k] = \int_G \sum_{i=1}^n A_i(x) \left[\left| \frac{\partial u_k}{\partial x_i} \right|^{\alpha+1} + \alpha \left| \frac{u_k}{v} \frac{\partial v}{\partial x_i} \right|^{\alpha+1} - (\alpha+1) \frac{\partial u_k}{\partial x_i} \left| \frac{u_k}{v} \frac{\partial v}{\partial x_i} \right|^{\alpha-1} \frac{u_k}{v} \frac{\partial v}{\partial x_i} \right] dx$$

$$\geq 0 \tag{8.2.9}$$

in view of Lemma 8.1.1 of Section 8.1. We first show that $\lim_{k\to\infty} M[u_k] = M[u] = 0$. Since $A_i(x)$ and $C(x)$ are bounded on $\overline{G}$, there exists a constant $K_1 > 0$ satisfying

$$\big|M[u_k] - M[u]\big| \leq K_1 \int_G \sum_{i=1}^n \left| \left| \frac{\partial u_k}{\partial x_i} \right|^{\alpha+1} - \left| \frac{\partial u}{\partial x_i} \right|^{\alpha+1} \right| dx + K_1 \int_G \big| |u_k|^{\alpha+1} - |u|^{\alpha+1} \big| \, dx. \tag{8.2.10}$$

The mean value theorem implies that

$$\left| \left| \frac{\partial u_k}{\partial x_i} \right|^{\alpha+1} - \left| \frac{\partial u}{\partial x_i} \right|^{\alpha+1} \right| \leq (\alpha+1) \left(\left| \frac{\partial u_k}{\partial x_i} \right| + \left| \frac{\partial u}{\partial x_i} \right| \right)^{\alpha} \left| \frac{\partial u_k}{\partial x_i} - \frac{\partial u}{\partial x_i} \right|. \tag{8.2.11}$$

Using Hölder's inequality, from (8.2.11) we see that

$$\begin{aligned} &\int_G \sum_{i=1}^n \left| \left| \frac{\partial u_k}{\partial x_i} \right|^{\alpha+1} - \left| \frac{\partial u}{\partial x_i} \right|^{\alpha+1} \right| dx \\ &\leq (\alpha+1) \sum_{i=1}^n \left(\int_G \left(\left| \frac{\partial u_k}{\partial x_i} \right| + \left| \frac{\partial u}{\partial x_i} \right| \right)^{\alpha+1} dx \right)^{\frac{\alpha}{\alpha+1}} \times \\ &\qquad \times \left(\int_G \left| \frac{\partial}{\partial x_i} \big(u_k - u \big) \right|^{\alpha+1} dx \right)^{\frac{1}{\alpha+1}} \\ &\leq n(\alpha+1) \big(\|u_k\| + \|u\| \big)^{\alpha} \|u_k - u\|. \end{aligned} \tag{8.2.12}$$

Analogously we have

$$\int_G \big| |u_k|^{\alpha+1} - |u|^{\alpha+1} \big| \, dx \leq (\alpha+1) \big(\|u_k\| + \|u\| \big)^{\alpha} \|u_k - u\|. \tag{8.2.13}$$

Combining (8.2.10), (8.2.12) and (8.2.13), we obtain

$$\big|M[u_k] - M[u]\big| \leq K_2 \big(\|u_k\| + \|u\| \big)^{\alpha} \|u_k - u\|$$

for some positive constant K_2 depending only on K_1, α and n, from which we see that $\lim_{k\to\infty} M[u_k] = M[u]$. It follows from (8.2.9) that $M[u] \geq 0$, which together with (8.2.7) implies $M[u] = 0$.

Let B be an arbitrary ball with $\overline{B} \subset G$ and define

$$Q_B[w] = \int_B \sum_{i=1}^n A_i(x) \left[\left| \frac{\partial w}{\partial x_i} \right|^{\alpha+1} + \alpha \left| \frac{w}{v} \frac{\partial v}{\partial x_i} \right|^{\alpha+1} - (\alpha+1) \frac{\partial w}{\partial x_i} \left| \frac{w}{v} \frac{\partial v}{\partial x_i} \right|^{\alpha-1} \frac{w}{v} \frac{\partial v}{\partial x_i} \right] dx \quad (8.2.14)$$

for $w \in C^1(G; \mathbb{R})$. We easily see that

$$0 \leq Q_B[u_k] \leq Q_G[u_k] = M[u_k], \qquad (8.2.15)$$

where $Q_G[u_k]$ denotes the right hand side of (8.2.14) with $w = u_k$ and with B replaced by G. It is easily checked that $\lim_{k\to\infty} Q_B[u_k] = Q_B[u]$, and letting $k \to \infty$ in (8.2.15), we find that $Q_B[u] = 0$. Since $A_i(x) > 0$ in B $(i = 1, 2, ..., n)$, we observe that

$$\left| \frac{\partial u}{\partial x_i} \right|^{\alpha+1} + \alpha \left| \frac{u}{v} \frac{\partial v}{\partial x_i} \right|^{\alpha+1} - (\alpha+1) \frac{\partial u}{\partial x_i} \left| \frac{u}{v} \frac{\partial v}{\partial x_i} \right|^{\alpha-1} \frac{u}{v} \frac{\partial v}{\partial x_i} \equiv 0 \text{ in } B \ (i=1, 2, ..., n).$$

It follows Lemma 8.1.1 implies that

$$\frac{\partial u}{\partial x_i} \equiv \frac{u}{v} \frac{\partial v}{\partial x_i} \quad \text{in } B \ (i = 1, 2, ..., n)$$

or

$$v \frac{\partial}{\partial x_i} \left(\frac{u}{v} \right) \equiv 0 \quad \text{in } B \ (i = 1, 2, ..., n).$$

Hence, we observe that $u/v = k_0$ in B for some constant k_0. Since B is an arbitrary ball with $\overline{B} \subset G$, we conclude that $u/v = k_0$ in G, where $k_0 \neq 0$ in view of the hypothesis that u is nontrivial, and therefore v is a constant multiple of u in G. The proof is complete. □

Corollary 8.2.1 (Sturmian comparison theorem). *Let $\partial G \in C^1$. If there exists a nontrivial solution $u \in \mathcal{D}_{p_\alpha}(G)$ of $p_\alpha[u] = 0$ for which $u = 0$ on ∂G and* (8.2.6) *holds, then every solution $v \in \mathcal{D}_{P_\alpha}(G)$ of $P_\alpha[v] = 0$ must vanish at some point of G unless v is a constant multiple of u.*

Proof. The hypothesis $V[u] \geq 0$ implies that

$$M[u] \leq \int_G \left[\sum_{i=1}^n a_i(x) \left| \frac{\partial u}{\partial x_i} \right|^{\alpha+1} - c(x)|u|^{\alpha+1} \right] dx = 0$$

in view of (8.2.4). The conclusion follows from Theorem 8.2.3. □

Remark 8.2.1. In case $M[u] < 0$ in Theorem 8.2.3 [or $V[u] > 0$ in Corollary 8.2.1], then every solution $v \in \mathcal{D}_{P_\alpha}(G)$ of $P_\alpha[v] = 0$ must vanish at some point of G.

Remark 8.2.2. In the case where $a_i(x) \geq A_i(x)$ in G $(i = 1, 2, ..., n)$ and $C(x) \geq c(x)$ in G, the condition (8.2.6) is satisfied.

Next we study the oscillation character of

$$P_\alpha[v] = \sum_{i=1}^{n} \frac{\partial}{\partial x_i}\left(A_i(x)\left|\frac{\partial v}{\partial x_i}\right|^{\alpha-1}\frac{\partial v}{\partial x_i}\right) + C(x)|v|^{\alpha-1}v = 0 \qquad (8.2.16)$$

in Ω, where $\alpha > 0$ and Ω is an exterior domain in $\mathbb{R}^n$, that is, $\Omega \supset \{x \in \mathbb{R}^n;\ |x| \geq r_0\}$ for some $r_0 > 0$. It is assumed that $A_i(x) \in C(\Omega; (0, \infty))$ $(i = 1, 2, ..., n)$ and $C(x) \in C(\Omega; \mathbb{R})$.

The domain $\mathcal{D}_{P_\alpha}(\Omega)$ of P_α is defined to be the set of all functions v of class $C^1(\Omega; \mathbb{R})$ with the property that $A_i(x)\left|\frac{\partial v}{\partial x_i}\right|^{\alpha-1}\frac{\partial v}{\partial x_i} \in C^1(\Omega; \mathbb{R})$ $(i = 1, 2, ..., n)$.

A solution $v \in \mathcal{D}_{P_\alpha}(\Omega)$ of (8.2.16) is said to be *oscillatory* in Ω if it has a zero in Ω_r for any $r > 0$, where

$$\Omega_r = \Omega \cap \{x \in \mathbb{R}^n;\ |x| > r\}.$$

Let

$$A_0(x) = n \max_{1 \leq i \leq n} A_i(x)$$

and let $\bar{A}_0(r)$ and $\bar{C}(r)$ denote the spherical means of $A_0(x)$ and $C(x)$, respectively (*cf.* Section 8.1).

Theorem 8.2.4. *If the half-linear ordinary differential equation*

$$(r^{n-1}\bar{A}_0(r)|y'|^{\alpha-1}y')' + r^{n-1}\bar{C}(r)|y|^{\alpha-1}y = 0 \qquad (8.2.17)$$

is oscillatory at $r = \infty$, *then every solution* $v \in \mathcal{D}_{P_\alpha}(\Omega)$ *of the half-linear elliptic equation* (8.2.16) *is oscillatory in* Ω.

Proof. Let $\{r_k\}_{k=1}^{\infty}$ be the sequence of the zeros of a nontrivial solution $y(r)$ of (8.2.17) such that $r_0 \leq r_1 < r_2 < \cdots$, $\lim_{k\to\infty} r_k = \infty$. Letting

$$G_k = \{x \in \mathbb{R}^n;\ r_k < |x| < r_{k+1}\} \quad (k = 1, 2, ...)$$

and $u(x) = y(|x|)$, we see that

$$\begin{aligned} M_{G_k}[u] &= \int_{G_k} \left[\sum_{i=1}^n A_i(x) \left| \frac{\partial u}{\partial x_i} \right|^{\alpha+1} - C(x)|u|^{\alpha+1} \right] dx \\ &\le \int_{G_k} \left[\max_{1\le i\le n} A_i(x) \sum_{i=1}^n \left| \frac{\partial u}{\partial x_i} \right|^{\alpha+1} - C(x)|u|^{\alpha+1} \right] dx \\ &\le \omega_n \int_{r_k}^{r_{k+1}} \left[\bar{A}_0(r)|y'(r)|^{\alpha+1} - \bar{C}(r)|y(r)|^{\alpha+1} \right] r^{n-1} dr \\ &= -\omega_n \int_{r_k}^{r_{k+1}} \Big[\left(r^{n-1}\bar{A}_0(r)|y'(r)|^{\alpha-1}y'(r) \right)' \\ &\qquad + r^{n-1}\bar{C}(r)|y(r)|^{\alpha-1}y(r) \Big] y(r)\, dr \\ &= 0, \end{aligned}$$

where ω_n denotes the surface area of the unit sphere S_1. It follows from Theorem 8.2.3 that v has a zero on each $\overline{G_k}$ $(k = 1, 2, ...)$, that is, v is oscillatory in Ω. The proof is complete. □

Corollary 8.2.2. *Assume that there exist the constants* $A_i > 0$ $(i = 1, 2, ..., n)$ *and* $C_0 > 0$ *such that*

$$A_i(x) \le A_i \ (i = 1, 2, ..., n), \qquad C(x) \ge C_0 \quad \text{in } \Omega.$$

Then every solution $v \in \mathcal{D}_{P_\alpha}(\Omega)$ *of* (8.2.16) *is oscillatory in* Ω.

Proof. Arguing as in the proof of Theorem 8.2.4, we observe that

$$\begin{aligned} M_{G_k}[u] &\le -\omega_n \int_{r_k}^{r_{k+1}} \left[\left(r^{n-1}A_0|y'|^{\alpha-1}y' \right)' + r^{n-1}C_0|y|^{\alpha-1}y \right] y\, dr \\ &= 0, \end{aligned}$$

where $A_0 = n \max_{1\le i\le n} A_i$, $\{r_k\}$ is the sequence of the zeros of a nontrivial solution y of

$$\left(r^{n-1}A_0|y'|^{\alpha-1}y' \right)' + r^{n-1}C_0|y|^{\alpha-1}y = 0 \tag{8.2.18}$$

satisfying $r_1 < r_2 < \cdots$, $\lim_{k\to\infty} r_k = \infty$, and $u = y(|x|)$. Here we used the fact that (8.2.18) is oscillatory at $r = \infty$ (see Example 8.1.1 of Section 8.1). Therefore, every solution $v \in \mathcal{D}_{P_\alpha}(\Omega)$ of (8.2.16) has a zero on each $\overline{G_k}$ $(k = 1, 2, ...)$, that is, v is oscillatory in Ω. This completes the proof. □

Remark 8.2.3. Applying Lemma 8.1.2 in Section 8.1 to (8.2.17), we can establish oscillation results, but we omit them (*cf.* Theorems 8.1.5 and 8.1.6 of Section 8.1).

8.3 Half-Linear Damped Elliptic Equations

In this section we consider the half-linear elliptic operator P_α defined by

$$P_\alpha[v] = \nabla \cdot \left(A(x)|\nabla v|^{\alpha-1}\nabla v\right) + (\alpha+1)|\nabla v|^{\alpha-1}B(x) \cdot \nabla v + C(x)|v|^{\alpha-1}v$$

which has the first order term $(\alpha+1)|\nabla v|^{\alpha-1}B(x) \cdot \nabla v$ called damping term. We note that in case $\alpha = 1$ the half-linear elliptic equation $P_\alpha[v] = 0$ reduces to a non-self-adjoint linear elliptic equation.

Let G be a bounded domain in $\mathbb{R}^n$ with piecewise smooth boundary ∂G, and assume that $\alpha > 0$ is a constant, $A(x) \in C(\overline{G}; (0, \infty))$, $B(x) \in C(\overline{G}; \mathbb{R}^n)$ and $C(x) \in C(\overline{G}; \mathbb{R})$.

The domain $\mathcal{D}_{P_\alpha}(G)$ of P_α is defined to be the set of all functions v of class $C^1(\overline{G}; \mathbb{R})$ with the property that $A(x)|\nabla v|^{\alpha-1}\nabla v \in C^1(G; \mathbb{R}^n) \cap C(\overline{G}; \mathbb{R}^n)$.

Theorem 8.3.1 (Picone identity). *If $v \in \mathcal{D}_{P_\alpha}(G)$, $v \neq 0$ in G, then the following Picone identity holds for any $u \in C^1(G; \mathbb{R})$:*

$$\begin{aligned}
&-\nabla \cdot \left(u\varphi(u)\frac{A(x)|\nabla v|^{\alpha-1}\nabla v}{\varphi(v)}\right) \\
&\quad = -A(x)\left|\nabla u - \frac{u}{A(x)}B(x)\right|^{\alpha+1} + C(x)|u|^{\alpha+1} \\
&\qquad + A(x)\left[\left|\nabla u - \frac{u}{A(x)}B(x)\right|^{\alpha+1} + \alpha\left|\frac{u}{v}\nabla v\right|^{\alpha+1}\right. \\
&\qquad\qquad \left. -(\alpha+1)\left(\nabla u - \frac{u}{A(x)}B(x)\right) \cdot \Phi\left(\frac{u}{v}\nabla v\right)\right] \\
&\qquad - \frac{u\varphi(u)}{\varphi(v)}P_\alpha[v],
\end{aligned} \tag{8.3.1}$$

where $\varphi(s) = |s|^{\alpha-1}s$ $(s \in \mathbb{R})$ and $\Phi(\xi) = |\xi|^{\alpha-1}\xi$ $(\xi \in \mathbb{R}^n)$.

Proof. It can be shown that the following identity holds:

$$\begin{aligned}
&-\nabla \cdot \left(u\varphi(u)\frac{A(x)|\nabla v|^{\alpha-1}\nabla v}{\varphi(v)}\right) \\
&\quad = C(x)|u|^{\alpha+1} \\
&\qquad + A(x)\left[\alpha\left|\frac{u}{v}\nabla v\right|^{\alpha+1} - (\alpha+1)(\nabla u) \cdot \Phi\left(\frac{u}{v}\nabla v\right)\right] \\
&\qquad - \frac{u\varphi(u)}{\varphi(v)}\left[\nabla \cdot \left(A(x)|\nabla v|^{\alpha-1}\nabla v\right) + C(x)|v|^{\alpha-1}v\right]
\end{aligned} \tag{8.3.2}$$

(see (8.1.9) of Section 8.1, or Kusano Jaroš and Yoshida [151, p.385]). Since

$$\frac{u\varphi(u)}{\varphi(v)}(\alpha+1)B(x)\cdot\Phi(\nabla v) = (\alpha+1)u\varphi\left(\frac{u}{v}\right)B(x)\cdot\Phi(\nabla v)$$
$$= (\alpha+1)uB(x)\cdot\Phi\left(\frac{u}{v}\nabla v\right),$$

the identity (8.3.2) can be written in the form

$$\begin{aligned}
&-\nabla\cdot\left(u\varphi(u)\frac{A(x)|\nabla v|^{\alpha-1}\nabla v}{\varphi(v)}\right)\\
&= -A(x)\left|\nabla u - \frac{u}{A(x)}B(x)\right|^{\alpha+1} + C(x)|u|^{\alpha+1}\\
&\quad + A(x)\left[\left|\nabla u - \frac{u}{A(x)}B(x)\right|^{\alpha+1} + \alpha\left|\frac{u}{v}\nabla v\right|^{\alpha+1}\right.\\
&\qquad\left. -(\alpha+1)\left(\nabla u - \frac{u}{A(x)}B(x)\right)\cdot\Phi\left(\frac{u}{v}\nabla v\right)\right]\\
&\quad - \frac{u\varphi(u)}{\varphi(v)}\left[\nabla\cdot\left(A(x)|\nabla v|^{\alpha-1}\nabla v\right) + (\alpha+1)B(x)\cdot\Phi(\nabla v) + C(x)|v|^{\alpha-1}v\right]
\end{aligned}$$

which is the desired Picone identity (8.3.1). □

Theorem 8.3.2. *Assume that there exists a nontrivial function* $u \in C^1(\overline{G};\mathbb{R})$ *such that* $u = 0$ *on* ∂G *and*

$$M_G[u] := \int_G \left[A(x)\left|\nabla u - \frac{u}{A(x)}B(x)\right|^{\alpha+1} - C(x)|u|^{\alpha+1}\right] dx \le 0. \quad (8.3.3)$$

Then every solution $v \in \mathcal{D}_{P_\alpha}(G)$ *of* $P_\alpha[v] = 0$ *must vanish at some point of* $\overline{G}$.

Proof. Suppose to the contrary that there exists a solution $v \in \mathcal{D}_{P_\alpha}(G)$ of $P_\alpha[v] = 0$ such that $v \neq 0$ on $\overline{G}$. It follows from Theorem 8.3.1 that the Picone identity (8.3.1) holds for the nontrivial function u, and integrating

(8.3.1) over G yields

$$\begin{aligned}0 = &-M_G[u]\\ &+\int_G A(x)\left[\left|\nabla u - \frac{u}{A(x)}B(x)\right|^{\alpha+1} + \alpha\left|\frac{u}{v}\nabla v\right|^{\alpha+1}\right.\\ &\qquad\left. -(\alpha+1)\left(\nabla u - \frac{u}{A(x)}B(x)\right)\cdot\Phi\left(\frac{u}{v}\nabla v\right)\right]dx\\ \geq &\int_G A(x)\left[\left|\nabla u - \frac{u}{A(x)}B(x)\right|^{\alpha+1} + \alpha\left|\frac{u}{v}\nabla v\right|^{\alpha+1}\right.\\ &\qquad\left. -(\alpha+1)\left(\nabla u - \frac{u}{A(x)}B(x)\right)\cdot\Phi\left(\frac{u}{v}\nabla v\right)\right]dx. \quad (8.3.4)\end{aligned}$$

It is easily checked that

$$\nabla u - \frac{u}{A(x)}B(x) - \frac{u}{v}\nabla v = v\nabla\left(\frac{u}{v}\right) - \frac{u}{A(x)}B(x) = v\left[\nabla\left(\frac{u}{v}\right) - \frac{B(x)}{A(x)}\frac{u}{v}\right].$$

In case

$$\nabla\left(\frac{u}{v}\right) - \frac{B(x)}{A(x)}\frac{u}{v} \equiv 0 \quad \text{in } G,$$

then it follows from a result of Jaroš, Kusano and Yoshida [118, Lemma] that

$$\frac{u}{v} = C_0\, e^{\alpha(x)} \quad \text{on } \overline{G}$$

for some constant C_0 and some continuous function $\alpha(x)$. Since $u = 0$ on ∂G, we see that $C_0 = 0$, which contradicts the fact that u is nontrivial. Hence we obtain

$$\nabla\left(\frac{u}{v}\right) - \frac{B(x)}{A(x)}\frac{u}{v} \not\equiv 0 \quad \text{in } G$$

and therefore

$$\nabla u - \frac{u}{A(x)}B(x) \not\equiv \frac{u}{v}\nabla v \quad \text{in } G.$$

We see from a result of Kusano, Jaroš and Yoshida [151, Lemma 2.1] that

$$\begin{aligned}&\int_G A(x)\left[\left|\nabla u - \frac{u}{A(x)}B(x)\right|^{\alpha+1} + \alpha\left|\frac{u}{v}\nabla v\right|^{\alpha+1}\right.\\ &\qquad\left. -(\alpha+1)\left(\nabla u - \frac{u}{A(x)}B(x)\right)\cdot\Phi\left(\frac{u}{v}\nabla v\right)\right]dx > 0\end{aligned}$$

which together with (8.3.4) yields a contradiction. □

Let Ω be an exterior domain in $\mathbb{R}^n$, that is, $\Omega \supset \{x \in \mathbb{R}^n;\ |x| \geq r_0\}$ for some $r_0 > 0$. We investigate oscillations of the half-linear elliptic equation

$$P_\alpha[v] = 0 \quad \text{in } \Omega, \tag{8.3.5}$$

where $\alpha > 0$ is a constant, $A(x) \in C(\Omega; (0, \infty))$, $B(x) \in C(\Omega; \mathbb{R}^n)$ and $C(x) \in C(\Omega; \mathbb{R})$.

The domain $\mathcal{D}_{P_\alpha}(\Omega)$ of P_α is defined to be the set of all functions v of class $C^1(\Omega; \mathbb{R})$ with the property that $A(x)|\nabla v|^{\alpha-1}\nabla v \in C^1(\Omega; \mathbb{R}^n)$.

A solution $v \in \mathcal{D}_{P_\alpha}(\Omega)$ of (8.3.5) is said to be *oscillatory* in Ω if it has a zero in Ω_r for any $r > 0$, where

$$\Omega_r = \Omega \cap \{x \in \mathbb{R}^n;\ |x| > r\}.$$

Theorem 8.3.3. *Assume that for any $r > 0$ there exists a bounded and piecewise smooth domain G with $\overline{G} \subset \Omega_r$. If there is a nontrivial function $u \in C^1(\overline{G}; \mathbb{R})$ such that $u = 0$ on ∂G and $M_G[u] \leq 0$, where M_G is defined by (8.3.3), then every solution $v \in \mathcal{D}_{P_\alpha}(\Omega)$ of (8.3.5) is oscillatory in Ω.*

Proof. Let $r > 0$ be an arbitrary number. It follows from Theorem 8.3.2 that every solution $v \in \mathcal{D}_{P_\alpha}(\Omega)$ of (8.3.5) has a zero on $\overline{G} \subset \Omega_r$, that is, every solution v of (8.3.5) is oscillatory in Ω. □

Lemma 8.3.1. *Let $0 < \alpha < 1$. Then we obtain the inequality*

$$|\nabla u - uW(x)|^{\alpha+1} \leq \frac{|\nabla u|^{\alpha+1}}{1-\alpha} + \frac{|W(x)|^{\alpha+1}}{1-\alpha}|u|^{\alpha+1} \tag{8.3.6}$$

for any function $u \in C^1(G; \mathbb{R})$ and any n-vector function $W(x) \in C(G; \mathbb{R}^n)$.

Proof. The following inequality holds:

$$|\nabla u|^{\alpha+1} + \alpha|\nabla u - uW(x)|^{\alpha+1} - (\alpha+1)\,(\nabla u) \cdot \Phi\,(\nabla u - uW(x)) \geq 0$$

(*cf.* Kusano, Jaroš and Yoshida [151, Lemma 2.1]). Hence we have

$$\begin{aligned}&|\nabla u|^{\alpha+1} + \alpha|\nabla u - uW(x)|^{\alpha+1}\\ &\quad -(\alpha+1)\,(\nabla u - uW(x) + uW(x)) \cdot \Phi\,(\nabla u - uW(x)) \geq 0\end{aligned}$$

and therefore

$$\begin{aligned}&|\nabla u|^{\alpha+1} + \alpha|\nabla u - uW(x)|^{\alpha+1}\\ &\quad -(\alpha+1)\left[|\nabla u - uW(x)|^{\alpha+1} + uW(x) \cdot \Phi\,(\nabla u - uW(x))\right] \geq 0\end{aligned}$$

or

$$|\nabla u|^{\alpha+1} - (\alpha+1)uW(x) \cdot \Phi\,(\nabla u - uW(x)) \geq |\nabla u - uW(x)|^{\alpha+1}. \tag{8.3.7}$$

Using Schwarz's inequality and Young's inequality, we find that

$$\begin{aligned}
&|(\alpha+1)uW(x)\cdot\Phi\left(\nabla u - uW(x)\right)| \\
&\le (\alpha+1)|uW(x)||\nabla u - uW(x)|^{\alpha} \\
&\le (\alpha+1)\left[\frac{|uW(x)|^{\alpha+1}}{\alpha+1} + \frac{|\nabla u - uW(x)|^{\alpha+1}}{\frac{\alpha+1}{\alpha}}\right] \\
&= |uW(x)|^{\alpha+1} + \alpha|\nabla u - uW(x)|^{\alpha+1}. \qquad (8.3.8)
\end{aligned}$$

Combining (8.3.7) with (8.3.8) yields the following

$$\begin{aligned}
|\nabla u - uW(x)|^{\alpha+1} &\le |\nabla u|^{\alpha+1} + |(\alpha+1)uW(x)\cdot\Phi\left(\nabla u - uW(x)\right)| \\
&\le |\nabla u|^{\alpha+1} + |uW(x)|^{\alpha+1} + \alpha|\nabla u - uW(x)|^{\alpha+1}
\end{aligned}$$

and hence

$$(1-\alpha)|\nabla u - uW(x)|^{\alpha+1} \le |\nabla u|^{\alpha+1} + |W(x)|^{\alpha+1}|u|^{\alpha+1}$$

which is equivalent to (8.3.6). The proof is complete. □

Theorem 8.3.4. *Let $0 < \alpha < 1$. Assume that for any $r > 0$ there exist a bounded and piecewise smooth domain G with $\overline{G} \subset \Omega_r$ and a nontrivial function $u \in C^1(\overline{G};\mathbb{R})$ such that $u = 0$ on ∂G and*

$$\int_G \left[\frac{A(x)}{1-\alpha}|\nabla u|^{\alpha+1} - \left\{C(x) - \frac{|B(x)|^{\alpha+1}}{(1-\alpha)A(x)^{\alpha}}\right\}|u|^{\alpha+1}\right] dx \le 0.$$

Then every solution $v \in \mathcal{D}_{P_\alpha}(\Omega)$ of (8.3.5) is oscillatory in Ω.

Proof. Using Lemma 8.3.1 with $W(x) = B(x)/A(x)$, we obtain

$$A(x)\left|\nabla u - \frac{u}{A(x)}B(x)\right|^{\alpha+1} \le \frac{A(x)}{1-\alpha}|\nabla u|^{\alpha+1} + \frac{|B(x)|^{\alpha+1}}{(1-\alpha)A(x)^{\alpha}}|u|^{\alpha+1}$$

and therefore

$$M_G[u] \le \int_G \left[\frac{A(x)}{1-\alpha}|\nabla u|^{\alpha+1} - \left\{C(x) - \frac{|B(x)|^{\alpha+1}}{(1-\alpha)A(x)^{\alpha}}\right\}|u|^{\alpha+1}\right] dx \le 0.$$

The conclusion follows from Theorem 8.3.3. □

Lemma 8.3.2. *Let $E(x) \in C(G;(0,\infty))$ satisfy $E(x) > \alpha$. Then we obtain the inequality*

$$|\nabla u - uW(x)|^{\alpha+1} \le \frac{E(x)}{E(x)-\alpha}|\nabla u|^{\alpha+1} + \frac{|E(x)W(x)|^{\alpha+1}}{E(x)-\alpha}|u|^{\alpha+1} \qquad (8.3.9)$$

for any function $u \in C^1(G;\mathbb{R})$ and any n-vector function $W(x) \in C(G;\mathbb{R}^n)$.

Proof. Proceeding as in the proof of Lemma 8.3.1, we see that the inequality (8.3.7) holds. Applying Schwarz's inequality and Young's inequality, we have

$$\begin{aligned} &|(\alpha+1)uW(x)\cdot\Phi(\nabla u - uW(x))| \\ &= \frac{1}{E(x)}(\alpha+1)|uE(x)W(x)||\nabla u - uW(x)|^{\alpha} \\ &\le \frac{1}{E(x)}\Big(|uE(x)W(x)|^{\alpha+1} + \alpha|\nabla u - uW(x)|^{\alpha+1}\Big). \end{aligned} \tag{8.3.10}$$

Combining (8.3.7) with (8.3.10) yields the following

$$|\nabla u - uW(x)|^{\alpha+1} \le |\nabla u|^{\alpha+1} + \frac{|E(x)W(x)|^{\alpha+1}}{E(x)}|u|^{\alpha+1} + \frac{\alpha}{E(x)}|\nabla u - uW(x)|^{\alpha+1}$$

and therefore

$$\left(1-\frac{\alpha}{E(x)}\right)|\nabla u - uW(x)|^{\alpha+1} \le |\nabla u|^{\alpha+1} + \frac{|E(x)W(x)|^{\alpha+1}}{E(x)}|u|^{\alpha+1}$$

which is equivalent to (8.3.9). The proof is complete. □

Theorem 8.3.5. *Let $A(x) > \alpha$ in Ω. Assume that for any $r > 0$ there exist a bounded and piecewise smooth domain G with $\overline{G} \subset \Omega_r$ and a nontrivial function $u \in C^1(\overline{G};\mathbb{R})$ such that $u = 0$ on ∂G and*

$$\int_G \left[\frac{A(x)^2}{A(x)-\alpha}|\nabla u|^{\alpha+1} - \left\{C(x) - \frac{A(x)}{A(x)-\alpha}|B(x)|^{\alpha+1}\right\}|u|^{\alpha+1}\right] dx \le 0.$$

Then every solution $v \in \mathcal{D}_{P_\alpha}(\Omega)$ of (8.3.5) is oscillatory in Ω.

Proof. The inequality (8.3.9) with $E(x) = A(x)$ and $W(x) = B(x)/A(x)$ implies

$$A(x)\left|\nabla u - \frac{u}{A(x)}B(x)\right|^{\alpha+1} \le \frac{A(x)^2}{A(x)-\alpha}|\nabla u|^{\alpha+1} + \frac{A(x)}{A(x)-\alpha}|B(x)|^{\alpha+1}|u|^{\alpha+1}.$$

Hence, it can be shown that

$$\begin{aligned} M_G[u] &\le \int_G \left[\frac{A(x)^2}{A(x)-\alpha}|\nabla u|^{\alpha+1} - \left\{C(x) - \frac{A(x)}{A(x)-\alpha}|B(x)|^{\alpha+1}\right\}|u|^{\alpha+1}\right] dx \\ &\le 0 \end{aligned}$$

and consequently the conclusion follows from Theorem 8.3.3. □

We recall that $\overline{Q(x)}(r)$ denotes the spherical mean of $Q(x)$ over the sphere $S_r = \{x \in \mathbb{R}^n;\ |x| = r\}$ (*cf.* Section 8.1).

Theorem 8.3.6. *Let $0 < \alpha < 1$. If the half-linear ordinary differential equation*

$$\left(r^{n-1}\overline{\left\{\frac{A(x)}{1-\alpha}\right\}}(r)\,|y'|^{\alpha-1}y'\right)' + r^{n-1}\overline{\left\{C(x) - \frac{|B(x)|^{\alpha+1}}{(1-\alpha)A(x)^{\alpha}}\right\}}(r)\,|y|^{\alpha-1}y = 0 \quad (8.3.11)$$

is oscillatory at $r = \infty$, then every solution $v \in \mathcal{D}_{P_\alpha}(\mathbb{R}^n)$ of (8.3.5) is oscillatory in $\mathbb{R}^n$.

Proof. Let $\{r_k\}$ be the sequence of the zeros of a nontrivial solution $y(r)$ of (8.3.11) such that $r_1 < r_2 < \cdots$, $\lim_{k\to\infty} r_k = \infty$. Letting

$$G_k = \{x \in \mathbb{R}^n;\ r_k < |x| < r_{k+1}\}\ (k = 1, 2, ...)$$

and $u(x) = y(|x|)$, we find that

$$\begin{aligned} M_{G_k}[u] &\le \int_{G_k} \left[\frac{A(x)}{1-\alpha}|\nabla u|^{\alpha+1} - \left\{C(x) - \frac{|B(x)|^{\alpha+1}}{(1-\alpha)A(x)^{\alpha}}\right\}|u|^{\alpha+1}\right] dx \\ &= \omega_n \int_{r_k}^{r_{k+1}} \left[\overline{\left\{\frac{A(x)}{1-\alpha}\right\}}(r)\,|y'(r)|^{\alpha+1} \right. \\ &\qquad \left. - \overline{\left\{C(x) - \frac{|B(x)|^{\alpha+1}}{(1-\alpha)A(x)^{\alpha}}\right\}}(r)\,|y(r)|^{\alpha+1}\right] r^{n-1}dr \\ &= 0, \end{aligned}$$

where ω_n denotes the surface area of the unit sphere S_1. Theorem 8.3.4 implies that every solution $v \in \mathcal{D}_{P_\alpha}(\mathbb{R}^n)$ of (8.3.5) is oscillatory in $\mathbb{R}^n$. The proof is complete. □

Analogously we have the following theorem.

Theorem 8.3.7. *Let $A(x) > \alpha$ in $\mathbb{R}^n$. If the half-linear ordinary differential equation*

$$\left(r^{n-1}\overline{\left\{\frac{A(x)^2}{A(x)-\alpha}\right\}}(r)\,|y'|^{\alpha-1}y'\right)' + r^{n-1}\overline{\left\{C(x) - \frac{A(x)}{A(x)-\alpha}|B(x)|^{\alpha+1}\right\}}(r)\,|y|^{\alpha-1}y = 0$$

is oscillatory at $r = \infty$, then every solution $v \in \mathcal{D}_{P_\alpha}(\mathbb{R}^n)$ of (8.3.5) is oscillatory in $\mathbb{R}^n$.

Theorem 8.3.8. *Let $0 < \alpha < 1$. If there are positive constants A_0 and C_0 satisfying*

$$\frac{A(x)}{1-\alpha} \le A_0, \quad C(x) - \frac{|B(x)|^{\alpha+1}}{(1-\alpha)A(x)^\alpha} \ge C_0,$$

then every solution $v \in \mathcal{D}_{P_\alpha}(\mathbb{R}^n)$ of (8.3.5) is oscillatory in $\mathbb{R}^n$.

Proof. The conclusion follows by taking into account the inequality

$$M_G[u] \le \int_G \left[A_0|\nabla u|^{\alpha+1} - C_0|u|^{\alpha+1}\right] dx$$

and the fact that

$$\left(r^{n-1}A_0\,|y'|^{\alpha-1}y'\right)' + r^{n-1}C_0\,|y|^{\alpha-1}y = 0$$

is oscillatory at $r = \infty$ (see Example 8.1.1 of Section 8.1). □

Similarly we obtain the following.

Theorem 8.3.9. *Let $A(x) > \alpha$ in $\mathbb{R}^n$. If there are positive constants A_0 and C_0 satisfying*

$$\frac{A(x)^2}{A(x)-\alpha} \le A_0, \quad C(x) - \frac{A(x)}{A(x)-\alpha}|B(x)|^{\alpha+1} \ge C_0,$$

then every solution $v \in \mathcal{D}_{P_\alpha}(\mathbb{R}^n)$ of (8.3.5) is oscillatory in $\mathbb{R}^n$.

Example 8.3.1. We consider the half-linear elliptic equation

$$\nabla \cdot (3|\nabla v|\nabla v) + 3|\nabla v|\left(\frac{\partial v}{\partial x_1} + 2\frac{\partial v}{\partial x_2}\right) + 38|v|v = 0 \tag{8.3.12}$$

for $x = (x_1, x_2) \in \mathbb{R}^2$. Here $n = \alpha = 2$, $A(x) = 3$, $B(x) = (1,2)$, $C(x) = 38$. It is easily seen that

$$\frac{A(x)^2}{A(x)-\alpha} = 9,$$

$$C(x) - \frac{A(x)}{A(x)-\alpha}|B(x)|^{\alpha+1} = 38 - 3(\sqrt{5})^3 \ge 1$$

and therefore we can take $A_0 = 9$ and $C_0 = 1$. It follows from Theorem 8.3.9 that every solution v of (8.3.12) is oscillatory in $\mathbb{R}^2$.

Remark 8.3.1. More general half-linear elliptic equations of the form

$$\sum_{i=1}^{n} \frac{\partial}{\partial x_i}\left(\left(A_i(x)\right)^2|\nabla_A v|^{\alpha-1}\frac{\partial v}{\partial x_i}\right) + (\alpha+1)|\nabla_A v|^{\alpha-1}B(x)\cdot\nabla_A v$$
$$+C(x)|v|^{\alpha-1}v = 0$$

were studied by Yoshida [322], where

$$\nabla_A v = \left(A_1(x)\frac{\partial v}{\partial x_1}, ..., A_n(x)\frac{\partial v}{\partial x_n}\right).$$

8.4 Forced Superlinear Elliptic Equations

In 1993 El-Sayed [74] investigated oscillations of the linear ordinary differential equation

$$(p(t)y')' + q(t)y = f(t)$$

by using a Sturmian comparison theorem. Nasr [219] studied forced oscillations of nonlinear differential equations of the form

$$y'' + q(t)|y|^{\beta}\operatorname{sgn} y = f(t) \quad (\beta > 1).$$

More general ordinary differential equations

$$(p(t)y')' + q(t)|y|^{\beta}\operatorname{sgn} y = f(t) \quad (\beta > 1),$$
$$(p(t)|y'|^{\alpha}\operatorname{sgn} y')' + q(t)|y|^{\beta}\operatorname{sgn} y = f(t) \quad (\beta \geq \alpha > 0)$$

were treated by Jaroš, Kusano and Yoshida [112, 115].

Forced oscillations of the superlinear elliptic equation

$$\sum_{i,j=1}^{n} \frac{\partial}{\partial x_i}\left(A_{ij}(x)\frac{\partial v}{\partial x_j}\right) + C(x)|v|^{\beta-1}v = f(x) \quad (\beta > 1)$$

were treated by Jaroš, Kusano and Yoshida [113], and non-self-adjoint cases were studied by Jaroš, Kusano and Yoshida [118]. Oscillation results for the superlinear elliptic equation with forcing term

$$\nabla \cdot (A(x)|\nabla v|^{\alpha-1}\nabla v) + C(x)|v|^{\beta-1}v = f(x) \quad (\beta > \alpha > 0)$$

were derived by Jaroš, Kusano and Yoshida [115]. For more general equations we refer to Yoshida [321, 323].

In this section we establish a Picone-type inequality for the superlinear elliptic operator P defined by

$$P[v] = \nabla \cdot \left(A(x)|\nabla v|^{\alpha-1}\nabla v\right) + (\alpha+1)|\nabla v|^{\alpha-1}B(x)\cdot\nabla v + C(x)|v|^{\beta-1}v$$

and derive oscillation results for the superlinear elliptic equation

$$P[v] = f(x), \tag{8.4.1}$$

where α, β are constants satisfying $\beta > \alpha > 0$.

We assume that $A(x) \in C(\overline{G};(0,\infty))$, $B(x) \in C(\overline{G};\mathbb{R}^n)$, $C(x) \in C(\overline{G};[0,\infty))$ and $f(x) \in C(\overline{G};\mathbb{R})$.

The domain $\mathcal{D}_P(G)$ of P is defined to be the set of all functions v of class $C^1(\overline{G};\mathbb{R})$ with the property that $A(x)|\nabla v|^{\alpha-1}\nabla v \in C^1(G;\mathbb{R}^n) \cap C(\overline{G};\mathbb{R}^n)$.

Theorem 8.4.1 (Picone-type inequality). *If $v \in \mathcal{D}_P(G)$, $v \neq 0$ in G and $vf(x) \leq 0$ in G, then we obtain the following Picone-type inequality for any $u \in C^1(G;\mathbb{R})$:*

$$
\begin{aligned}
&-\nabla \cdot \left(u\varphi(u) \frac{A(x)\Phi(\nabla v)}{\varphi(v)} \right) \\
&\geq -A(x) \left| \nabla u - \frac{u}{A(x)} B(x) \right|^{\alpha+1} + \frac{\beta}{\alpha} \left(\frac{\beta-\alpha}{\alpha} \right)^{\frac{\alpha-\beta}{\beta}} C(x)^{\frac{\alpha}{\beta}} |f(x)|^{\frac{\beta-\alpha}{\beta}} |u|^{\alpha+1} \\
&\quad + A(x) \left[\left| \nabla u - \frac{u}{A(x)} B(x) \right|^{\alpha+1} + \alpha \left| \frac{u}{v} \nabla v \right|^{\alpha+1} \right. \\
&\qquad \left. -(\alpha+1) \left(\nabla u - \frac{u}{A(x)} B(x) \right) \cdot \Phi \left(\frac{u}{v} \nabla v \right) \right] \\
&\quad - \frac{u\varphi(u)}{\varphi(v)} \Big(P[v] - f(x) \Big),
\end{aligned}
\tag{8.4.2}
$$

where $\varphi(s) = |s|^{\alpha-1}s$ ($s \in \mathbb{R}$) and $\Phi(\xi) = |\xi|^{\alpha-1}\xi$ ($\xi \in \mathbb{R}^n$).

Proof. The following identity holds:

$$
\begin{aligned}
&-\nabla \cdot \left(u\varphi(u) \frac{A(x)\Phi(\nabla v)}{\varphi(v)} \right) \\
&= -A(x) \left| \nabla u - \frac{u}{A(x)} B(x) \right|^{\alpha+1} \\
&\quad + A(x) \left[\left| \nabla u - \frac{u}{A(x)} B(x) \right|^{\alpha+1} + \alpha \left| \frac{u}{v} \nabla v \right|^{\alpha+1} \right. \\
&\qquad \left. -(\alpha+1) \left(\nabla u - \frac{u}{A(x)} B(x) \right) \cdot \Phi \left(\frac{u}{v} \nabla v \right) \right] \\
&\quad - \frac{u\varphi(u)}{\varphi(v)} \Big(\nabla \cdot (A(x)|\nabla v|^{\alpha-1}\nabla v) + (\alpha+1)|\nabla v|^{\alpha-1} B(x) \cdot \nabla v \Big)
\end{aligned}
\tag{8.4.3}
$$

(see Theorem 8.3.1 of Section 8.3 or Yoshida [320, Theorem 1.1]). It is easy to see that

$$
\begin{aligned}
&\nabla \cdot (A(x)|\nabla v|^{\alpha-1}\nabla v) + (\alpha+1)|\nabla v|^{\alpha-1} B(x) \cdot \nabla v \\
&= P[v] - f(x) + f(x) - C(x)|v|^{\beta-1}v
\end{aligned}
$$

and therefore

$$
\begin{aligned}
&\frac{u\varphi(u)}{\varphi(v)}\Big(\nabla\cdot(A(x)|\nabla v|^{\alpha-1}\nabla v)+(\alpha+1)|\nabla v|^{\alpha-1}B(x)\cdot\nabla v\Big)\\
&=\frac{u\varphi(u)}{\varphi(v)}\Big(P[v]-f(x)+f(x)-C(x)|v|^{\beta-1}v\Big)\\
&=\frac{u\varphi(u)}{\varphi(v)}\Big(P[v]-f(x)\Big)-u\varphi(u)\left(C(x)\frac{|v|^{\beta-1}v}{\varphi(v)}-\frac{f(x)}{\varphi(v)}\right)\\
&=\frac{u\varphi(u)}{\varphi(v)}\Big(P[v]-f(x)\Big)-|u|^{\alpha+1}\left(C(x)|v|^{\beta-\alpha}-\frac{f(x)}{|v|^{\alpha-1}v}\right). \qquad (8.4.4)
\end{aligned}
$$

It can be shown that the inequality

$$
\begin{aligned}
C(x)|v|^{\beta-\alpha}-\frac{f(x)}{|v|^{\alpha-1}v}&=C(x)|v|^{\beta-\alpha}+\frac{|f(x)|}{|v|^{\alpha}}\\
&\geq\frac{\beta}{\alpha}\left(\frac{\beta-\alpha}{\alpha}\right)^{\frac{\alpha-\beta}{\beta}}C(x)^{\frac{\alpha}{\beta}}|f(x)|^{\frac{\beta-\alpha}{\beta}} \qquad (8.4.5)
\end{aligned}
$$

holds (see, for example, Jaroš, Kusano and Yoshida [115, p.55], [116, p.712]). Combining (8.4.3)–(8.4.5) yields the desired Picone-type inequality (8.4.2). The proof is complete. □

Theorem 8.4.2. *If there exists a nontrivial function* $u\in C^1(\overline{G};\mathbb{R})$ *such that* $u=0$ *on* ∂G *and*

$$
\begin{aligned}
M_G[u]:=\int_G\Bigg[&A(x)\left|\nabla u-\frac{u}{A(x)}B(x)\right|^{\alpha+1}\\
&-\frac{\beta}{\alpha}\left(\frac{\beta-\alpha}{\alpha}\right)^{\frac{\alpha-\beta}{\beta}}C(x)^{\frac{\alpha}{\beta}}|f(x)|^{\frac{\beta-\alpha}{\beta}}|u|^{\alpha+1}\Bigg]\,dx\leq 0,
\end{aligned}
$$

then every solution $v\in\mathcal{D}_P(G)$ *of* (8.4.1) *satisfying* $vf(x)\leq 0$ *must vanish at some point of* $\overline{G}$.

Proof. Suppose to the contrary that there is a solution $v\in\mathcal{D}_P(G)$ of (8.4.1) satisfying $vf(x)\leq 0$ and $v\neq 0$ on $\overline{G}$. Theorem 8.4.1 implies that the Picone-type inequality (8.4.2) holds for the nontrivial function u. Integrating (8.4.2) over G and proceeding as in the proof of Theorem 8.3.2 of Section 8.3, we are led to a contradiction. □

Corollary 8.4.1. *Assume that* $f(x)\geq 0$ [*or* $f(x)\leq 0$] *in* G. *If there is a nontrivial function* $u\in C^1(\overline{G};\mathbb{R})$ *such that* $u=0$ *on* ∂G *and* $M_G[u]\leq 0$, *then* (8.4.1) *has no negative* [*or positive*] *solution on* $\overline{G}$.

Proof. Suppose that (8.4.1) has a negative [or positive] solution v on $\overline{G}$. It is easy to see that $vf(x) \le 0$ in G. Therefore it follows from Theorem 8.4.2 that v must vanish at some point of $\overline{G}$. This is a contradiction and the proof is complete. □

Theorem 8.4.3. *Assume that G is divided into two subdomains G_1 and G_2 by an $(n-1)$-dimensional piecewise smooth hypersurface in such a way that*

$$f(x) \ge 0 \quad \text{in } G_1 \quad \text{and} \quad f(x) \le 0 \quad \text{in } G_2.$$

If there are nontrivial functions $u_k \in C^1(\overline{G_k};\mathbb{R})$ $(k = 1,2)$ such that $u_k = 0$ on ∂G_k and

$$M_{G_k}[u_k] = \int_{G_k} \left[A(x) \left| \nabla u_k - \frac{u_k}{A(x)} B(x) \right|^{\alpha+1} - \frac{\beta}{\alpha}\left(\frac{\beta-\alpha}{\alpha}\right)^{\frac{\alpha-\beta}{\beta}} C(x)^{\frac{\alpha}{\beta}} |f(x)|^{\frac{\beta-\alpha}{\beta}} |u_k|^{\alpha+1} \right] dx \le 0, \tag{8.4.6}$$

then every solution $v \in \mathcal{D}_P(G)$ of (8.4.1) has a zero on $\overline{G}$.

Proof. Suppose that there exists a solution $v \in \mathcal{D}_P(G)$ of (8.4.1) which has no zero on $\overline{G}$. Then, either $v > 0$ on $\overline{G}$ or $v < 0$ on $\overline{G}$. If $v > 0$ on $\overline{G}$, then $v > 0$ on $\overline{G_2}$, and therefore $vf(x) \le 0$ in G_2. It follows from Corollary 8.4.1 that (8.4.1) has no positive solution $\overline{G_2}$. This is a contradiction. In the case where $v < 0$ on $\overline{G}$, a similar argument leads us to a contradiction. The proof is complete. □

Now we establish oscillation criteria for (8.4.1) in an exterior domain Ω in $\mathbb{R}^n$, that is, $\Omega \supset \{x \in \mathbb{R}^n;\ |x| \ge r_0\}$ for some $r_0 > 0$. It is assumed that $A(x) \in C(\Omega;(0,\infty))$, $B(x) \in C(\Omega;\mathbb{R}^n)$, $C(x) \in C(\Omega;\mathbb{R})$ and $f(x) \in C(\Omega;\mathbb{R})$.

The domain $\mathcal{D}_P(\Omega)$ of P is defined to be the set of all functions $v \in C^1(\Omega;\mathbb{R})$ with the property that $A(x)|\nabla v|^{\alpha-1}\nabla v \in C^1(\Omega;\mathbb{R}^n)$.

A solution $v \in \mathcal{D}_P(\Omega)$ of (8.4.1) is said to be *oscillatory* in Ω if it has a zero in Ω_r for any $r > 0$, where

$$\Omega_r = \Omega \cap \{x \in \mathbb{R}^n;\ |x| > r\}.$$

Theorem 8.4.4. *Assume that for any $r > 0$ there exists a bounded and piecewise smooth domain G with $\overline{G} \subset \Omega_r$, which can be divided into two* *subdomains G_1 and G_2 by an $(n-1)$-dimensional piecewise smooth hypersurface in such a way that $f(x) \geq 0$ in G_1 and $f(x) \leq 0$ in G_2. Furthermore, assume that $C(x) \geq 0$ in G and that there are nontrivial functions $u_k \in C^1(\overline{G_k};\mathbb{R})$ such that $u_k = 0$ on ∂G_k and $M_{G_k}[u_k] \leq 0$ $(k = 1, 2)$, where M_{G_k} are defined by (8.4.6). Then every solution $v \in \mathcal{D}_P(\Omega)$ of (8.4.1) is oscillatory in Ω.*

Proof. For any $r > 0$ there exists a bounded domain G as mentioned in the hypotheses of Theorem 8.4.4. Theorem 8.4.3 implies that every solution v of (8.4.1) has a zero on $\overline{G} \subset \Omega_r$, that is, v is oscillatory in Ω. □

Example 8.4.1. We consider the forced superlinear elliptic equation

$$\nabla\cdot(|\nabla v|^2\nabla v) + 4|\nabla v|^2\left(\frac{\partial v}{\partial x_1} + \frac{\partial v}{\partial x_2}\right)$$
$$+K(\sin x_1)(\sin x_2)|v|^{\beta-1}v = (\cos x_1)\sin x_2, \quad (x_1, x_2) \in \Omega, \quad (8.4.7)$$

where β and K are the constants satisfying $\beta > 3$, $K > 0$, and Ω is an unbounded domain in $\mathbb{R}^2$ containing a horizontal strip such that

$$[2\pi, \infty) \times [0, \pi] \subset \Omega.$$

Here $n = 2, \alpha = 3, A(x) = 1, B(x) = (1, 1)$, $C(x) = K(\sin x_1)\sin x_2$ and $f(x) = (\cos x_1)\sin x_2$. For any fixed $j \in \mathbb{N}$ we consider the rectangle

$$G^{(j)} = (2j\pi, (2j+1)\pi) \times (0, \pi)$$

which is divided into two subdomains

$$G_1^{(j)} = (2j\pi, (2j + (1/2))\pi) \times (0, \pi),$$
$$G_2^{(j)} = ((2j + (1/2))\pi, (2j+1)\pi) \times (0, \pi)$$

by the vertical line $x_1 = (2j+(1/2))\pi$. It is easily seen that $f(x) \geq 0$ in $G_1^{(j)}$, $f(x) \leq 0$ in $G_2^{(j)}$ and $C(x) \geq 0$ in $G^{(j)}$. Letting $u_k = (\sin 2x_1)\sin x_2$ $(k = 1, 2)$, we find that $u_k = 0$ on $\partial G_k^{(j)}$ $(k = 1, 2)$. A simple computation shows that

$$M_{G_k^{(j)}}[u_k]$$
$$= \int_{G_k^{(j)}} \Bigg[|\nabla u_k - (u_k, u_k)|^4$$
$$-\frac{\beta}{3}\left(\frac{\beta-3}{3}\right)^{\frac{3-\beta}{\beta}} (K(\sin x_1)\sin x_2)^{\frac{3}{\beta}} \times |(\cos x_1)\sin x_2|^{\frac{\beta-3}{\beta}} |u_k|^4\Bigg] dx$$
$$= \frac{261}{128}\pi^2 - \frac{128}{15}K^{3/\beta}\frac{\beta}{3}\left(\frac{\beta-3}{3}\right)^{\frac{3-\beta}{\beta}} B\left(\frac{5}{2} + \frac{3}{2\beta}, 3 - \frac{3}{2\beta}\right),$$

where $B(s,t)$ is the beta function. If $K > 0$ is chosen so large that

$$K \geq \left[\frac{3915}{16384} \pi^2 \left(\frac{\beta}{3} \left(\frac{\beta - 3}{3} \right)^{\frac{3-\beta}{\beta}} B\left(\frac{5}{2} + \frac{3}{2\beta}, 3 - \frac{3}{2\beta} \right) \right)^{-1} \right]^{\frac{\beta}{3}},$$

then $M_{G_k^{(j)}}[u_k] \leq 0$ hold for $k = 1, 2$. It follows from Theorem 8.4.3 that every solution v of (8.4.7) is oscillatory in Ω for all sufficiently large $K > 0$.

8.5 Superlinear-Sublinear Elliptic Equations

Oscillations of the superlinear-sublinear elliptic equation

$$\sum_{i,j=1}^{n} \frac{\partial}{\partial x_i} \left(A_{ij}(x) \frac{\partial v}{\partial x_j} \right) + C(x)|v|^{\beta-1}v + D(x)|v|^{\gamma-1}v = 0$$

were investigated by Jaroš, Kusano and Yoshida [113] by using a Picone-type inequality, where $\beta > 1$ and $0 < \gamma < 1$. More general non-self-adjoint equations of the form

$$\sum_{i,j=1}^{n} \frac{\partial}{\partial x_i} \left(A_{ij}(x) \frac{\partial v}{\partial x_j} \right) + 2\sum_{i=1}^{n} B_i(x) \frac{\partial v}{\partial x_i} + C(x)|v|^{\beta-1}v + D(x)|v|^{\gamma-1}v = 0$$

were studied by Jaroš, Kusano and Yoshida [118]. Oscillation results for the super-sublinear elliptic equation

$$\nabla \cdot (A(x)|\nabla v|^{\alpha-1}\nabla v) + C(x)|v|^{\beta-1}v + D(x)|v|^{\gamma-1}v = 0$$

were obtained by Jaroš, Kusano and Yoshida [116], where $\beta > \alpha$ and $0 < \gamma < \alpha$.

In this section we establish a Picone-type inequality for the elliptic operator P defined by

$$P[v] = \nabla \cdot (A(x)|\nabla v|^{\alpha-1}\nabla v) + (\alpha+1)|\nabla v|^{\alpha-1}B(x) \cdot \nabla v$$
$$+C(x)|v|^{\beta-1}v + D(x)|v|^{\gamma-1}v$$

and present oscillation results for the superlinear-sublinear elliptic equation

$$P[v] = 0, \tag{8.5.1}$$

where $\beta > \alpha$ and $0 < \gamma < \alpha$.

We assume that $A(x) \in C(\overline{G}; (0,\infty))$, $B(x) \in C(\overline{G}; \mathbb{R}^n)$, $C(x) \in C(\overline{G}; [0,\infty))$ and $D(x) \in C(\overline{G}; [0,\infty))$.

The domain $\mathcal{D}_P(G)$ of P is defined to be the set of all functions v of class $C^1(\overline{G}; \mathbb{R})$ with the property that $A(x)|\nabla v|^{\alpha-1}\nabla v \in C^1(G; \mathbb{R}^n) \cap C(\overline{G}; \mathbb{R}^n)$.

Theorem 8.5.1 (Picone-type inequality). *If $v \in \mathcal{D}_P(G)$ and $v \neq 0$ in G, then we obtain the following Picone-type inequality for any $u \in C^1(G;\mathbb{R})$:*

$$\begin{aligned}
&-\nabla \cdot \left(u\varphi(u) \frac{A(x)\Phi(\nabla v)}{\varphi(v)} \right) \\
&\geq -A(x) \left| \nabla u - \frac{u}{A(x)} B(x) \right|^{\alpha+1} \\
&\quad + \frac{\beta-\gamma}{\alpha-\gamma} \left(\frac{\beta-\alpha}{\alpha-\gamma} \right)^{\frac{\alpha-\beta}{\beta-\gamma}} C(x)^{\frac{\alpha-\gamma}{\beta-\gamma}} D(x)^{\frac{\beta-\alpha}{\beta-\gamma}} |u|^{\alpha+1} \\
&\quad + A(x) \Bigg[\left| \nabla u - \frac{u}{A(x)} B(x) \right|^{\alpha+1} + \alpha \left| \frac{u}{v} \nabla v \right|^{\alpha+1} \\
&\qquad\qquad - (\alpha+1) \left(\nabla u - \frac{u}{A(x)} B(x) \right) \cdot \Phi\left(\frac{u}{v} \nabla v \right) \Bigg] \\
&\quad - \frac{u\varphi(u)}{\varphi(v)} P[v],
\end{aligned} \tag{8.5.2}$$

where $\varphi(s) = |s|^{\alpha-1}s$ $(s \in \mathbb{R})$ and $\Phi(\xi) = |\xi|^{\alpha-1}\xi$ $(\xi \in \mathbb{R}^n)$.

Proof. First we note that the identity (8.4.3) of Section 8.4 holds for any $v \in \mathcal{D}_P(G)$ with $v \neq 0$ in G and $u \in C^1(G;\mathbb{R})$. It is easily checked that

$$\begin{aligned}
&\nabla \cdot \left(A(x)|\nabla v|^{\alpha-1}\nabla v \right) + (\alpha+1)|\nabla v|^{\alpha-1} B(x) \cdot \nabla v \\
&= P[v] - C(x)|v|^{\beta-1}v - D(x)|v|^{\gamma-1}v
\end{aligned}$$

and therefore

$$\begin{aligned}
&\frac{u\varphi(u)}{\varphi(v)} \left(\nabla \cdot \left(A(x)|\nabla v|^{\alpha-1}\nabla v \right) + (\alpha+1)|\nabla v|^{\alpha-1} B(x) \cdot \nabla v \right) \\
&= \frac{u\varphi(u)}{\varphi(v)} \left(P[v] - C(x)|v|^{\beta-1}v - D(x)|v|^{\gamma-1}v \right) \\
&= \frac{u\varphi(u)}{\varphi(v)} P[v] - |u|^{\alpha+1} \left(C(x)|v|^{\beta-\alpha} + \frac{D(x)}{|v|^{\alpha-\gamma}} \right).
\end{aligned} \tag{8.5.3}$$

We obtain the following inequality:

$$C(x)|v|^{\beta-\alpha} + \frac{D(x)}{|v|^{\alpha-\gamma}} \geq \frac{\beta-\gamma}{\alpha-\gamma} \left(\frac{\beta-\alpha}{\alpha-\gamma} \right)^{\frac{\alpha-\beta}{\beta-\gamma}} C(x)^{\frac{\alpha-\gamma}{\beta-\gamma}} D(x)^{\frac{\beta-\alpha}{\beta-\gamma}} \tag{8.5.4}$$

(see Jaroš, Kusano and Yoshida [116, p.717]). Combining (8.4.3), (8.5.3) and (8.5.4) yields the desired Picone-type inequality (8.5.2). □

Theorem 8.5.2. *If there is a nontrivial function $u \in C^1(\overline{G}; \mathbb{R})$ such that $u = 0$ on ∂G and*

$$\begin{aligned} M_G[u] := \int_G \Bigg[A(x) \left| \nabla u - \frac{u}{A(x)} B(x) \right|^{\alpha+1} \\ - \frac{\beta-\gamma}{\alpha-\gamma} \left(\frac{\beta-\alpha}{\alpha-\gamma} \right)^{\frac{\alpha-\beta}{\beta-\gamma}} C(x)^{\frac{\alpha-\gamma}{\beta-\gamma}} D(x)^{\frac{\beta-\alpha}{\beta-\gamma}} |u|^{\alpha+1} \Bigg] dx \\ \leq 0, \end{aligned} \tag{8.5.5}$$

then every solution $v \in \mathcal{D}_P(G)$ of (8.5.1) *vanishes at some point of $\overline{G}$.*

Proof. Suppose to the contrary that there exists a solution $v \in \mathcal{D}_P(G)$ of (8.5.1) such that $v \neq 0$ on $\overline{G}$. It follows from Theorem 8.5.1 that the Picone-type inequality (8.5.2) holds for the nontrivial function u. Integrating (8.5.2) over G and arguing as in the proof of Theorem 8.3.2 of Section 8.3, we find that a contradiction yields. This completes the proof. □

Now we derive oscillation results for (8.5.1) in an exterior domain Ω in $\mathbb{R}^n$, that is, $\Omega \supset \{x \in \mathbb{R}^n;\ |x| \geq r_0\}$ for some $r_0 > 0$. It is assumed that $A(x) \in C(\Omega; (0, \infty))$, $B(x) \in C(\Omega; \mathbb{R}^n)$, $C(x) \in C(\Omega; [0, \infty))$ and $D(x) \in C(\Omega; [0, \infty))$.

The domain $\mathcal{D}_P(\Omega)$ of P is defined to be the set of all functions $v \in C^1(\Omega; \mathbb{R})$ with the property that $A(x)|\nabla v|^{\alpha-1}\nabla v \in C^1(\Omega; \mathbb{R}^n)$.

A solution $v \in \mathcal{D}_P(\Omega)$ of (8.5.1) is said to be *oscillatory* in Ω if it has a zero in Ω_r for any $r > 0$, where

$$\Omega_r = \Omega \cap \{x \in \mathbb{R}^n;\ |x| > r\}.$$

Theorem 8.5.3. *Assume that for any $r > 0$ there is a bounded and piecewise smooth domain G with $\overline{G} \subset \Omega_r$. If there exists a nontrivial function $u \in C^1(\overline{G}; \mathbb{R})$ such that $u = 0$ on ∂G and $M_G[u] \leq 0$, where M_G is defined by* (8.5.5), *then every solution $v \in \mathcal{D}_P(\Omega)$ of* (8.5.1) *is oscillatory in Ω.*

Proof. Let $r > 0$ be an arbitrary number. It follows from Theorem 8.5.2 that every solution $v \in \mathcal{D}_P(\Omega)$ of (8.5.1) has a zero on $\overline{G} \subset \Omega_r$, that is, every solution v of (8.5.1) is oscillatory in Ω. □

Theorem 8.5.4. *Let $0 < \alpha < 1$. Assume that for any $r > 0$ there are a bounded and piecewise smooth domain G with $\overline{G} \subset \Omega_r$ and a nontrivial function $u \in C^1(\overline{G}; \mathbb{R})$ such that $u = 0$ on ∂G and*

$$\int_G \left[\frac{A(x)}{1-\alpha} |\nabla u|^{\alpha+1} - \left\{ H(x) - \frac{|B(x)|^{\alpha+1}}{(1-\alpha)A(x)^{\alpha}} \right\} |u|^{\alpha+1} \right] dx \leq 0, \tag{8.5.6}$$

where

$$H(x) = \frac{\beta-\gamma}{\alpha-\gamma}\left(\frac{\beta-\alpha}{\alpha-\gamma}\right)^{\frac{\alpha-\beta}{\beta-\gamma}} C(x)^{\frac{\alpha-\gamma}{\beta-\gamma}} D(x)^{\frac{\beta-\alpha}{\beta-\gamma}}.$$

Then every solution $v \in \mathcal{D}_P(\Omega)$ of (8.5.1) is oscillatory in Ω.

Proof. The inequality (8.3.6) of Lemma 8.3.1 with $W(x) = B(x)/A(x)$ yields the inequality

$$A(x)\left|\nabla u - \frac{u}{A(x)}B(x)\right|^{\alpha+1} \le \frac{A(x)}{1-\alpha}|\nabla u|^{\alpha+1} + \frac{|B(x)|^{\alpha+1}}{(1-\alpha)A(x)^{\alpha}}|u|^{\alpha+1}.$$

Hence we see from (8.5.6) that

$$M_G[u] \le \int_G \left[\frac{A(x)}{1-\alpha}|\nabla u|^{\alpha+1} - \left\{H(x) - \frac{|B(x)|^{\alpha+1}}{(1-\alpha)A(x)^{\alpha}}\right\}|u|^{\alpha+1}\right] dx \le 0.$$

The conclusion follows from Theorem 8.5.3. □

Theorem 8.5.5. *Let $A(x) > \alpha$ in Ω. Assume that for any $r > 0$ there are a bounded and piecewise smooth domain G with $\overline{G} \subset \Omega_r$ and a nontrivial function $u \in C^1(\overline{G};\mathbb{R})$ such that $u = 0$ on ∂G and*

$$\int_G \left[\frac{A(x)^2}{A(x)-\alpha}|\nabla u|^{\alpha+1} - \left\{H(x) - \frac{A(x)}{A(x)-\alpha}|B(x)|^{\alpha+1}\right\}|u|^{\alpha+1}\right] dx \le 0. \tag{8.5.7}$$

Then every solution $v \in \mathcal{D}_P(\Omega)$ of (8.5.1) is oscillatory in Ω.

Proof. The inequality (8.3.9) of Lemma 8.3.2 with $E(x) = A(x)$ and $W(x) = B(x)/A(x)$ yields the inequality

$$A(x)\left|\nabla u - \frac{u}{A(x)}B(x)\right|^{\alpha+1} \le \frac{A(x)^2}{A(x)-\alpha}|\nabla u|^{\alpha+1} + \frac{A(x)}{A(x)-\alpha}|B(x)|^{\alpha+1}|u|^{\alpha+1}.$$

Using the above inequality and (8.5.7), we observe that $M_G[u] \le 0$. The conclusion follows from Theorem 8.5.3. □

Theorem 8.5.6. *Let $0 < \alpha < 1$. If the half-linear ordinary differential equation*

$$\left(r^{n-1}\overline{\left\{\frac{A(x)}{1-\alpha}\right\}}(r)\,|y'|^{\alpha-1}y'\right)' + r^{n-1}\overline{\left\{H(x) - \frac{|B(x)|^{\alpha+1}}{(1-\alpha)A(x)^{\alpha}}\right\}}(r)\,|y|^{\alpha-1}y = 0$$

is oscillatory at $r = \infty$, then every solution $v \in \mathcal{D}_P(\mathbb{R}^n)$ of (8.5.1) is oscillatory in $\mathbb{R}^n$.

Proof. Proceeding as in the proof of Theorem 8.3.6 of Section 8.3, we observe that the conclusion follows from Theorem 8.5.4. □

Analogously we obtain the following:

Theorem 8.5.7. *Let $A(x) > \alpha$ in $\mathbb{R}^n$. If the half-linear ordinary differential equation*

$$\left(r^{n-1} \overline{\left\{ \frac{A(x)^2}{A(x)-\alpha} \right\}}(r)\, |y'|^{\alpha-1} y' \right)' + r^{n-1} \overline{\left\{ H(x) - \frac{A(x)}{A(x)-\alpha} |B(x)|^{\alpha+1} \right\}}(r)\, |y|^{\alpha-1} y = 0$$

is oscillatory at $r = \infty$, then every solution $v \in \mathcal{D}_P(\mathbb{R}^n)$ of (8.5.1) is oscillatory in $\mathbb{R}^n$.

Theorem 8.5.8. *Let $0 < \alpha < 1$. If there are positive constants A_0 and H_0 satisfying*

$$\frac{A(x)}{1-\alpha} \le A_0, \quad H(x) - \frac{|B(x)|^{\alpha+1}}{(1-\alpha)A(x)^\alpha} \ge H_0,$$

then every solution $v \in \mathcal{D}_P(\mathbb{R}^n)$ of (8.5.1) is oscillatory in $\mathbb{R}^n$.

Theorem 8.5.9. *Let $A(x) > \alpha$ in $\mathbb{R}^n$. If there are positive constants A_0 and H_0 satisfying*

$$\frac{A(x)^2}{A(x)-\alpha} \le A_0, \quad H(x) - \frac{A(x)}{A(x)-\alpha} |B(x)|^{\alpha+1} \ge H_0,$$

then every solution $v \in \mathcal{D}_P(\mathbb{R}^n)$ of (8.5.1) is oscillatory in $\mathbb{R}^n$.

The proofs of Theorems 8.5.8 and 8.5.9 follow from the inequality

$$M_G[u] \le \int_G \left[A_0 |\nabla u|^{\alpha+1} - H_0 |u|^{\alpha+1} \right] dx$$

and the fact that

$$\left(r^{n-1} A_0\, |y'|^{\alpha-1} y' \right)' + r^{n-1} H_0\, |y|^{\alpha-1} y = 0$$

is oscillatory at $r = \infty$ (see Example 8.1.1 of Section 8.1).

Example 8.5.1. We consider the superlinear-sublinear equation

$$\nabla \cdot (3|\nabla v| \nabla v) + 3|\nabla v| \left(\frac{\partial v}{\partial x_1} + \frac{\partial v}{\partial x_2} \right) + 32|v|^3 v + 32|v|^{-1/2} v = 0 \quad (8.5.8)$$

for $x = (x_1, x_2) \in \mathbb{R}^2$. Here $n = 2$, $\alpha = 2$, $\beta = 4$, $\gamma = 1/2$, $A(x) = 3$, $B(x) = (1, 1)$, $C(x) = D(x) = 32$. An easy computation yields

$$\frac{A(x)^2}{A(x) - \alpha} = 9,$$

$$H(x) - \frac{A(x)}{A(x) - \alpha}|B(x)|^{\alpha+1} = 16 - 6\sqrt{2} > 0$$

and hence we can take $A_0 = 9$ and $H_0 = 16 - 6\sqrt{2}$. Theorem 8.5.9 implies that every solution v of (8.5.8) is oscillatory in $\mathbb{R}^2$.

8.6 Quasilinear Parabolic Equations

In 1962 McNabb [196] established criteria for unboundedness of solutions of linear parabolic equations on the basis of Picone identity. His results were extended by Dunninger [72], Kusano and Narita [155] to parabolic differential inequalities, and by Chan [46], Chan and Young [47,48], Kobayashi and Yoshida [133], Kuks [149] to time-dependent matrix differential inequalities. All of them also contain the results about zeros of solutions or singularities of matrix solutions.

Recently Jaroš, Kusano and Yoshida [113] established Picone-type inequalities which connect a linear elliptic operator with an associated superlinear-sublinear elliptic operator. Extending Picone-type inequalities to parabolic equations with time-dependent coefficients, Jaroš, Kusano and Yoshida [114] studied oscillatory behavior and unboundedness of solutions of superlinear-sublinear parabolic equations of the form

$$\frac{\partial v}{\partial t} - \left[\sum_{i,j=1}^{n} \frac{\partial}{\partial x_i}\left(A_{ij}(x,t)\frac{\partial v}{\partial x_j}\right) + C(x,t)|v|^{\beta-1}v + D(x,t)|v|^{\gamma-1}v\right] = 0$$

in a cylindrical domain $G \times (0, \infty) \subset \mathbb{R}^{n+1}$. We note that Jaroš, Kusano and Yoshida [117] studied the quasilinear parabolic equation

$$\frac{\partial v}{\partial t} - \left[\nabla \cdot (A(x,t)|\nabla v|^{\alpha-1}\nabla v) + C(x,t)|v|^{\alpha-1}v\right] = 0,$$

where $\alpha > 0$ is a constant.

In this section we deal with the quasilinear parabolic equation

$$\frac{\partial v}{\partial t} - P[v] = 0, \quad (x,t) \in \Omega := G \times (0, \infty), \tag{8.6.1}$$

where G is a bounded domain in $\mathbb{R}^n$ with piecewise smooth boundary ∂G and

$$P[v] = \nabla \cdot (A(x,t)|\nabla v|^{\alpha-1}\nabla v) + C(x,t)|v|^{\beta-1}v + D(x,t)|v|^{\gamma-1}v. \tag{8.6.2}$$

We investigate oscillations of solutions of (8.6.1), and unboundedness of solutions is also obtained as corollaries.

It is assumed that $A(x,t) \in C(\overline{\Omega};(0,\infty))$, $C(x,t) \in C(\overline{\Omega};[0,\infty))$, $D(x,t) \in C(\overline{\Omega};[0,\infty))$, and α,β,γ are constants such that $\beta > \alpha$, $0 < \gamma < \alpha$.

The domain $\mathcal{D}_P(\Omega)$ of P is defined to be the set of all functions v of class $C^1(\overline{\Omega};\mathbb{R})$ with the property that $A(x,t)|\nabla v|^{\alpha-1}\nabla v \in C^1(\Omega;\mathbb{R}^n)\cap C(\overline{\Omega};\mathbb{R}^n)$.

Definition 8.6.1. By a *solution* of Eq. (8.6.1) we mean a function $v \in \mathcal{D}(\Omega)$ which satisfies (8.6.1).

Definition 8.6.2. A solution v of (8.6.1) is said to be *oscillatory* on $\overline{\Omega}$ if v has a zero on $\overline{G}\times[t,\infty)$ for any $t > 0$. Otherwise, v is called *nonoscillatory* on $\overline{\Omega}$.

Associated with (8.6.2) we consider the half-linear elliptic operator p defined by

$$p[u] = \nabla\cdot\left(a(x)|\nabla u|^{\alpha-1}\nabla u\right) + c(x)|u|^{\alpha-1}u,$$

where $a(x) \in C(\overline{G};(0,\infty))$ and $c(x) \in C(\overline{G};\mathbb{R})$.

The domain $\mathcal{D}_p(G)$ of p is defined to be the set of all functions u of class $C^1(\overline{G};\mathbb{R})$ with the property that $a(x)|\nabla u|^{\alpha-1}\nabla u \in C^1(G;\mathbb{R}^n)\cap C(\overline{G};\mathbb{R}^n)$.

Theorem 8.6.1 (Picone-type inequality). *Assume that $u \in \mathcal{D}_p(G)$, $v \in \mathcal{D}_P(\Omega)$ and $v \neq 0$ in $G\times I$, where I is any interval contained in $(0,\infty)$. Then we have the Picone-type inequality:*

$$\begin{aligned}
&\nabla\cdot\left(\frac{u}{\varphi(v)}[\varphi(v)a(x)\Phi(\nabla u) - \varphi(u)A(x,t)\Phi(\nabla v)]\right)\\
&\ge \left(a(x)-A(x,t)\right)|\nabla u|^{\alpha+1} + \left(H(x,t)-c(x)\right)|u|^{\alpha+1}\\
&\quad + A(x,t)\left[|\nabla u|^{\alpha+1} + \alpha\left|\frac{u}{v}\nabla v\right|^{\alpha+1} - (\alpha+1)(\nabla u)\cdot\Phi\left(\frac{u}{v}\nabla v\right)\right]\\
&\quad + \frac{u}{\varphi(v)}\left(\varphi(v)p[u] - \varphi(u)P[v]\right),\quad (x,t)\in G\times I,
\end{aligned}\tag{8.6.3}$$

where $\varphi(s) = |s|^{\alpha-1}s$ $(s\in\mathbb{R})$, $\Phi(\xi) = |\xi|^{\alpha-1}\xi$ $(\xi\in\mathbb{R}^n)$ and

$$H(x,t) = \frac{\beta-\gamma}{\alpha-\gamma}\left(\frac{\beta-\alpha}{\alpha-\gamma}\right)^{\frac{\alpha-\beta}{\beta-\gamma}} C(x,t)^{\frac{\alpha-\gamma}{\beta-\gamma}} D(x,t)^{\frac{\beta-\alpha}{\beta-\gamma}}.$$

Proof. The identity (8.1.4) with $p_\alpha[u]$ replaced by $p[u]$ is valid:

$$\begin{aligned}
&\nabla\cdot\left(\frac{u}{\varphi(v)}[\varphi(v)a(x)\Phi(\nabla u)]\right)\\
&\qquad = a(x)|\nabla u|^{\alpha+1} - c(x)|u|^{\alpha+1} + \frac{u}{\varphi(v)}\left(\varphi(v)p[u]\right).
\end{aligned}\tag{8.6.4}$$

By the same arguments as were used in the proof of Theorem 8.5.1, we obtain the following inequality:

$$\begin{aligned}
&-\nabla\cdot\left(u\varphi(u)\frac{A(x,t)\Phi(\nabla v)}{\varphi(v)}\right)\\
&\geq -A(x,t)\,|\nabla u|^{\alpha+1} + H(x,t)\,|u|^{\alpha+1}\\
&\quad +A(x,t)\left[|\nabla u|^{\alpha+1} + \alpha\left|\frac{u}{v}\nabla v\right|^{\alpha+1} - (\alpha+1)(\nabla u)\cdot\Phi\left(\frac{u}{v}\nabla v\right)\right]\\
&\quad -\frac{u\varphi(u)}{\varphi(v)}P[v] \qquad (8.6.5)
\end{aligned}$$

which is the inequality (8.5.2) with $A(x) = A(x,t)$, $B(x) = 0$, $C(x) = C(x,t)$, $D(x) = D(x,t)$. Combining (8.6.4) with (8.6.5) yields the desired Picone-type inequality (8.6.3). □

The following notation will be used:

$$V[u](t) = \int_G \left[(a(x) - A(x,t))|\nabla u|^{\alpha+1} + \left(H(x,t) - c(x)\right)|u|^{\alpha+1}\right] dx,$$

$$M[u](t) = \int_G \left[A(x,t)|\nabla u|^{\alpha+1} - H(x,t)|u|^{\alpha+1}\right] dx.$$

Theorem 8.6.2. *Assume that there is a nontrivial function $u \in \mathcal{D}_p(G)$ such that*

$$\begin{aligned}
&p[u] = 0 \quad \textit{in } G,\\
&u = 0 \quad \textit{on } \partial G,\\
&\lim_{t\to\infty}\int_T^t V[u](s)\,ds = \infty \quad \textit{for any } T > 0.
\end{aligned}$$

In case $0 < \alpha \leq 1$, every solution $v \in \mathcal{D}_P(\Omega)$ of (8.6.1) which is nonoscillatory on $\overline{\Omega}$ satisfies

$$\lim_{t\to\infty}\int_G |u|^{\alpha+1}\theta(|v|)\,dx = \infty, \qquad (8.6.6)$$

where

$$\theta(s) = \begin{cases} \log s & (\textit{if} \quad \alpha = 1)\\ s^{-\alpha+1} & (\textit{if} \quad 0 < \alpha < 1). \end{cases}$$

In case $\alpha > 1$, every solution $v \in \mathcal{D}_P(\Omega)$ of (8.6.1) is oscillatory on $\overline{\Omega}$.

Proof. Let $0 < \alpha \le 1$ and $v \in \mathcal{D}_P(\Omega)$ be a solution of (8.6.1) which is nonoscillatory on $\overline{\Omega}$. Then there is a number $t_0 > 0$ such that $v \neq 0$ on $\overline{G} \times [t_0, \infty)$. Integrating the Picone-type inequality (8.6.3) over G, we see that

$$\begin{aligned} 0 &\ge V[u](t) - \int_G \frac{u\varphi(u)}{\varphi(v)} P[v]\, dx \\ &= V[u](t) - \int_G |u|^{\alpha+1} \frac{1}{|v|^{\alpha-1}v} \frac{\partial v}{\partial t}\, dx, \quad t \ge t_0 \end{aligned} \tag{8.6.7}$$

in view of the fact that

$$A(x,t)\left[|\nabla u|^{\alpha+1} + \alpha\left|\frac{u}{v}\nabla v\right|^{\alpha+1} - (\alpha+1)(\nabla u)\cdot \Phi\left(\frac{u}{v}\nabla v\right)\right] \ge 0 \text{ on } \overline{G}\times[t_0,\infty)$$

(see Lemma 8.1.1 of Section 8.1). It is easy to check that

$$\frac{1}{|v|^{\alpha-1}v}\frac{\partial v}{\partial t} = \begin{cases} \dfrac{\partial}{\partial t}\log|v| & (\alpha = 1) \\ \dfrac{\partial}{\partial t}\left(\dfrac{1}{-\alpha+1}|v|^{-\alpha+1}\right) & (\alpha \neq 1) \end{cases}$$

and therefore (8.6.7) implies

$$\frac{d}{dt}\left(\int_G |u|^{\alpha+1}\log|v|\, dx\right) \ge V[u](t) \quad (\alpha = 1), \tag{8.6.8}$$

$$\frac{d}{dt}\left(\frac{1}{-\alpha+1}\int_G |u|^{\alpha+1}|v|^{-\alpha+1}\, dx\right) \ge V[u](t) \quad (\alpha \neq 1) \tag{8.6.9}$$

for $t \ge t_0$. We integrate (8.6.8) and (8.6.9) over $[t_0, T]$ to obtain

$$\Theta(T) - \Theta(t_0) \ge \int_{t_0}^T V[u](s)\, ds \quad (\alpha = 1), \tag{8.6.10}$$

$$\frac{1}{-\alpha+1}(\Theta(T) - \Theta(t_0)) \ge \int_{t_0}^T V[u](s)\, ds \quad (\alpha \neq 1), \tag{8.6.11}$$

where

$$\Theta(t) = \int_G |u|^{\alpha+1}\theta(|v|)\, dx. \tag{8.6.12}$$

In case $0 < \alpha \le 1$, we observe, using (8.6.10) and (8.6.11), that

$$\lim_{T\to\infty} \Theta(T) = \infty$$

which is equivalent to (8.6.6).

Let $\alpha > 1$. Suppose to the contrary that there is a nonoscillatory solution $v \in \mathcal{D}_P(\Omega)$ on $\overline{\Omega}$ of (8.6.1). Arguing as in the proof of the first statement, we see that (8.6.11) holds. Since $-\alpha + 1 < 0$, from (8.6.11) it follows that

$$\frac{1}{\alpha - 1}\Theta(t_0) \geq \int_{t_0}^{T} V[u](s)\,ds.$$

The right hand side of the above inequality tends to ∞ as $T \to \infty$, and therefore a contradiction yields. This completes the proof. □

Corollary 8.6.1. *Let $0 < \alpha \leq 1$ and assume that there is a nontrivial function $u \in \mathcal{D}_p(G)$ such that*

$$p[u] = 0 \quad \text{in } G,$$
$$u = 0 \quad \text{on } \partial G,$$
$$\lim_{t\to\infty} \int_T^t V[u](s)\,ds = \infty \quad \text{for any } T > 0.$$

Then every bounded solution $v \in \mathcal{D}_P(\Omega)$ of (8.6.1) *is oscillatory on $\overline{\Omega}$.*

Proof. Let $v \in \mathcal{D}_P(\Omega)$ be any bounded solution of (8.6.1). We easily see that $\theta(|v|)$ is bounded from above, and so is $\int_G |u|^{\alpha+1}\theta(|v|)\,dx$. Then (8.6.6) does not hold, hence Theorem 8.6.2 implies that the (bounded) solution v is oscillatory on $\overline{\Omega}$. □

Corollary 8.6.2. *Let $0 < \alpha \leq 1$ and assume that the same hypotheses as those of Theorem* 8.6.2 *hold. If $v \in \mathcal{D}_P(\Omega)$ is a solution of* (8.6.1) *which is nonoscillatory on $\overline{\Omega}$, then v is unbounded in Ω.*

Proof. Since v is nonoscillatory on $\overline{\Omega}$, it follows from Theorem 8.6.2 that v satisfies the condition (8.6.6). Hence, $|v|$ cannot be bounded from above in Ω, that is, v is unbounded in Ω. □

Theorem 8.6.3. *Assume that there exists a nontrivial function $u \in C^1(\overline{G};\mathbb{R})$ such that $u = 0$ on ∂G and*

$$\lim_{t\to\infty} \int_T^t M[u](s)\,ds = -\infty \quad \text{for any } T > 0. \tag{8.6.13}$$

In case $0 < \alpha \leq 1$, every solution $v \in \mathcal{D}_P(\Omega)$ of (8.6.1) *which is nonoscillatory on $\overline{\Omega}$ satisfies* (8.6.6). *In case $\alpha > 1$, every solution $v \in \mathcal{D}_P(\Omega)$ of* (8.6.1) *is oscillatory on $\overline{\Omega}$.*

Proof. Let $0 < \alpha \leq 1$ and assume that $v \in \mathcal{D}_P(\Omega)$ is a solution of (8.6.1) which is nonoscillatory on $\overline{\Omega}$. Then $v \neq 0$ on $\overline{G} \times [t_0, \infty)$ for some $t_0 > 0$.

In the proof of Theorem 8.6.2 we used the Picone-type inequality (8.6.3). Integrating (8.6.5) over G instead of (8.6.3), we obtain

$$\begin{aligned} 0 &\leq M[u](t) + \int_G |u|^{\alpha+1} \frac{1}{|v|^{\alpha-1}v} P[v]\, dx \\ &= M[u](t) + \int_G |u|^{\alpha+1} \frac{1}{|v|^{\alpha-1}v} \frac{\partial v}{\partial t}\, dx, \quad t \geq t_0. \end{aligned}$$

Proceeding as in the proof of Theorem 8.6.2, we observe that

$$\Theta(T) - \Theta(t_0) \geq -\int_{t_0}^{T} M[u](s)\, ds \quad (\alpha = 1),$$

$$\frac{1}{-\alpha+1}(\Theta(T) - \Theta(t_0)) \geq -\int_{t_0}^{T} M[u](s)\, ds \quad (\alpha \neq 1),$$

where $\Theta(t)$ is given by (8.6.12). Using the same arguments as in the proof of Theorem 8.6.2, we find that v satisfies (8.6.6). The case where $\alpha > 1$ can be handled by an argument similar to that of Theorem 8.6.2. The proof is complete. □

Corollary 8.6.3. *Let $0 < \alpha \leq 1$ and assume that there is a nontrivial function $u \in C^1(\overline{G}; \mathbb{R})$ satisfying* (8.6.13) *and the boundary condition $u = 0$ on ∂G. Then every bounded solution $v \in \mathcal{D}_P(\Omega)$ of* (8.6.1) *is oscillatory on $\overline{\Omega}$.*

Corollary 8.6.4. *Let $0 < \alpha \leq 1$ and assume that there is a nontrivial function $u \in C^1(\overline{G}; \mathbb{R})$ satisfying* (8.6.13) *and the boundary condition $u = 0$ on ∂G. If $v \in \mathcal{D}_P(\Omega)$ is a solution of* (8.6.1) *which is nonoscillatory on $\overline{\Omega}$, then v is unbounded in Ω.*

Corollaries 8.6.3 and 8.6.4 follow from Theorem 8.6.3, and the proofs of them are quite similar to those of Corollaries 8.6.1 and 8.6.2, respectively, and will be omitted.

Example 8.6.1. We consider the quasilinear parabolic equation

$$\frac{\partial v}{\partial t} - \left[\frac{\partial}{\partial x}\left(A_0 \left|\frac{\partial v}{\partial x}\right|^{\alpha-1} \frac{\partial v}{\partial x}\right) + C_0 v^3 + C_0 v^{1/3}\right] = 0,$$

$$(x,t) \in (-1,1) \times (0,\infty), \quad (8.6.14)$$

where A_0 and C_0 are positive constants. Here $n = 1$, $A(x,t) = A_0 > 0$, $C(x,t) = D(x,t) = C_0 > 0$, $\beta = 3$, $\gamma = 1/3$, $G = (-1,1)$ and $\Omega = (-1,1) \times (0,\infty)$. We consider two cases where $\alpha = 2$ or $\alpha = 1/2$. First we treat the case where $\alpha = 2$. Choosing $u = 1 - x^2$, we observe that $u(-1) = u(1) = 0$. It is easily verified that

$$H(x,t) = H_0 = \frac{8}{5}\left(\frac{5}{3}\right)^{3/8} C_0.$$

An easy calculation yields

$$\begin{aligned} M[u](t) &= \int_{-1}^{1} \left[A_0|u'(x)|^3 - H_0|u(x)|^3\right] dx \\ &= 4A_0 - \frac{32}{35}H_0. \end{aligned}$$

If $A_0 < (8/35)H_0$, then the condition (8.6.13) is satisfied, and therefore Theorem 8.6.3 implies that every solution v of (8.6.14) with $\alpha = 2$ is oscillatory on $\overline{\Omega}$.

Next we deal with the case where $\alpha = 1/2$. Choosing $u = 1 - x^2$, we find that $u(-1) = u(1) = 0$ and

$$\begin{aligned} M[u](t) &= \int_{-1}^{1} \left[A_0|u'(x)|^{3/2} - H_0|u(x)|^{3/2}\right] dx \\ &= \frac{8}{5}\sqrt{2}A_0 - \frac{3}{8}\pi H_0. \end{aligned}$$

If $A_0 < (15/128)\sqrt{2}\pi H_0$, we see that the condition (8.6.13) is satisfied, and therefore Theorem 8.6.3 implies that every solution v of (8.6.14) with $\alpha = 1/2$ which is nonoscillatory on $\overline{\Omega}$ satisfies

$$\lim_{t\to\infty} \int_{-1}^{1} \left(1 - x^2\right)^{3/2} |v|^{1/2}\, dx = \infty.$$

Example 8.6.2. We consider the quasilinear parabolic equation

$$\frac{\partial v}{\partial t} - \left[\frac{\partial}{\partial x}\left(A_0\left|\frac{\partial v}{\partial x}\right|\frac{\partial v}{\partial x}\right) + \frac{1}{2}e^{-4t}v^5 + \frac{1}{2}e^{(2/3)t}v^{1/3}\right] = 0 \qquad (8.6.15)$$

for $(x,t) \in (0,\pi) \times (0,\infty)$, where A_0 is a positive constant. Here $n = 1$, $\alpha = 2$, $\beta = 5$, $\gamma = 1/3$, $A(x,t) = A_0 > 0$, $C(x,t) = (1/2)e^{-4t}$; $D(x,t) = (1/2)e^{(2/3)t}$, $G = (0,\pi)$ and $\Omega = (0,\pi) \times (0,\infty)$. Choosing $u = \sin x$, we find that $u(0) = u(\pi) = 0$,

$$H(x,t) = H_0(t) = \frac{7}{5}\left(\frac{5}{9}\right)^{9/14} e^{-t}$$

and

$$\begin{aligned} M[u](t) &= \int_0^{\pi} \left[A_0 |u'(x)|^3 - H(x,t)|u(x)|^3 \right] dx \\ &= 2A_0 \int_0^{\pi/2} \cos^3 x\, dx - 2H_0(t) \int_0^{\pi/2} \sin^3 x\, dx \\ &= \frac{4}{3}(A_0 - H_0(t)). \end{aligned}$$

Hence, the condition (8.6.13) is violated. Then there exists a nonoscillatory solution $v = e^t$ of (8.6.15).

Example 8.6.3. We consider the quasilinear parabolic equation

$$\frac{\partial v}{\partial t} - \left[\frac{\partial}{\partial x} \left(A_0 \left| \frac{\partial v}{\partial x} \right|^{-1/2} \frac{\partial v}{\partial x} \right) + C_0 v^3 + C_0 v^{1/5} \right] = 0 \qquad (8.6.16)$$

for $(x,t) \in (0,\pi/2) \times (0,\infty)$, where A_0 and C_0 are positive constants. Here $n = 1$, $\alpha = 1/2$, $\beta = 3$, $\gamma = 1/5$, $A(x,t) = A_0 > 0$, $C(x,t) = D(x,t) = C_0 > 0$, $G = (0,\pi/2)$ and $\Omega = (0,\pi/2) \times (0,\infty)$. Letting $u = x \cos x$, we see that $u(0) = u(\pi/2) = 0$ and

$$M[u](t) = \int_0^{\pi/2} \left[A_0 |u'(x)|^{3/2} - H_0 |u(x)|^{3/2} \right] dx,$$

where

$$H_0 = \frac{28}{3} \left(\frac{3}{25} \right)^{25/28} C_0.$$

If C_0 is sufficiently large, then $M[u](t)$ is a negative constant, and therefore the condition (8.6.13) is satisfied. From Corollary 8.6.3 it follows that every bounded solution v of (8.6.16) is oscillatory on $\overline{\Omega}$.

8.7 Parabolic Systems

Beginning with the work of McNabb [196], unboundedness of solutions has been studied by numerous authors. We refer the reader to Dunninger [72], Jaroš, Kusano and Yoshida [114, 117] for scalar parabolic equations, and to Chan [46], Chan and Young [47, 48], Kuks [149], Kusano and Narita [155] for parabolic systems.

We are concerned with matrix solutions of the time-dependent differential system of parabolic type

$$\frac{\partial W}{\partial t} - P[W] = 0 \quad \text{in } \Omega := G \times (0,\infty), \qquad (8.7.1)$$

where G is a bounded domain in $\mathbb{R}^n$ with piecewise smooth boundary ∂G and

$$P[W] = \sum_{i,j=1}^{n} \frac{\partial}{\partial x_i}\left(G_{ij}(x,t)\frac{\partial W}{\partial x_j}\right) + H(x,t)W.$$

Definition 8.7.1. Let $(a_{ij}(x))$ be a $k \times \ell$ matrix function in a domain (or closed domain) $E \subset \mathbb{R}^n$. Then the following holds:

$$(a_{ij}(x)) = (\tilde{a}_1(x), \tilde{a}_2(x), ..., \tilde{a}_\ell(x)),$$

where $\tilde{a}_j(x) = (a_{1j}(x), a_{2j}(x), ..., a_{kj}(x))^T$. A matrix function $(a_{ij}(x))$ is said to be an $\mathbb{R}^{\ell\times k}$*-valued function of class* $C^r(E)$ if each component $a_{ij}(x)$ is of class $C^r(E;\mathbb{R})$ $(i = 1,2,...,k;\ j = 1,2,...,\ell)$, where r is a nonnegative integer. We use the notation:

$$C^r(E;\mathbb{R}^{\ell\times k}) = \{(a_{ij}(x));\ a_{ij}(x) \in C^r(E;\mathbb{R})\ (i = 1,2,...,k;\ j = 1,2,...,\ell)\}.$$

It is assumed that:

(H8.7-1) $G_{ij}(x,t)$ $(i,j = 1,2,...,n)$ and $H(x,t)$ are $m \times m$ real symmetric matrix functions;

(H8.7-2) $G_{ij}(x,t) \in C^1(\overline{\Omega};\mathbb{R}^{m\times m})$ $(i,j=1,2,...,n)$, $H(x,t) \in C(\overline{\Omega};\mathbb{R}^{m\times m})$;

(H8.7-3) $G_{ij}(x,t) = G_{ji}(x,t)$ $(i,j = 1,2,...,n)$ and the $mn \times mn$ matrix $\mathscr{G} = (G_{ij}(x,t))_{i,j=1}^{n}$ is positive definite in Ω.

The domain $\mathcal{D}_P(\Omega)$ of P is defined to be the set of all $m \times m$ matrix functions $W \in C^2(\Omega;\mathbb{R}^{m\times m}) \cap C^1(\overline{\Omega};\mathbb{R}^{m\times m})$.

Definition 8.7.2. A function $v : \overline{\Omega} \longrightarrow \mathbb{R}$ is said to be *oscillatory* on $\overline{\Omega}$ if v has a zero on $\overline{G} \times [t,\infty)$ for any $t > 0$. Otherwise, v is called *nonoscillatory* on $\overline{\Omega}$.

Definition 8.7.3. An $m \times m$ matrix $W(x,t) \in C^1(\tilde{\Omega};\mathbb{R}^{m\times m})$, $\tilde{\Omega} \subset \Omega$, is said to be *prepared* in $\tilde{\Omega}$ with respect to P if the matrices

$$\sum_{j=1}^{n} W(x,t)^T G_{ij}(x,t)\frac{\partial W}{\partial x_j}(x,t) \quad (i = 1,2,...,n)$$

are symmetric in $\tilde{\Omega}$, where the superscript T denotes the transpose.

Theorem 8.7.1 (Picone identity). *Let* $W \in \mathcal{D}_P(\Omega)$ *and let* $\det W \neq 0$ *in* $G \times I$, *where* I *is any interval contained in* $(0,\infty)$. *If* W *is prepared* *in $G \times I$ with respect to P, then the following Picone identity holds for any m-column vector $u \in C^1(G;\mathbb{R}^{1\times m})$:*

$$\sum_{i,j=1}^{n}\left(W\frac{\partial}{\partial x_i}(W^{-1}u)\right)^T G_{ij}(x,t)\left(W\frac{\partial}{\partial x_j}(W^{-1}u)\right)$$
$$+\sum_{i,j=1}^{n}\frac{\partial}{\partial x_i}\left(u^T G_{ij}(x,t)\frac{\partial W}{\partial x_j}W^{-1}u\right)$$
$$=\sum_{i,j=1}^{n}\left(\frac{\partial u}{\partial x_i}\right)^T G_{ij}(x,t)\frac{\partial u}{\partial x_j}-u^T H(x,t)u+u^T P[W]W^{-1}u. \tag{8.7.2}$$

Proof. In the case where $G_{ij}(x,t)=G_{ij}(x)$, the Picone identity (8.7.2) was established (see, for example, Kusano and Yoshida [156, p.172]). Differentiations appearing in (8.7.2) are only partial differentiations with respect to x_i, and so we can consider t as a parameter. Hence, we conclude that the identity (8.7.2) holds. □

Lemma 8.7.1. *Assume that $W \in \mathcal{D}_P(\Omega)$ is symmetric and nonsingular in $G\times[t_0,\infty)$ for some $t_0>0$. If $\frac{\partial}{\partial t}\log W$ commutes with $\log W$ in $G\times[t_0,\infty)$, then we obtain*

$$\frac{\partial W}{\partial t}W^{-1}=\frac{\partial}{\partial t}\log W=\frac{\partial}{\partial t}\Big(\mathrm{Re}\log W\Big)\quad \text{in } G\times[t_0,\infty), \tag{8.7.3}$$

where $\log W$ denotes the principal value of logarithm of W and Re *means the real part.*

Proof. Since

$$W=\exp(\log W)=\sum_{j=0}^{\infty}\frac{1}{j\,!}(\log W)^j,$$

we have

$$\begin{aligned}\frac{\partial W}{\partial t}&=\sum_{j=0}^{\infty}\frac{1}{j\,!}\frac{\partial}{\partial t}(\log W)^j\\&=\sum_{j=1}^{\infty}\frac{1}{j\,!}\,j\left(\frac{\partial}{\partial t}\log W\right)(\log W)^{j-1}\\&=\left(\frac{\partial}{\partial t}\log W\right)\exp(\log W)\\&=\left(\frac{\partial}{\partial t}\log W\right)W.\end{aligned}$$

Hence we obtain

$$\frac{\partial W}{\partial t} W^{-1} = \frac{\partial}{\partial t} \log W.$$

Since W is a real symmetric matrix, there exists an orthogonal matrix S such that $S^{-1}WS = J$, where

$$J = \begin{pmatrix} \lambda_1 & & & 0 \\ & \lambda_2 & & \\ & & \ddots & \\ 0 & & & \lambda_m \end{pmatrix},$$

λ_i $(i = 1, 2, ..., n)$ being the eigenvalues of W. It can be shown that

$$\begin{aligned} \log W &= \log (SJS^{-1}) = S(\log J)S^{-1} \\ &= S \begin{pmatrix} \log \lambda_1 & & & 0 \\ & \log \lambda_2 & & \\ & & \ddots & \\ 0 & & & \log \lambda_m \end{pmatrix} S^{-1}, \end{aligned}$$

where

$$\begin{aligned} \log \lambda_i &= \log |\lambda_i| + \sqrt{-1} \arg \lambda_i \quad (0 \le \arg \lambda_i < 2\pi) \\ &= \begin{cases} \log |\lambda_i| & \text{if } \lambda_i > 0 \\ \log |\lambda_i| + \sqrt{-1}\pi & \text{if } \lambda_i < 0. \end{cases} \end{aligned}$$

Therefore we have

$$\begin{aligned} \frac{\partial}{\partial t} \log W &= \frac{\partial}{\partial t} \left(S \begin{pmatrix} \log |\lambda_1| & & & 0 \\ & \log |\lambda_2| & & \\ & & \ddots & \\ 0 & & & \log |\lambda_m| \end{pmatrix} S^{-1} \right) \\ &= \frac{\partial}{\partial t} \Big(\text{Re} \log W \Big). \end{aligned}$$

□

Theorem 8.7.2. *Assume that* (H8.7-1)–(H8.7-3) *hold, and that there is a nontrivial m-column vector function* $u \in C^1(\overline{G}; \mathbb{R}^{1\times m})$ *such that* $u = 0$ *on* ∂G *and*

$$\lim_{t\to\infty} \int_T^t M[u](s)\, ds = -\infty \quad \textit{for any } T > 0, \tag{8.7.4}$$

where

$$M[u](t) = \int_G \left[\sum_{i,j=1}^{n} \left(\frac{\partial u}{\partial x_i} \right)^T G_{ij}(x,t) \frac{\partial u}{\partial x_j} - u^T H(x,t) u \right] dx.$$

Let $W \in \mathcal{D}_P(\Omega)$ *be a solution of* (8.7.1) *such that:*

(i) W *is symmetric in* Ω;
(ii) W *is prepared in* Ω *with respect to* P;
(iii) $\det W$ *is nonoscillatory on* $\overline{\Omega}$, *that is,* $\det W \neq 0$ *on* $\overline{G} \times [t_0, \infty)$ *for some* $t_0 > 0$;
(iv) $\frac{\partial}{\partial t} \log W$ *commutes with* $\log W$ *in* $G \times [t_0, \infty)$.

Then the following condition holds:

$$\lim_{t \to \infty} \int_G u^T \Big(\mathrm{Re} \log W \Big) u \, dx = \infty. \tag{8.7.5}$$

Proof. The hypotheses (i) and (ii) imply that the identity (8.7.2) holds in $G \times [t_0, \infty)$. Integrating (8.7.2) over G and taking account of (H8.7-3) yield

$$\begin{aligned} 0 &\leq M[u](t) + \int_G u^T P[W] W^{-1} u \, dx \\ &= M[u](t) + \int_G u^T \frac{\partial W}{\partial t} W^{-1} u \, dx, \quad t \geq t_0 \end{aligned}$$

which implies

$$-M[u](t) \leq \int_G u^T \frac{\partial W}{\partial t} W^{-1} u \, dx, \quad t \geq t_0.$$

Using Lemma 8.7.1, we obtain

$$-M[u](t) \leq \frac{d}{dt} \int_G u^T \Big(\mathrm{Re} \log W \Big) u \, dx, \quad t \geq t_0. \tag{8.7.6}$$

Integrating (8.7.6) over $[t_0, t]$, we have

$$-\int_{t_0}^t M[u](s) \, ds \leq z(t) - z(t_0),$$

where

$$z(t) = \int_G u^T \Big(\mathrm{Re} \log W \Big) u \, dx.$$

It follows from the hypothesis (8.7.4) that

$$\lim_{t \to \infty} z(t) = \infty$$

which is equivalent to (8.7.5). □

Theorem 8.7.3. *Assume that* (H8.7-1)–(H8.7-3) *hold, and that there is a nontrivial* m*-column vector function* $u \in C^1(\overline{G}; \mathbb{R}^{1 \times m})$ *satisfying* (8.7.4) *and the boundary condition* $u = 0$ *on* ∂G. *Let* $W \in \mathcal{D}_P(\Omega)$ *be a solution of* (8.7.1) *which satisfies the hypotheses* (i)–(iv) *of Theorem* 8.7.2. *Then,* $\|W\|$ *is unbounded in* Ω, *where*

$$\|W\| = \left(\mathrm{tr} \; W^T W \right)^{1/2},$$

$\mathrm{tr} \; W$ *being the trace of* W.

Proof. It is easy to see that

$$\left| \int_G u^T \Big(\text{Re} \log W \Big) u \, dx \right| \le K \|\text{Re} \log W\| \tag{8.7.7}$$

for some positive constant K. As was shown in the proof of Lemma 8.7.1, $\text{Re} \log W$ can be written in the form

$$\text{Re} \log W = S(\log \tilde{J}) S^{-1},$$

where S is an orthogonal matrix such that $S^{-1} W S = J$ and

$$\tilde{J} = \begin{pmatrix} |\lambda_1| & & & 0 \\ & |\lambda_2| & & \\ & & \ddots & \\ 0 & & & |\lambda_m| \end{pmatrix}.$$

Hence we obtain

$$\|\text{Re} \log W\| \le \|S\| \cdot \|\log \tilde{J}\| \cdot \|S^{-1}\| = m \, \|\log \tilde{J}\|. \tag{8.7.8}$$

We easily see that

$$\|\tilde{J}\| = \|J\| \le \|S^{-1}\| \cdot \|W\| \cdot \|S\| = m \, \|W\|. \tag{8.7.9}$$

Assume that $\|W\|$ is bounded. Then $\|\tilde{J}\|$ is bounded from (8.7.9), and therefore $\|\log \tilde{J}\|$ is also bounded. The inequality (8.7.8) implies that $\|\text{Re} \log W\|$ is bounded. In view of (8.7.7), we find that $\int_G u^T (\text{Re} \log W) u \, dx$ is bounded. Hence, the condition (8.7.5) means that $\|W\|$ is unbounded. □

The following corollary is an immediate consequence of Theorem 8.7.3.

Corollary 8.7.1. *Assume that* (H8.7-1)–(H8.7-3) *hold, and that there is a nontrivial m-column vector function $u \in C^1(\overline{G}; \mathbb{R}^{1 \times m})$ satisfying* (8.7.4) *and the boundary condition $u = 0$ on ∂G. Let $W \in \mathcal{D}_P(\Omega)$ be a solution of* (8.7.1) *which satisfies the hypotheses* (i)–(ii) *of Theorem* 8.7.2. *If $\|W\|$ is bounded in Ω, then either that* $\det W$ *is oscillatory on $\overline{\Omega}$, or* (*if* $\det W$ *is nonoscillatory on $\overline{\Omega}$*) *that* (iv) *of Theorem* 8.7.2 *does not hold.*

We now consider the comparison operator L defined by

$$L[u] = \sum_{i,j=1}^{n} \frac{\partial}{\partial x_i} \left(A_{ij}(x) \frac{\partial u}{\partial x_j} \right) + C(x) u, \tag{8.7.10}$$

where $A_{ij}(x)$ $(i, j = 1, 2, ..., n)$ and $C(x)$ satisfy the following hypotheses:

(H8.7-4) $A_{ij}(x)$ $(i,j=1,2,...,n)$ and $C(x)$ are $m \times m$ real symmetric matrix functions;

(H8.7-5) $A_{ij}(x) \in C^1(\overline{G};\mathbb{R}^{m\times m})$ $(i,j=1,2,...,n)$ and $C(x) \in C(\overline{G};\mathbb{R}^{m\times m})$;

(H8.7-6) $A_{ij}(x) = A_{ji}(x)$ $(i,j=1,2,...,n)$ and the $mn \times mn$ matrix $\mathscr{A} = \left(A_{ij}(x)\right)_{i,j=1}^{n}$ is positive definite in G.

The domain $\mathcal{D}_L(G)$ of L is defined to be the set of all m-column vector functions $u \in C^2(G;\mathbb{R}^{1\times m}) \cap C^1(\overline{G};\mathbb{R}^{1\times m})$.

Theorem 8.7.4. *Assume that* (H8.7-1)–(H8.7-6) *hold, and that there is a nontrivial m-column vector function $u \in \mathcal{D}_L(G)$ such that:*

$$L[u] = 0 \quad \text{in } G, \tag{8.7.11}$$

$$u = 0 \quad \text{on } \partial G, \tag{8.7.12}$$

$$\lim_{t\to\infty} \int_T^t V[u](s)\,ds = \infty \quad \text{for any } T > 0, \tag{8.7.13}$$

where

$$V[u](t) = \int_G \Bigg[\sum_{i,j=1}^{n} \left(\frac{\partial u}{\partial x_i}\right)^T \Big(A_{ij}(x) - G_{ij}(x,t)\Big)\frac{\partial u}{\partial x_j} + u^T \Big(H(x,t) - C(x)\Big)u\Bigg]\,dx.$$

Let $W \in \mathcal{D}_P(\Omega)$ be a solution of (8.7.1) *which satisfies the hypotheses* (i)–(iv) *of Theorem* 8.7.2. *Then, the condition* (8.7.5) *holds.*

Proof. Proceeding as in the proof of Theorem 8.7.1, we observe that the following Picone identity

$$\begin{aligned}
&\sum_{i,j=1}^{n} \frac{\partial}{\partial x_i}\left(u^T A_{ij}(x)\frac{\partial u}{\partial x_j} - u^T G_{ij}(x,t)\frac{\partial W}{\partial x_j}W^{-1}u\right)\\
&= \sum_{i,j=1}^{n}\left(\frac{\partial u}{\partial x_i}\right)^T \Big(A_{ij}(x) - G_{ij}(x,t)\Big)\frac{\partial u}{\partial x_j} + u^T\Big(H(x,t) - C(x)\Big)u\\
&\quad + \sum_{i,j=1}^{n}\left(W\frac{\partial}{\partial x_i}(W^{-1}u)\right)^T G_{ij}(x,t)\left(W\frac{\partial}{\partial x_j}(W^{-1}u)\right)\\
&\quad + u^T L[u] - u^T P[W]W^{-1}u
\end{aligned}$$

holds in $G \times [t_0,\infty)$. Integrating the above identity over G and taking account of the hypotheses, we obtain the inequality

$$0 \geq V[u](t) - \int_G u^T P[W]W^{-1}u\,dx, \quad t \geq t_0$$

or

$$V[u](t) \le \int_G u^T \frac{\partial W}{\partial t} W^{-1} u \, dx, \quad t \ge t_0.$$

Arguing as in the proof of Theorem 8.7.2, we conclude that the condition (8.7.5) holds. □

Theorem 8.7.5. *Assume that* (H8.7-1)–(H8.7-6) *hold, and that there is a nontrivial m-column vector function $u \in \mathcal{D}_L(G)$ satisfying* (8.7.11)–(8.7.13). *Let $W \in \mathcal{D}_P(\Omega)$ be a solution of* (8.7.1) *which satisfies the hypotheses* (i)–(iv) *of Theorem* 8.7.2. *Then, $\|W\|$ is unbounded in Ω.*

Proof. By the same arguments as were used in Theorem 8.7.3, we conclude that the conclusion follows from Theorem 8.7.4. □

We obtain the analogue of Corollary 8.7.1.

Corollary 8.7.2. *Assume that* (H8.7-1)–(H8.7-6) *hold, and that there is a nontrivial m-column vector function $u \in \mathcal{D}_L(G)$ satisfying* (8.7.11)–(8.7.13). *Let $W \in \mathcal{D}_P(\Omega)$ be a solution of* (8.7.1) *which satisfies the hypotheses* (i)–(ii) *of Theorem* 8.7.2. *If $\|W\|$ is bounded in Ω, then either that* $\det W$ *is oscillatory on $\overline{\Omega}$, or (if* $\det W$ *is nonoscillatory on $\overline{\Omega}$) that* (iv) *of Theorem* 8.7.2 *does not hold.*

Example 8.7.1. We consider the matrix differential system

$$\frac{\partial W}{\partial t} - \left(\alpha \frac{\partial^2 W}{\partial x^2} + \beta W \right) = 0, \quad (x,t) \in (0,\pi) \times (0,\infty), \tag{8.7.14}$$

where α and β are positive constants with $\alpha < \beta$. Here $n = 1$, $G_{11}(x,t) = \alpha \mathrm{I}_m$ ($\mathrm{I}_m : m \times m$ identity matrix), $H = \beta \mathrm{I}_m$, $G = (0,\pi)$ and $\Omega = (0,\pi) \times (0,\infty)$. Letting

$$u = \begin{pmatrix} \sin x \\ \sin x \\ \vdots \\ \sin x \end{pmatrix},$$

we see that $u(0) = u(\pi) = 0$ and

$$\begin{aligned} M[u](t) &= \int_0^\pi \left[\alpha \left(\frac{\partial u}{\partial x} \right)^T \frac{\partial u}{\partial x} - \beta u^T u \right] dx \\ &= \int_0^\pi \left[\alpha m \cos^2 x - \beta m \sin^2 x \right] dx \\ &= \frac{\pi}{2} m(\alpha - \beta) < 0. \end{aligned}$$

Hence we find that

$$\lim_{t\to\infty} \int_T^t M[u](s)\, ds = -\infty$$

for any $T > 0$. It follows from Theorem 8.7.2 that if W is a solution of (8.7.14) satisfying (i)–(iv) of Theorem 8.7.2, then (8.7.5) holds. One such solution is $W = e^{\beta t}\mathrm{I}_m$. In fact, it is clear that (i)–(iv) hold for $W = e^{\beta t}\mathrm{I}_m$, and that

$$\begin{aligned}
&\lim_{t\to\infty} \int_0^\pi u^T \Big(\mathrm{Re}\ \log W\Big) u\, dx \\
&= \lim_{t\to\infty} \int_0^\pi \beta t m \sin^2 x\, dx \\
&= \lim_{t\to\infty} \frac{\pi}{2}\beta m t = \infty.
\end{aligned}$$

Example 8.7.2. We consider the matrix differential system

$$\frac{\partial W}{\partial t} - \left(\alpha \frac{\partial^2 W}{\partial x^2} + \beta W\right) = 0, \quad (x,t) \in (-1,1)\times(0,\infty), \tag{8.7.15}$$

where α and β are positive constants satisfying $\alpha < (5/2)\beta$. Here $n = 1$, $G_{11}(x,t) = \alpha \mathrm{I}_m$, $H = \beta\,\mathrm{I}_m$, $G = (-1,1)$ and $\Omega = (-1,1)\times(0,\infty)$. We let

$$u = \begin{pmatrix} 1-x^2 \\ 1-x^2 \\ \vdots \\ 1-x^2 \end{pmatrix}$$

and find that $u(-1) = u(1) = 0$ and

$$\begin{aligned}
M[u](t) &= \int_{-1}^1 \left[\alpha \left(\frac{\partial u}{\partial x}\right)^T \frac{\partial u}{\partial x} - \beta u^T u\right] dx \\
&= \int_{-1}^1 \left[4\alpha m x^2 - \beta m (1-x^2)^2\right] dx \\
&= m\left(\frac{8}{3}\alpha - \frac{16}{15}\beta\right) < 0.
\end{aligned}$$

Hence it is easily seen that

$$\lim_{t\to\infty} \int_T^t M[u](s)\, ds = -\infty$$

for any $T > 0$. Theorem 8.7.3 implies that if W is a solution of (8.7.15) satisfying (i)–(iv) of Theorem 8.7.2, then $\|W\|$ is unbounded in $(-1,1)\times(0,\infty)$. For example, $W = e^{\beta t}\mathrm{I}_m$ is such a solution. In fact, we see that $\|W\| = \sqrt{m}\, e^{\beta t}$.

8.8 Applications to Riccati Method

In 1980 Noussair and Swanson [225] utilized Riccati method to investigate oscillations of elliptic equations of second order. Usami [275] extended the Wintner Theorem [283] to half-linear elliptic equations by using Riccati inequality. Riccati techniques were employed to obtain various oscillation results (for example, Wintner-type criteria, Kamenev-type criteria, Philos-type criteria) for quasilinear elliptic equations including half-linear equations, see Došlý and Mařík [70], Mařík [187–195], Xu [289, 290], Xu and Xing [291].

We deal with the half-linear damped elliptic equation

$$P_\alpha[v] = 0 \quad \text{in } \Omega, \tag{8.8.1}$$

where $\alpha > 0$ is a constant, Ω is an exterior domain which includes $\{x \in \mathbb{R}^n;\ |x| \geq r_0\}$ for some $r_0 > 0$, and $P_\alpha[v]$ is defined by

$$P_\alpha[v] = \nabla \cdot \left(A(x)|\nabla v|^{\alpha-1}\nabla v\right) + |\nabla v|^{\alpha-1}B(x) \cdot \nabla v + C(x)|v|^{\alpha-1}v.$$

It is assumed that $A(x) \in C(\Omega; (0, \infty))$, $B(x) \in C(\Omega; \mathbb{R}^n)$ and $C(x) \in C(\Omega; \mathbb{R})$.

The domain $\mathcal{D}_{P_\alpha}(\Omega)$ of P_α is defined to be the set of all functions v of class $C^1(\Omega; \mathbb{R})$ with the property that $A(x)|\nabla v|^{\alpha-1}\nabla v \in C^1(\Omega; \mathbb{R}^n)$.

A solution $v \in \mathcal{D}_{P_\alpha}(\Omega)$ of (8.8.1) is said to be *oscillatory* in Ω if it has a zero in Ω_r for any $r > 0$, where

$$\Omega_r = \Omega \cap \{x \in \mathbb{R}^n;\ |x| > r\}.$$

We use the notation:

$$A(r, s) = \{x \in \mathbb{R}^n;\ r < |x| < s\},$$
$$A(r, \infty) = \{x \in \mathbb{R}^n;\ |x| > r\},$$
$$A[r, \infty) = \{x \in \mathbb{R}^n;\ |x| \geq r\}.$$

Since Ω is an exterior domain in $\mathbb{R}^n$, we see that

$$\Omega_{r_1} = A(r_1, \infty)$$

for some large $r_1 \geq r_0$.

Lemma 8.8.1. *If $v \in \mathcal{D}_{P_\alpha}(\Omega)$ is a solution of* (8.8.1) *and $v \neq 0$ on $A[r_2, \infty)$ for some $r_2 > r_1$, then we obtain the Riccati-type equation*

$$\nabla \cdot W(x) + C(x) + \alpha A(x)^{-1/\alpha}|W(x)|^{1+(1/\alpha)} + \langle W(x), A(x)^{-1}B(x)\rangle = 0 \quad \text{on } A[r_2, \infty), \tag{8.8.2}$$

where $\langle U, V\rangle$ denotes the scalar product of U, $V \in \mathbb{R}^n$ and

$$W(x) = \frac{A(x)|\nabla v|^{\alpha-1}\nabla v}{|v|^{\alpha-1}v}. \tag{8.8.3}$$

Proof. First we note that the Picone identity (8.3.1) of Section 8.3 with $(\alpha+1)B(x)$ replaced by $B(x)$ holds on $A[r_2,\infty)$. Letting $u=1$ and $P_\alpha[v]=0$ in (8.3.1), we have

$$-\nabla\cdot\left(\frac{A(x)|\nabla v|^{\alpha-1}\nabla v}{|v|^{\alpha-1}v}\right)$$
$$= C(x)+\alpha A(x)\left|\frac{\nabla v}{v}\right|^{\alpha+1}+B(x)\cdot\left(\frac{|\nabla v|^{\alpha-1}\nabla v}{|v|^{\alpha-1}v}\right), \tag{8.8.4}$$

where the dot $\cdot$ denotes the scalar product. From (8.8.3) we see that

$$|W(x)| = A(x)\left|\frac{\nabla v}{v}\right|^{\alpha}$$

and therefore

$$\left|\frac{\nabla v}{v}\right|^{\alpha+1}=\left(\frac{|W(x)|}{A(x)}\right)^{\frac{\alpha+1}{\alpha}}=\frac{|W(x)|^{1+(1/\alpha)}}{A(x)^{1+(1/\alpha)}}.$$

Hence we obtain

$$A(x)\left|\frac{\nabla v}{v}\right|^{\alpha+1}=A(x)^{-1/\alpha}|W(x)|^{1+(1/\alpha)}. \tag{8.8.5}$$

It is easy to check that

$$B(x)\cdot\left(\frac{|\nabla v|^{\alpha-1}\nabla v}{|v|^{\alpha-1}v}\right)=\langle A(x)^{-1}B(x),W(x)\rangle. \tag{8.8.6}$$

Combining (8.8.4)–(8.8.6) yields the desired equation (8.8.2). □

Lemma 8.8.2. *If $v\in\mathcal{D}_{P_\alpha}(\Omega)$ is a solution of (8.8.1) and $v\neq 0$ on $A[r_2,\infty)$ for some $r_2>r_1$, then we obtain*

$$\nabla\cdot(\psi(x)W(x))+\psi(x)C(x)+\alpha\psi(x)A(x)^{-1/\alpha}|W(x)|^{1+(1/\alpha)}$$
$$+\langle W(x),\psi(x)A(x)^{-1}B(x)-\nabla\psi(x)\rangle=0 \tag{8.8.7}$$

on $A[r_2,\infty)$ for any $\psi(x)\in C^1(A[r_2,\infty);\mathbb{R})$.

Proof. We easily see that

$$\nabla\cdot(\psi(x)W(x))=\psi(x)\nabla\cdot W(x)+\langle W(x),\nabla\psi(x)\rangle. \tag{8.8.8}$$

Combining (8.8.2) with (8.8.8) yields the desired equation (8.8.7). □

Lemma 8.8.3. *If $v\in\mathcal{D}_{P_\alpha}(\Omega)$ is a solution of (8.8.1) and $v\neq 0$ on $A[r_2,\infty)$ for some $r_2>r_1$, then we obtain the following Riccati-type inequality*

$$\nabla\cdot(\psi(x)W(x))+C_\psi(x)+\frac{\alpha}{\alpha+1}d(x)|W(x)|^{1+(1/\alpha)}\le 0 \tag{8.8.9}$$

on $A[r_2, \infty)$ *for any* $\psi(x) \in C^1(A[r_2, \infty); (0, \infty))$, *where*

$$d(x) = \frac{\alpha+1}{2}\psi(x)A(x)^{-1/\alpha},$$

$$C_\psi(x) = \psi(x)C(x) - \frac{1}{\alpha+1}d(x)^{-\alpha}\psi(x)^{\alpha+1}\left|\frac{B(x)}{A(x)} - \frac{\nabla\psi(x)}{\psi(x)}\right|^{\alpha+1}.$$

Proof. It is easily seen that

$$\begin{aligned}&\alpha\psi(x)A(x)^{-1/\alpha}|W(x)|^{1+(1/\alpha)} + \langle W(x), \psi(x)A(x)^{-1}B(x) - \nabla\psi(x)\rangle \\ &= d(x)\left(2\frac{\alpha}{\alpha+1}|W(x)|^{1+(1/\alpha)}\right) \\ &\qquad + \langle W(x), \psi(x)A(x)^{-1}B(x) - \nabla\psi(x)\rangle. \qquad (8.8.10)\end{aligned}$$

Applying Young's inequality, we obtain

$$\begin{aligned}&|\langle W(x), \psi(x)A(x)^{-1}B(x) - \nabla\psi(x)\rangle| \\ &= |\langle d(x)^{\alpha/(\alpha+1)}W(x), d(x)^{-\alpha/(\alpha+1)}(\psi(x)A(x)^{-1}B(x) - \nabla\psi(x))\rangle| \\ &\le \frac{\alpha}{\alpha+1}d(x)|W(x)|^{1+(1/\alpha)} + \frac{1}{\alpha+1}d(x)^{-\alpha}|\psi(x)A(x)^{-1}B(x) - \nabla\psi(x)|^{\alpha+1}\end{aligned}$$

and therefore

$$\begin{aligned}&\langle W(x), \psi(x)A(x)^{-1}B(x) - \nabla\psi(x)\rangle \\ &\ge -\frac{\alpha}{\alpha+1}d(x)|W(x)|^{1+(1/\alpha)} \\ &\qquad -\frac{1}{\alpha+1}d(x)^{-\alpha}|\psi(x)A(x)^{-1}B(x) - \nabla\psi(x)|^{\alpha+1}. \quad (8.8.11)\end{aligned}$$

Combining (8.8.10) with (8.8.11) yields

$$\begin{aligned}&\alpha\psi(x)A(x)^{-1/\alpha}|W(x)|^{1+(1/\alpha)} + \langle W(x), \psi(x)A(x)^{-1}B(x) - \nabla\psi(x)\rangle \\ &\ge \frac{\alpha}{\alpha+1}d(x)|W(x)|^{1+(1/\alpha)} \\ &\qquad -\frac{1}{\alpha+1}d(x)^{-\alpha}\psi(x)^{\alpha+1}\left|\frac{B(x)}{A(x)} - \frac{\nabla\psi(x)}{\psi(x)}\right|^{\alpha+1}. \quad (8.8.12)\end{aligned}$$

From (8.8.7) and (8.8.12) it follows that (8.8.9) holds. □

Lemma 8.8.4. *If* $v \in \mathcal{D}_{P_\alpha}(\Omega)$ *is a solution of* (8.8.1) *such that* $v \neq 0$ *on* $A[r_2, \infty)$ *for some* $r_2 > r_1$, *then we obtain the following differential inequality*

$$Y'(r) + \int_{S_r} C_\psi(x)\, dS + \frac{\alpha}{\alpha+1}\Psi(r)^{-1/\alpha}|Y(r)|^{1+(1/\alpha)} \le 0 \qquad (8.8.13)$$

for $r \ge r_2$, where

$$S_r = \{x \in \mathbb{R}^n;\ |x| = r\},$$
$$\Psi(r) = \int_{S_r} d(x)^{-\alpha}\psi(x)^{\alpha+1}\,dS,$$
$$Y(r) = \int_{S_r} \psi(x)\langle W(x), \nu(x)\rangle\,dS,$$

$\nu(x)$ being the unit exterior normal vector x/r on S_r.

Proof. Integrating (8.8.9) over S_r, we see that

$$\int_{S_r} \nabla\cdot(\psi(x)W(x))\,dS + \int_{S_r} C_\psi(x)\,dS + \frac{\alpha}{\alpha+1}\int_{S_r} d(x)|W(x)|^{1+(1/\alpha)}dS \le 0, \quad r \ge r_2. \tag{8.8.14}$$

It follows from a result of Noussair and Swanson [223, Lemma 1] that

$$\int_{S_r} \nabla\cdot(\psi(x)W(x))\,dS = \frac{d}{dr}\int_{S_r} \psi(x)\langle W(x), \nu(x)\rangle\,dS = Y'(r). \tag{8.8.15}$$

Using Hölder's inequality, we obtain

$$\begin{aligned} |Y(r)| &\le \int_{S_r} \psi(x)|W(x)|\,dS \\ &= \int_{S_r} \psi(x)d(x)^{-\frac{\alpha}{\alpha+1}}\left(d(x)^{\frac{\alpha}{\alpha+1}}|W(x)|\right)dS \\ &\le \left(\int_{S_r} d(x)^{-\alpha}\psi(x)^{\alpha+1}\,dS\right)^{\frac{1}{\alpha+1}}\left(\int_{S_r} d(x)|W(x)|^{\frac{\alpha+1}{\alpha}}\,dS\right)^{\frac{\alpha}{\alpha+1}} \end{aligned}$$

and hence

$$\int_{S_r} d(x)|W(x)|^{\frac{\alpha+1}{\alpha}}\,dS \ge \Psi(r)^{-1/\alpha}|Y(r)|^{1+(1/\alpha)}. \tag{8.8.16}$$

Combining (8.8.14)–(8.8.16), we obtain the desired inequality (8.8.13). □

Theorem 8.8.1. *If there is a function $\psi(x) \in C^1(A[r_1,\infty);(0,\infty))$ such that the Riccati-type inequality (8.8.13) has no solution on $[r,\infty)$ for all large r, then every solution $v \in \mathcal{D}_{P_\alpha}(\Omega)$ of (8.8.1) is oscillatory in Ω.*

Proof. Suppose to the contrary that there is a nonoscillatory solution $v \in \mathcal{D}_{P_\alpha}(\Omega)$ of (8.8.1). Let $v \ne 0$ on $A[r_2,\infty)$ for some $r_2 > r_1$. Then Lemma 8.8.4 implies that (8.8.13) holds for $r \ge r_2$. This contradicts the hypothesis and completes the proof. □

Now we consider the differential inequality

$$y'(r) + \frac{1}{\beta}\frac{1}{p(r)}|y(r)|^{\beta} \leq -q(r), \tag{8.8.17}$$

where $\beta > 1$, $p(r) \in C([r_1, \infty); (0, \infty))$ and $q(r) \in C([r_1, \infty); \mathbb{R})$.

The following lemma was established by Usami [275, Proposition 3].

Lemma 8.8.5. *If there exists a function $\theta(r) \in C^1([r_1, \infty); (0, \infty))$ such that*

$$\int_{r_2}^{\infty} \left(\frac{p(r)|\theta'(r)|^{\beta}}{\theta(r)}\right)^{1/(\beta-1)} dr < \infty,$$

$$\int_{r_2}^{\infty} \frac{1}{p(r)(\theta(r))^{\beta-1}}\, dr = \infty,$$

$$\int_{r_2}^{\infty} \theta(r) q(r)\, dr = \infty$$

for some $r_2 > r_1$, then (8.8.17) has no solution on $[r, \infty)$ for all large r.

Theorem 8.8.2. *If there is a function $\psi(x) \in C^1(A[r_1, \infty); (0, \infty))$ such that*

$$\int_{r_2}^{\infty} \frac{1}{\Psi(r)^{1/\alpha}}\, dr = \infty, \tag{8.8.18}$$

$$\int_{r_2}^{\infty} r^{n-1}\overline{C_{\psi}}(r)\, dr = \infty \tag{8.8.19}$$

for some $r_2 > r_1$, then every solution $v \in \mathcal{D}_{P_\alpha}(\Omega)$ of (8.8.1) is oscillatory in Ω, where $\overline{C_{\psi}}(r)$ denotes the spherical mean of $C_{\psi}(x)$ over $S_{\bar{r}}$.

Proof. The conclusion follows by combining Theorem 8.8.1 with Lemma 8.8.5 with $\theta(r) = 1$. □

An important special case of (8.8.1) is the following:

$$\nabla \cdot \left(|\nabla v|^{\alpha-1}\nabla v\right) + C(x)|v|^{\alpha-1}v = 0 \quad \text{in } \Omega \tag{8.8.20}$$

which was studied by Usami [275], Xu and Xing [291].

Corollary 8.8.1. (Usami [275, Theorem 4]) *If there exists a function $\theta(r) \in C^1([r_1, \infty); (0, \infty))$ such that*

$$\int_{r_2}^{\infty} \frac{1}{(r^{n-1}\theta(r))^{1/\alpha}}\, dr = \infty,$$

$$\int_{r_2}^{\infty} r^{n-1}\theta(r)\bar{C}(r)\, dr = \infty,$$

$$\int_{r_2}^{\infty} r^{n-1}\frac{|\theta'(r)|^{\alpha+1}}{\theta(r)^{\alpha}}\, dr < \infty$$

for some $r_2 > r_1$, then every solution $v \in \mathcal{D}_{P_\alpha}(\Omega)$ of (8.8.20) is oscillatory in Ω, where $\bar{C}(r)$ denotes the spherical mean of $C(x)$ over S_r.

Proof. In Theorem 8.8.2 we let $A(x) = 1$, $B(x) = 0$ and $\psi(x) = \theta(|x|)$. Then we see that

$$\int_{r_2}^{\infty} \frac{1}{\Psi(r)^{1/\alpha}}\,dr = \frac{\alpha+1}{2}\int_{r_2}^{\infty} \frac{1}{(\omega_n r^{n-1}\theta(r))^{1/\alpha}}\,dr = \infty,$$

$$\int_{r_2}^{\infty} r^{n-1}\overline{C_\psi}(r)\,dr = \int_{r_2}^{\infty} r^{n-1}\theta(r)\bar{C}(r)\,dr - \frac{1}{\alpha+1}\left(\frac{\alpha+1}{2}\right)^{-\alpha}\int_{r_2}^{\infty} r^{n-1}\frac{|\theta'(r)|^{\alpha+1}}{\theta(r)^{\alpha}}\,dr = \infty,$$

where ω_n denotes the surface area of the unit sphere S_1. The conclusion follows from Theorem 8.8.2. □

Next we derive Kamenev-type oscillation criteria for (8.8.1) using Lemma 8.8.4 (see Kamenev [122]).

Let

$$D = \{(r,s) \in \mathbb{R}^2;\ r \geq s \geq r_1\},$$
$$D_0 = \{(r,s) \in \mathbb{R}^2;\ r > s \geq r_1\}.$$

We consider the kernel function $H(r,s)$, which is defined, continuous and sufficiently smooth on D, so that the following conditions are satisfied:

(K$_1$) $H(r,s) \geq 0$ and $H(r,r) = 0$ for $r \geq s \geq r_1$;

(K$_2$) there exists a constant $k_0 > 0$ such that

$$\lim_{r\to\infty} \frac{H(r,s)}{H(r,r_1)} = k_0 \quad \text{for all } s \geq r_1;$$

(K$_3$) $\dfrac{\partial H}{\partial s}(r,s) \leq 0$, $-\dfrac{\partial H}{\partial s}(r,s) = h(r,s)H(r,s)$ for $(r,s) \in D_0$, where $h(r,s) \in C(D_0;\mathbb{R})$

(*cf.* Kong [134], Philos [239]).

Let $\rho(s) \in C^1([r_1,\infty);(0,\infty))$, and define an integral operator A_τ^ρ by

$$A_\tau^\rho(y;r) = \int_\tau^r H(r,s)y(s)\rho(s)\,ds, \quad r \geq \tau \geq r_1,$$

where $y \in C([\tau,\infty);\mathbb{R})$. It is easily seen that A_τ^ρ is linear and positive, and in fact satisfies the following:

(A$_1$) $A_\tau^\rho(k_1y_1 + k_2y_2;r) = k_1A_\tau^\rho(y_1;r) + k_2A_\tau^\rho(y_2;r)$ for $k_1,k_2 \in \mathbb{R}$;

(A$_2$) $A_\tau^\rho(y;r) \geq 0$ for $y \geq 0$;

(A$_3$) $A_\tau^\rho(y';r) = -H(r,\tau)y(\tau)\rho(\tau) + A_\tau^\rho([h-\rho^{-1}\rho']y;r)$

(see Wong [284]).

Lemma 8.8.6. *If*

$$\limsup_{r\to\infty} \frac{1}{H(r,r_1)} A_{r_1}^\rho \left(q - \frac{\beta-1}{\beta} p^{1/(\beta-1)} \left| h - \frac{\rho'}{\rho} \right|^{\frac{\beta}{\beta-1}} ; r \right) = \infty, \quad (8.8.21)$$

then (8.8.17) *has no solution on* $[r,\infty)$ *for all large* r.

Proof. Let (8.8.17) have a solution $y(r)$ which is defined on $[r_2,\infty)$ for some $r_2 > r_1$. Applying the operator A_τ^ρ to (8.8.17), we obtain

$$A_\tau^\rho(y';r) + A_\tau^\rho\left(\frac{1}{\beta p}|y|^\beta;r\right) \le -A_\tau^\rho(q;r), \quad r \ge \tau \ge r_2. \quad (8.8.22)$$

Since

$$\begin{aligned} A_\tau^\rho(y';r) &= -H(r,\tau)y(\tau)\rho(\tau) + A_\tau^\rho([h-\rho^{-1}\rho']y;r) \\ &\ge -H(r,\tau)y(\tau)\rho(\tau) - A_\tau^\rho(|h-\rho^{-1}\rho'||y|;r), \end{aligned} \quad (8.8.23)$$

a combination of (8.8.22) and (8.8.23) yields

$$A_\tau^\rho(q;r) \le H(r,\tau)y(\tau)\rho(\tau) + A_\tau^\rho(|h-\rho^{-1}\rho'||y|;r) - A_\tau^\rho\left(\frac{1}{\beta p}|y|^\beta;r\right). \quad (8.8.24)$$

Using Young's inequality, we derive

$$\begin{aligned} |h-\rho^{-1}\rho'||y| &= \left|h - \frac{\rho'}{\rho}\right| \left(\frac{1}{p}\right)^{-1/\beta} \left(\frac{1}{p}\right)^{1/\beta} |y| \\ &\le \frac{\beta-1}{\beta} p^{1/(\beta-1)} \left| h - \frac{\rho'}{\rho}\right|^{\frac{\beta}{\beta-1}} + \frac{1}{\beta}\frac{1}{p}|y|^\beta \end{aligned}$$

and hence

$$\begin{aligned} A_\tau^\rho(|h-\rho^{-1}\rho'||y|;r) \le A_\tau^\rho &\left(\frac{\beta-1}{\beta} p^{1/(\beta-1)} \left| h - \frac{\rho'}{\rho}\right|^{\frac{\beta}{\beta-1}} ; r \right) \\ &+ A_\tau^\rho\left(\frac{1}{\beta p}|y|^\beta;r\right). \end{aligned} \quad (8.8.25)$$

From (8.8.24) and (8.8.25) we see that

$$A_\tau^\rho\left(q - \frac{\beta-1}{\beta} p^{1/(\beta-1)} \left| h - \frac{\rho'}{\rho}\right|^{\frac{\beta}{\beta-1}} ; r \right) \le H(r,\tau)y(\tau)\rho(\tau). \quad (8.8.26)$$

Letting $\tau = r_2$ in (8.8.26) and then dividing by $H(r,r_2)$, we observe that

$$\frac{1}{H(r,r_2)} A_{r_2}^\rho \left(q - \frac{\beta-1}{\beta} p^{1/(\beta-1)} \left| h - \frac{\rho'}{\rho}\right|^{\frac{\beta}{\beta-1}} ; r \right) \le y(r_2)\rho(r_2)$$

which contradicts (8.8.21) in view of (K$_2$), (K$_3$). □

Theorem 8.8.3. *If there exist functions $\psi(x) \in C^1(A[r_1, \infty); (0, \infty))$ and $\rho(s) \in C^1([r_1, \infty); (0, \infty))$ such that*

$$\limsup_{r\to\infty} \frac{1}{H(r, r_1)} A^{\rho}_{r_1} \left(\int_{S_r} C_\psi(x)\, dS - \frac{1}{\alpha+1} \left| h - \frac{\rho'}{\rho} \right|^{\alpha+1} \Psi; r \right) = \infty, \tag{8.8.27}$$

then every solution $v \in \mathcal{D}_{P_\alpha}(\Omega)$ of (8.8.1) is oscillatory in Ω.

Proof. It follows from Lemma 8.8.6 that the hypothesis (8.8.27) implies that (8.8.13) has no solution on $[r, \infty)$ for all large r. The conclusion follows from Theorem 8.8.1. □

Corollary 8.8.2. *If there exists a function $\psi(x) \in C^1(A[r_1, \infty); (0, \infty))$ such that*

$$\limsup_{r\to\infty} \frac{1}{\left(\Theta(r)\right)^\lambda} \int_{r_1}^{r} \left(\Theta(r) - \Theta(s)\right)^\lambda s^{n-1} \overline{C_\psi}(s)\, ds = \infty \tag{8.8.28}$$

for some $\lambda > \alpha$, then every solution $v \in \mathcal{D}_{P_\alpha}(\Omega)$ of (8.8.1) is oscillatory in Ω, where

$$\Theta(r) = \int_{r_1}^{r} \Psi(s)^{-1/\alpha}\, ds.$$

Proof. Letting $\rho(s) = 1$ and

$$H(r, s) = \left(\Theta(r) - \Theta(s)\right)^\lambda, \quad (r, s) \in D,$$

we see that

$$\begin{aligned} &\frac{1}{H(r, r_1)} A^{\rho}_{r_1} \left(\int_{S_r} C_\psi(x)\, dS - \frac{1}{\alpha+1} |h|^{\alpha+1} \Psi; r \right) \\ &= \frac{1}{\left(\Theta(r)\right)^\lambda} \int_{r_1}^{r} H(r, s) \left(\int_{S_s} C_\psi(x)\, dS - \frac{1}{\alpha+1} |h(r, s)|^{\alpha+1} \Psi(s) \right) ds. \end{aligned} \tag{8.8.29}$$

It is easily verified that

$$h(r, s) = \lambda \left(\Theta(r) - \Theta(s)\right)^{-1} \Psi(s)^{-1/\alpha}$$

and that

$$\begin{aligned} \int_{r_1}^{r} H(r, s) |h(r, s)|^{\alpha+1} \Psi(s)\, ds &= \lambda^{\alpha+1} \int_{r_1}^{r} \left(\Theta(r) - \Theta(s)\right)^{\lambda-\alpha-1} \Psi(s)^{-1/\alpha}\, ds \\ &= \frac{\lambda^{\alpha+1}}{\lambda - \alpha} \left(\Theta(r)\right)^{\lambda-\alpha}. \end{aligned} \tag{8.8.30}$$

Combining (8.8.29) and (8.8.30), we arrive at

$$\begin{aligned}\frac{1}{H(r,r_1)}A^{\rho}_{r_1}\left(\int_{S_r} C_\psi(x)\,dS - \frac{1}{\alpha+1}|h|^{\alpha+1}\Psi;\,r\right)\\ = \frac{1}{(\Theta(r))^\lambda}\int_{r_1}^{r}(\Theta(r)-\Theta(s))^\lambda \omega_n s^{n-1}\overline{C_\psi}(s)\,ds\\ -\frac{\lambda^{\alpha+1}}{(\alpha+1)(\lambda-\alpha)}(\Theta(r))^{-\alpha}.\end{aligned} \tag{8.8.31}$$

We observe, using (8.8.28) and (8.8.31), that the conclusion follows from Theorem 8.8.3. □

Following the classical idea of Kamenev [122], we define $H(r,s)$ and $\rho(s)$ by

$$H(r,s) = (r-s)^\mu,\ \mu > 1,\ (r,s)\in D,$$
$$\rho(s) = s^\nu,\ \nu\in\mathbb{R}.$$

Corollary 8.8.3. *Let $\mu > 1$ and ν is a real number. If there exists a function $\psi(x)\in C^1(A[r_1,\infty);(0,\infty))$ such that*

$$\begin{aligned}\limsup_{r\to\infty}\frac{1}{r^\mu}\int_{r_1}^{r}\Big[\omega_n s^{\nu+n-1}(r-s)^\mu\overline{C_\psi}(s)\\ -\frac{1}{\alpha+1}s^{\nu-\alpha+1}|\nu r-(\mu+\nu)s|^{\alpha+1}(r-s)^{\mu-\alpha-1}\Psi(s)\Big]ds = \infty,\end{aligned} \tag{8.8.32}$$

then every solution $v\in\mathcal{D}_{P_\alpha}(\Omega)$ of (8.8.1) is oscillatory in Ω.

Proof. It is easy to see that $h(r,s) = \mu(r-s)^{-1}$, and that

$$\begin{aligned}\frac{1}{H(r,r_1)}A^{\rho}_{r_1}\left(\int_{S_r} C_\psi(x)\,dS - \frac{1}{\alpha+1}\left|h-\frac{\rho'}{\rho}\right|^{\alpha+1}\Psi;\,r\right)\\ = \frac{1}{(r-r_1)^\mu}\int_{r_1}^{r}(r-s)^\mu\Big(\omega_n s^{n-1}\overline{C_\psi}(s)\\ -\frac{1}{\alpha+1}|\mu(r-s)^{-1}-\nu s^{-1}|^{\alpha+1}\Psi(s)\Big)s^\nu\,ds\\ = \frac{1}{(r-r_1)^\mu}\int_{r_1}^{r}\Big[\omega_n s^{\nu+n-1}(r-s)^\mu\overline{C_\psi}(s)\\ -\frac{1}{\alpha+1}s^{\nu-\alpha-1}|\nu r-(\mu+\nu)s|^{\alpha+1}(r-s)^{\mu-\alpha-1}\Psi(s)\Big]ds.\end{aligned}$$

Hence, the conclusion follows from Theorem 8.8.3. □

Lemma 8.8.7. *Let $y(r)$ be a solution of (8.8.17) defined in an interval $[c, b)$. Then we obtain*

$$A_c^\rho\left(q - \frac{\beta-1}{\beta}p^{1/(\beta-1)}\left|h - \frac{\rho'}{\rho}\right|^{\frac{\beta}{\beta-1}}; b\right) \le H(b,c)y(c)\rho(c). \tag{8.8.33}$$

Proof. Application of the operator A_τ^ρ to (8.8.17) yields

$$A_\tau^\rho(y';r) + A_\tau^\rho\left(\frac{1}{\beta p}|y|^\beta; r\right) \le -A_\tau^\rho(q;r), \quad c \le \tau \le r < b.$$

Proceeding as in the proof of Lemma 8.8.6, we find that (8.8.26) holds for $c \le r < b$. Letting $\tau = c$ and $r \to b - 0$ in (8.8.26), we arrive at the desired inequality (8.8.33). □

In addition to the hypotheses (K$_1$)–(K$_3$) we suppose the following:

(K$_4$) $\dfrac{\partial H}{\partial r}(r,s) = \tilde{h}(r,s)H(r,s)$ for $(r,s) \in D_0$, where $\tilde{h}(r,s) \in C(D_0;\mathbb{R})$.

Lemma 8.8.8. *Let $y(r)$ be a solution of (8.8.17) defined in an interval $(a, c]$. Then we obtain*

$$\tilde{A}_a^\rho\left(q - \frac{\beta-1}{\beta}p^{1/(\beta-1)}\left|\tilde{h} + \frac{\rho'}{\rho}\right|^{\frac{\beta}{\beta-1}}; c\right) \le -H(c,a)y(c)\rho(c), \tag{8.8.34}$$

where

$$\tilde{A}_\tau^\rho(y;r) = \int_\tau^r H(s,\tau)y(s)\rho(s)\,ds, \quad r \ge \tau \ge r_1.$$

Proof. Applying the operator $\tilde{A}_\tau^\rho$ to (8.8.17), we have

$$\tilde{A}_\tau^\rho(y';r) + \tilde{A}_\tau^\rho\left(\frac{1}{\beta p}|y|; r\right) \le -\tilde{A}_\tau^\rho(q;r), \quad a < r \le c.$$

Noting that

$$\begin{aligned}\tilde{A}_\tau^\rho(y';r) &= H(r,\tau)y(r)\rho(r) - \tilde{A}_\tau^\rho([\tilde{h} + \rho^{-1}\rho']y;r)\\ &\ge H(r,\tau)y(r)\rho(r) - \tilde{A}_\tau^\rho(|\tilde{h} + \rho^{-1}\rho'||y|;r),\end{aligned}$$

and arguing as in the proof of Lemma 8.8.6, we observe that

$$\tilde{A}_\tau^\rho\left(q - \frac{\beta-1}{\beta}p^{1/(\beta-1)}\left|\tilde{h} + \frac{\rho'}{\rho}\right|^{\frac{\beta}{\beta-1}}; r\right) \le -H(r,\tau)y(r)\rho(r), \ a < \tau \le r \le c. \tag{8.8.35}$$

Letting $r = c$ and $\tau \to a + 0$ in (8.8.35), we obtain the desired inequality (8.8.34). □

Lemma 8.8.9. *If*

$$\frac{1}{H(b,c)} A_c^\rho \left(q - \frac{\beta-1}{\beta} p^{1/(\beta-1)} \left| h - \frac{\rho'}{\rho} \right|^{\frac{\beta}{\beta-1}} ; b \right) + \frac{1}{H(c,a)} \tilde{A}_a^\rho \left(q - \frac{\beta-1}{\beta} p^{1/(\beta-1)} \left| \tilde{h} + \frac{\rho'}{\rho} \right|^{\frac{\beta}{\beta-1}} ; c \right) > 0 \quad (8.8.36)$$

for some $c \in (a,b)$, *then there is no solution of* (8.8.17) *defined in* (a,b).

Proof. Suppose that there is a solution $y(r)$ of (8.8.17) defined in (a,b). From Lemmas 8.8.7 and 8.8.8 it follows that (8.8.33) and (8.8.34) hold, which imply that the left hand side of (8.8.36) is nonpositive. This is a contradiction and the proof is complete. □

Theorem 8.8.4. *If there exist functions* $\psi(x) \in C^1(A[r_1,\infty);(0,\infty))$ *and* $\rho(s) \in C^1([r_1,\infty);(0,\infty))$ *such that*

$$\frac{1}{H(b,c)} A_c^\rho \left(\int_{S_r} C_\psi(x)\, dS - \frac{1}{\alpha+1} \left| h - \frac{\rho'}{\rho} \right|^{\alpha+1} \Psi; b \right) + \frac{1}{H(c,a)} \tilde{A}_a^\rho \left(\int_{S_r} C_\psi(x)\, dS - \frac{1}{\alpha+1} \left| \tilde{h} + \frac{\rho'}{\rho} \right|^{\alpha+1} \Psi; c \right) > 0 \, (8.8.37)$$

for some $c \in (a,b)$, *then every solution* $v \in \mathcal{D}_{P_\alpha}(\Omega)$ *of* (8.8.1) *has at least one zero in* $A(a,b)$.

Proof. Suppose that there exists a solution $v \in \mathcal{D}_{P_\alpha}(\Omega)$ of (8.8.1) which has no zero in $A(a,b)$. Then there is a solution $Y(r)$ of (8.8.13) defined in (a,b). This contradicts Lemma 8.8.9. □

Theorem 8.8.5. *If there exist functions* $\psi(x) \in C^1(A[r_1,\infty);(0,\infty))$ *and* $\rho(s) \in C^1([r_1,\infty);(0,\infty))$ *such that for each* $\xi \geq r_1$

$$\limsup_{r\to\infty} A_\xi^\rho \left(\int_{S_r} C_\psi(x)\, dS - \frac{1}{\alpha+1} \left| h - \frac{\rho'}{\rho} \right|^{\alpha+1} \Psi; r \right) > 0, \quad (8.8.38)$$

$$\limsup_{r\to\infty} \tilde{A}_\xi^\rho \left(\int_{S_r} C_\psi(x)\, dS - \frac{1}{\alpha+1} \left| \tilde{h} + \frac{\rho'}{\rho} \right|^{\alpha+1} \Psi; r \right) > 0, \quad (8.8.39)$$

then every solution $v \in \mathcal{D}_{P_\alpha}(\Omega)$ *of* (8.8.1) *is oscillatory in* Ω.

Proof. For any $\xi \geq r_1$, let $\xi = a$. Then (8.8.38) implies that there is a constant $c > a$ such that

$$\tilde{A}_a^\rho \left(\int_{S_r} C_\psi(x)\, dS - \frac{1}{\alpha+1} \left| \tilde{h} + \frac{\rho'}{\rho} \right|^{\alpha+1} \Psi; c \right) > 0.$$

From (8.8.39) we see that there is a constant $b > c$ such that

$$A_c^\rho\left(\int_{S_r} C_\psi(x)\,dS - \frac{1}{\alpha+1}\left|h - \frac{\rho'}{\rho}\right|^{\alpha+1}\Psi;\, b\right) > 0.$$

It follows from Theorem 8.8.4 that every solution $v \in \mathcal{D}_{P_\alpha}(\Omega)$ of (8.8.1) has a zero in $A(a,b)$ for any $a \geq r_1$, that is, v is oscillatory in Ω. □

Corollary 8.8.4. *Let* $\lim_{r\to\infty} \Theta(r) = \infty$, *where* $\Theta(r)$ *is defined in Corollary 8.8.2, and let* $\lambda > \alpha$. *If there exists a function* $\psi(x) \in C^1(A[r_1,\infty);(0,\infty))$ *such that for each* $\xi \geq r_1$

$$\limsup_{r\to\infty} \frac{1}{(\Theta(r))^{\lambda-\alpha}} \int_\xi^r (\Theta(r)-\Theta(s))^\lambda \left(\int_{S_s} C_\psi(x)\,dS\right) ds > k_\lambda, \quad (8.8.40)$$

$$\limsup_{r\to\infty} \frac{1}{(\Theta(r))^{\lambda-\alpha}} \int_\xi^r (\Theta(s)-\Theta(\xi))^\lambda \left(\int_{S_s} C_\psi(x)\,dS\right) ds > k_\lambda, \quad (8.8.41)$$

then every solution $v \in \mathcal{D}_{P_\alpha}(\Omega)$ *of* (8.8.1) *is oscillatory in* Ω, *where*

$$k_\lambda = \frac{\lambda^{\alpha+1}}{(\alpha+1)(\lambda-\alpha)}.$$

Proof. We let $\rho(s) = 1$ and

$$H(r,s) = (\Theta(r)-\Theta(s))^\lambda, \quad (r,s) \in D.$$

Then it is easy to check that

$$h(r,s) = \lambda(\Theta(r)-\Theta(s))^{-1}\Psi(s)^{-1/\alpha}$$

and that

$$\int_\xi^r H(r,s)|h(r,s)|^{\alpha+1}\Psi(s)\,ds = \lambda^{\alpha+1}\int_\xi^r (\Theta(r)-\Theta(s))^{\lambda-\alpha-1}\Psi(s)^{-1/\alpha}\,ds$$
$$= \frac{\lambda^{\alpha+1}}{\lambda-\alpha}(\Theta(r)-\Theta(\xi))^{\lambda-\alpha}.$$

Hence we find that

$$A_\xi^\rho\left(\int_{S_r} C_\psi(x)\,dS - \frac{1}{\alpha+1}\left|h - \frac{\rho'}{\rho}\right|^{\alpha+1}\Psi;\, r\right)$$
$$= \int_\xi^r (\Theta(r)-\Theta(s))^\lambda\left(\int_{S_s} C_\psi(x)\,dS\right) ds - k_\lambda(\Theta(r)-\Theta(\xi))^{\lambda-\alpha}. \quad (8.8.42)$$

It is easily seen that

$$\tilde{h}(r,s) = \lambda(\Theta(r)-\Theta(s))^{-1}\Psi(r)^{-1/\alpha}$$

and that

$$\tilde{A}_{\xi}^{\rho}\left(\int_{S_r} C_{\psi}(x)\,dS - \frac{1}{\alpha+1}\left|\tilde{h}+\frac{\rho'}{\rho}\right|^{\alpha+1}\Psi;\, r\right)$$

$$= \int_{\xi}^{r} (\Theta(s)-\Theta(\xi))^{\lambda}\left(\int_{S_s} C_{\psi}(x)\,dS\right) ds - k_{\lambda}(\Theta(r)-\Theta(\xi))^{\lambda-\alpha}. \quad (8.8.43)$$

From (8.8.42) and (8.8.43) we find that the conclusion follows from Theorem 8.8.5. □

Corollary 8.8.5. *If there exists a function $\psi(x) \in C^1(A[r_1,\infty);(0,\infty))$ such that*

$$\limsup_{r\to\infty} \int_{\xi}^{r} \left[\omega_n s^{n-1}(r-s)^{\mu}\overline{C_{\psi}}(s) - \frac{1}{\alpha+1}\mu^{\alpha+1}(r-s)^{\mu-\alpha-1}\Psi(s)\right] ds > 0,$$

$$\limsup_{r\to\infty} \int_{\xi}^{r} \left[\omega_n s^{n-1}(s-\xi)^{\mu}\overline{C_{\psi}}(s) - \frac{1}{\alpha+1}\mu^{\alpha+1}(s-\xi)^{\mu-\alpha-1}\Psi(s)\right] ds > 0$$

for each $\xi \geq r_1$ and some $\mu > 1$, then every solution $v \in \mathcal{D}_{P_\alpha}(\Omega)$ of (8.8.1) is oscillatory in Ω.

Proof. Letting

$$H(r,s) = (r-s)^{\mu}, \quad \mu > 1, \ (r,s) \in D,$$
$$\rho(s) = 1,$$

we easily see that

$$h(r,s) = \tilde{h}(r,s) = \mu(r-s)^{-1}$$

and that

$$\int_{\xi}^{r} H(r,s)|h(r,s)|^{\alpha+1}\Psi(s)\,ds = \mu^{\alpha+1}\int_{\xi}^{r}(r-s)^{\mu-\alpha-1}\Psi(s)\,ds,$$

$$\int_{\xi}^{r} H(s,\xi)|\tilde{h}(s,\xi)|^{\alpha+1}\Psi(s)\,ds = \mu^{\alpha+1}\int_{\xi}^{r}(s-\xi)^{\mu-\alpha-1}\Psi(s)\,ds.$$

The conclusion follows from Theorem 8.8.5. □

8.9 Notes

Sturmian comparison theorems for (8.1.1) and (8.1.2) were first established by Dunninger [73] in 1995. The results of Section 8.1 are extracted from Kusano, Jaroš and Yoshida [151].

The results in Section 8.2 are based on Bognár and Došlý [32]. In particular, Theorems 8.2.1, 8.2.2 and Corollary 8.2.1 are due to Bognár and Došlý [32].

Section 8.3 is taken from Yoshida [320]. Lemmas 8.3.1 and 8.3.2 are extracted from Yoshida [322].

The Picone-type inequality and the oscillation results in Section 8.4 are extracted from Yoshida [321]. More general superlinear elliptic equations with forcing terms were considered by Yoshida [323].

In the case where $B(x) = 0$ the results in Section 8.5 were obtained by Jaroš, Kusano and Yoshida [116]. Section 8.5 is based on Yoshida [321].

Superlinear-sublinear parabolic equations are treated in Section 8.6, but the arguments used in Section 8.6 are similar to those of Jaroš, Kusano and Yoshida [117]. Section 8.6 is extracted from Yoshida [324].

Section 8.7 is based on Chan [46], Chan and Young [47], Kobayashi and Yoshida [133]. Parabolic systems considered in Section 8.7 are linear systems, whereas Chan [46], Chan and Young [47] deal with more general quasilinear parabolic systems. Fourth order quasilinear parabolic systems were treated by Chan and Young [48].

The results in Section 8.8 are based on Usami [275], Xu [289], Xu and Xing [291]. Riccati techniques were used by Mařík [195] to derive oscillation results for half-linear elliptic equations with damping terms. Lemma 8.8.5 and Corollary 8.8.1 are due to Usami [275]. Lemma 8.8.6, Theorem 8.8.3, Corollaries 8.8.2 and 8.8.3 are extracted from Xu and Xing [291] in which half-linear elliptic equations without damping terms are investigated. Lemmas 8.8.7–8.8.9, Theorems 8.8.4 and 8.8.5, Corollary 8.8.4 are taken from Xu [289] in which more general elliptic equations without damping terms are studied. For oscillations of quasilinear elliptic equations which include half-linear equations we mention the paper by Naito and Usami [214] which utilizes comparison principles to obtain oscillation results.

For half-linear second order differential equations we mention in particular the books of Došlý [69], Došlý and Řehák [71] in which oscillation theory is developed.

Symbols and Notation

Symbol	**Meaning**
$\mathbb{N}$	set of all positive integers
$\mathbb{Z}$	set of all integers
$\mathbb{R}$	set of all real numbers
$\mathbb{R}^n$	real n-dimensional Euclidean space
(a, b)	open interval $\{t \in \mathbb{R};\ a < t < b\}$
$[a, b]$	closed interval $\{t \in \mathbb{R};\ a \leq t \leq b\}$
$(a, b]$	interval $\{t \in \mathbb{R};\ a < t \leq b\}$
$[a, b)$	interval $\{t \in \mathbb{R};\ a \leq t < b\}$
$\lvert x \rvert = \left(\sum_{i=1}^{n} x_i^2 \right)^{1/2}$	Euclidean length of $x \in \mathbb{R}^n$
$A(r, s)$	annular domain $\{x \in \mathbb{R}^n;\ r < \lvert x \rvert < s\}$
$A(r, \infty)$	exterior domain $\{x \in \mathbb{R}^n;\ r < \lvert x \rvert\}$
$A[r, \infty)$	exterior domain $\{x \in \mathbb{R}^n;\ r \leq \lvert x \rvert\}$
$\langle a, b \rangle$	scalar product of $a \in \mathbb{R}^N$ and $b \in \mathbb{R}^N$
$a \cdot b$	scalar product of $a \in \mathbb{R}^N$ and $b \in \mathbb{R}^N$
$S_1 = \{x \in \mathbb{R}^n;\ \lvert x \rvert = 1\}$	unit sphere in $\mathbb{R}^n$
ω_n	surface area of unit sphere S_1, that is, $\omega_n = \dfrac{2\pi^{n/2}}{\Gamma(n/2)}$ ($\Gamma(s)$: gamma function)
$S_r = \{x \in \mathbb{R}^n;\ \lvert x \rvert = r\}$	$(n-1)$-dimensional sphere with radius r
∂G	boundary of a domain $G \subset \mathbb{R}^n$
$\overline{G}$	closure of a domain $G \subset \mathbb{R}^n$
$\lvert G \rvert$	volume of a domain $G \subset \mathbb{R}^n$, that is, $\lvert G \rvert = \int_G dx$

$C(E;\mathbb{R})$	space of real-valued continuous functions in $E \subset \mathbb{R}^m$
$C(E;[0,\infty))$	space of nonnegative-valued continuous functions in $E \subset \mathbb{R}^m$
$C(E;(0,\infty))$	space of positive-valued continuous functions in $E \subset \mathbb{R}^m$
$C(E;\mathbb{R}^N)$	space of $\mathbb{R}^N$-valued continuous functions in $E \subset \mathbb{R}^m$
$C^k(E;\mathbb{R})$	space of real-valued functions which are k-times continuously differentiable in $E \subset \mathbb{R}^m$
$C^k(E;\mathbb{R}^N)$	space of $\mathbb{R}^N$-valued functions which are k-times continuously differentiable in $E \subset \mathbb{R}^m$
$C^k(E;\mathbb{R}^{\ell\times j})$	space of $j \times \ell$ matrix functions whose components are of class $C^k(E;\mathbb{R})$
$\overline{f}(r),\ \overline{f(x)}(r)$	spherical mean of $f(x)$ over S_r, that is, $\dfrac{1}{\omega_n r^{n-1}} \displaystyle\int_{S_r} f(x)\,dS$
$[\theta(t)]_+ = \max\{0,\ \theta(t)\}$	positive part of $\theta(t)$
$[\theta(t)]_- = \max\{0,\ -\theta(t)\}$	negative part of $\theta(t)$
∇	nabla, that is, $\nabla = \left(\dfrac{\partial}{\partial x_1}, \dfrac{\partial}{\partial x_2}, ..., \dfrac{\partial}{\partial x_n}\right)$
$\nabla \cdot h = \displaystyle\sum_{i=1}^{n} \frac{\partial h_i}{\partial x_i}$	divergence of $h = (h_1, h_2, ..., h_n)$, that is, $\nabla \cdot h = \text{div}\ h$
∇_A	$\nabla_A = A(x) \cdot \nabla$, where $A(x) = \big(A_1(x), A_2(x), ..., A_n(x)\big)$
$\Delta = \displaystyle\sum_{i=1}^{n} \frac{\partial^2}{\partial x_i^2}$	Laplacian in $\mathbb{R}^n$
Δ^j	jth iterated Laplacian

Bibliography

[1] Adams, R. A. and Fournier, J. J. F. (2003). *Sobolev Spaces, Second Edition* (Academic Press).

[2] Agarwal, R. P., Bochner, M., Grace, S. R. and O'Regan, D. (2005). *Discrete Oscillation Theory* (Hindawi Publishing Corporation, New York).

[3] Agarwal, R. P., Bochner, M. and Li, W.-T. (2004). *Nonoscillation and Oscillation: Theory for Functional Differential Equations* (Marcel Dekker, Inc., New York).

[4] Agarwal, R. P., Grace, S. R. and O'Regan, D. (2000). *Oscillation Theory for Difference and Functional Differential Equations* (Kluwer Academic Publishers, Dordrecht).

[5] Agarwal, R. P., Grace, S. R. and O'Regan, D. (2002). *Oscillation Theory for Second Order Linear, Half-Linear, Superlinear and Sublinear Dynamic Equations* (Kluwer Academic Publishers, Dordrecht).

[6] Agarwal, R. P., Grace, S. R. and O'Regan, D. (2003). *Oscillation Theory for Second Order Dynamic Equations* (Taylor & Francis, Ltd., London).

[7] Agarwal, R. P., Meng, F. W. and Li, W. N. (2002). Oscillation of solutions of systems of neutral type partial functional differential equations, *Comput. Math. Appl.* **44**, pp. 777–786.

[8] Agmon, S. (1965). *Lectures on Elliptic Boundary Value Problems* (Van Nostrand Co., Inc.).

[9] Allegretto, W. (1974). Oscillation criteria for quasilinear equations, *Canad. J. Math.* **26**, pp. 931–947.

[10] Allegretto, W. (1976). Nonoscillation theory of elliptic equations of order $2n$, *Pacific J. Math.* **64**, pp. 1–16.

[11] Allegretto, W. (1976). Oscillation criteria for semilinear equations in general domains, *Canad. Math. Bull.* **19**, pp. 137–144.

[12] Allegretto, W. (1977). A Kneser theorem for higher order elliptic equations, *Canad. Math. Bull.* **20**, pp. 1–8.

[13] Allegretto, W. (1977). Nonoscillation criteria for elliptic equations in conical domains, *Proc. Amer. Math. Soc.* **63**, pp. 245–250.

[14] Allegretto, W. (2000). Sturm type theorems for solutions of elliptic nonlinear problems, *NoDEA Nonlinear Differential Equations Appl.* **7**, pp. 309–

321.

[15] Allegretto, W. (2001). Sturm theorems for degenerate elliptic equations, *Proc. Amer. Math. Soc.* **129**, pp. 3031–3035.

[16] Allegretto, W. and Huang, Y. X. (1998). A Picone's identity for the p-Laplacian and applications, *Nonlinear Anal.* **32**, pp. 819–830.

[17] Allegretto, W. and Huang, Y. X. (1999). Principal eigenvalues and Sturm comparison via Picone's identity, *J. Differential Equations* **156**, pp. 427–438.

[18] Altin, A. (1982). Comparison and oscillation theorems for singular ultrahyperbolic equations, *Comm. Fac. Sci. Univ. Ankara Sér. A_1 Math.* **31**, pp. 47–57.

[19] Angenent, S. (1991). Nodal properties of solutions of parabolic equations, *Rocky Mountain J. Math.* **21**, pp. 585–592.

[20] Aviles, P. (1982). A study of the singularities of solutions of a class of nonlinear elliptic partial differential equations, *Comm. Partial Differential Equations* **7**, pp. 609–643.

[21] Bainov, D. D., Kamont, Z. and Minchev, E. (1996). Monotone iterative methods for impulsive hyperbolic differential-functional equations, *J. Comput. Appl. Math.* **70**, pp. 329–347.

[22] Bainov, D. and Minchev, E. (1996). Oscillation of solutions of impulsive parabolic equations, *J. Comput. Appl. Math.* **69**, pp. 207–214.

[23] Bainov, D. and Minchev, E. (1996). Oscillation of solutions of impulsive nonlinear parabolic differential-difference equations, *Internat. J. Theoret. Phys.* **35**, pp. 207–215.

[24] Bainov, D. and Minchev, E. (1998). Forced oscillations of solutions of impulsive nonlinear parabolic differential-difference equations, *J. Korean Math. Soc.* **35**, pp. 881–890.

[25] Bainov, D. D. and Mishev, D. P. (1991). *Oscillation Theory for Neutral Differential Equations with Delay* (Adam Hilger, Ltd., Bristol).

[26] Bainov, D. D. and Mishev, D. P. (1995). *Oscillation Theory of Operator-Differential Equations* (World Scientific Publishing Co., Inc., River Edge, NJ).

[27] Ball, J. M. (1973). Stability theory for an extensible beam, *J. Differential Equations* **14**, pp. 399–418.

[28] Barański, F. (1976). The mean value theorem and oscillatory properties of certain elliptic equations in three dimensional space, *Comment. Math. Prace Mat.* **19**, pp. 13–14.

[29] Bateman, H. (1933). Logarithmic solutions of Bianchi's equation, *Proc. Nat. Acad. Sci. U.S.A.* **19**, pp. 852–854.

[30] Benchohra, M., Henderson, J. and Ntouyas, S. (2006). *Impulsive Differential Equations and Inclusions* (Hindawi Publishing Corporation, New York).

[31] Bianchi, L. (1895). Il metodo di Riemann esteso alla integrazione della equazione: $\dfrac{\partial^n u}{\partial x_1 \partial x_2 \cdots \partial x_n} = Mu$, *Atti Real. Accad. Lincei, Rend.* **4**, pp. 8–18.

[32] Bognár, G. and Došlý, O. (2003). The application of Picone-type identity for some nonlinear elliptic differential equations, *Acta Math. Univ. Comenian.* **72**, pp. 45–57.

[33] Borzymowski, A. (1980). A Goursat problem for a polyvibrating equation of D. Mangeron, *Funkcial. Ekvac.* **23**, pp. 1–16.

[34] Bouchekif, M. (1991). Comportement oscillatoire des solutions d'une équation elliptique, *Riv. Mat. Pura Appl.* **9**, pp. 69–79.

[35] Bouchekif, M. and Górowski, J. (1990). Propriétés oscillatoires des solutions de certaines équations elliptiques, *Facta Univ. Ser. Math. Inform.* **5**, pp. 67–79.

[36] Bugir, M. K. (1990). Oscillation properties of solutions of linear differential equations in space of constant curvature, *Differ. Equ.* **26**, pp. 1461–1466.

[37] Bugir, M. K. (1990). A remark on the conditions for the oscillation of solutions of nonlinear differential equations, *Ukrainian Math. J.* **42**, pp. 404–409.

[38] Bugir, M. K. (1992). A study of the nonoscillatory character of solutions of partial differential equations by the method of separation of variables, *Ukrainian Math. J.* **44**, pp. 278–283.

[39] Bugir, M. K. and Dobrotvor, I. G. (1988). *Oscillability of Solutions of Differential Equations in Spaces of Constant Curvature*, *Akad. Nauk Ukrain. SSR Inst. Mat. Preprint* No. 62.

[40] Bykov, Ya. V. and Kultaev, T. Ch. (1983). Oscillation of solutions of a class of parabolic equations, *Izv. Akad. Nauk Kirgiz. SSR* **6**, pp. 3–9.

[41] Bykova, L. Ya. and Marusich, A. I. (1989). Oscillation properties of solutions of a class of parabolic equations with delay, *Izv. Akad. Nauk Kirgiz. SSR* **2**, pp. 3–10.

[42] Cannon, J. R. (1984). *The One-Dimensional Heat Equation* (Addison-Wesley Publishing Company, Inc.).

[43] Cazenave, T. and Haraux, A. (1984). Propriétés oscillatoires des solutions de certaines équations des ondes semi-linéaires, *C. R. Acad. Sci. Paris Sér. I Math.* **298**, pp. 449–452.

[44] Cazenave, T. and Haraux, A. (1987). Oscillatory phenomena associated to semilinear wave equations in one spatial dimension, *Trans. Amer. Math. Soc.* **300**, pp. 207–233.

[45] Cazenave, T. and Haraux, A. (1988). Some oscillatory properties of the wave equation in several space dimensions, *J. Funct. Anal.* **76**, pp. 87–109.

[46] Chan, C. Y. (1982). Singular and unbounded matrix solutions for both time-dependent matrix and vector differential systems, *J. Math. Anal. Appl.* **87**, pp. 147–157.

[47] Chan, C. Y. and Young, E. C. (1973). Unboundedness of solutions and comparison theorems for time-dependent quasilinear differential matrix inequalities, *J. Differential Equations* **14**, pp. 195–201.

[48] Chan, C. Y. and Young, E. C. (1975). Singular matrix solutions for time-dependent fourth order quasilinear matrix differential inequalities, *J. Differential Equations* **18**, pp. 386–392.

[49] Chen, M. -P. and Zhang, B. G. (1995). Oscillation criteria for a class of

perturbed Schrödinger equations, *Hiroshima Math. J.* **25**, pp. 207–214.

[50] Cheng, S. S. (2003). *Partial Difference Equations* (Taylor & Francis, London).

[51] Cheng, S. S. and Zhang, B. G. (1994). Qualitative theory of partial difference equations (I): oscillation of nonlinear partial difference equations, *Tamkang J. Math.* **25**, pp. 279–288.

[52] Coddington, E. A. and Levinson, N. (1955). *Theory of Ordinary Differential Equations* (McGraw-Hill, New York).

[53] Conlan, J. and Diaz, J. B. (1963). Existence of solutions of an n-th order hyperbolic partial differential equation, *Contrib. Differential Equations* **2**, pp. 277–289.

[54] Courant, R. and Hilbert, D. (1966). *Methods of Mathematical Physics, Vol. I* (Interscience, New York).

[55] Cui, B. T. (1991). Oscillation theorems of nonlinear parabolic equations of neutral type, *Math. J. Toyama Univ.* **14**, pp. 113–123.

[56] Cui, B. T. (1992). Oscillation properties for parabolic equations of neutral type, *Comment. Math. Univ. Carolinae* **33**, pp. 581–588.

[57] Cui, B. T., Deng, F. Q., Li, W. N. and Liu, Y. Q. (2005). Oscillation problems for delay parabolic systems with impulses, *Dyn. Contin. Discrete Impuls. Syst. Ser. A Math. Anal.* **12**, pp. 67–76.

[58] Cui, B. T., Liu, Y. Q. and Deng, F. Q. (2003). Some oscillation problems for impulsive hyperbolic differential systems with several delays, *Appl. Math. Comput.* **146**, pp. 667–679.

[59] Deng, L. H. (2002). Oscillation criteria for certain hyperbolic functional differential equations with Robin boundary condition, *Indian J. Pure Appl. Math.* **33**, pp. 1137–1146.

[60] Deng, L. H. and Ge, W. G. (2001). Oscillation for certain delay hyperbolic equations satisfying the Robin boundary condition, *Indian J. Pure Appl. Math.* **32**, pp. 1269–1274.

[61] Deng, L. H. and Ge, W. G. (2001). Oscillation criteria of solutions for impulsive delay parabolic equation, *Acta Math. Sinica (Chin. Ser.)* **44**, pp. 501–506.

[62] Deng, L. H., Ge, W. G. and Wang, P. G. (2003). Oscillation of hyperbolic equations with continuous deviating arguments under the Robin boundary condition, *Soochow J. Math.* **29**, pp. 1–6.

[63] Deng, L. H., Tan, Y. M. and Yu, Y. H. (2002). Oscillation criteria of solutions for a class of impulsive parabolic differential equations, *Indian J. Pure Appl. Math.* **33**, pp. 1147–1153.

[64] Dobrotvor, I. G. (1984). Oscillation properties of the solutions of equations with a polyharmonic operator in the space E^n, *Ukrainian Math. J.* **36**, pp. 230–232.

[65] Domšlak, Ju. I. (1970). On the oscillation of solutions of vector differential equations, *Soviet Math. Dokl.* **11**, pp. 839–841.

[66] Domshlak, Yu. I. (1971). Oscillatory properties of solutions of vector differential equations, *Differ. Uravn.* **7**, pp. 961–969: *Differ. Equ.* **7**, pp. 728–734.

[67] Domshlak, Yu. I. and Tamoev, G. I. (1981). A vector analogue of the

Picone-Hartman-Wintner comparison theorem for operator-partial differential equations and its application to the investigation of ultrahyperbolic equations, *Izv. Akad. Nauk Azerbaidzhan. SSR Ser. Fiz.-Tekhn. Mat. Nauk* **2**, pp. 21–26.

[68] Došlý, O. (2002). The Picone identity for a class of partial differential equations, *Math. Bohem.* **127**, pp. 581–589.

[69] Došlý, O. (2004). *Half-linear Differential Equations, Handbook of Differential Equations: Ordinary Differential Equations, Volume 1* (Elsevier B. V., Amsterdam)

[70] Došlý, O. and Mařík, R. (2001). Nonexistence of positive solutions for PDE's with p-Laplacian, *Acta Math. Hungar.* **90**, pp. 89–107.

[71] Došlý, O. and Řehák, P. (2005). *Half-linear Differential Equations* (North-Holland Mathematics Studies, 202, Elsevier Science B.V., Amsterdam).

[72] Dunninger, D. R. (1969). Sturmian theorems for parabolic inequalities, *Rend. Accad. Sci. Fis. Mat. Napoli* **36**, pp. 406–410.

[73] Dunninger, D. R. (1995). A Sturm comparison theorem for some degenerate quasilinear elliptic operators, *Boll. Un. Mat. Ital. A (7)* **9**, pp. 117–121.

[74] El-Sayed, M. A. (1993). An oscillation criterion for forced second order linear differential equation, *Proc. Amer. Math. Soc.* **118**, pp. 814–817.

[75] Erbe, L. H., Freedman, H. I., Liu, X. Z. and Wu, J. H. (1991). Comparison principles for impulsive parabolic equations with applications to models of single species growth, *J. Austral. Math. Soc. Ser. B* **32**, pp. 382–400.

[76] Erbe, L. H., Kong, Q. and Zhang, B. G. (1995). *Oscillation Theory for Functional Differential Equations* (Marcel Dekker, Inc., New York).

[77] Evans, L. C. (1998). *Partial Differential Equations* (American Mathematical Society, Providence, Rhode Island).

[78] Feireisl, E. and Herrmann, L. (1992). Oscillations of a nonlinearly damped extensible beam, *Appl. Math.* **37**, pp. 469–478.

[79] Fitzgibbon, W. E. (1982). Global existence and boundedness of solutions to the extensible beam equations, *SIAM J. Math. Anal.* **13**, pp. 739–745.

[80] Frydrych, Z. (1975). On certain oscillatory properties for solutions of an equation with a biharmonic leading part, *Ann. Polon. Math.* **30**, pp. 225–228.

[81] Fu, X. and Liu, X. (1997). Oscillation criteria for impulsive hyperbolic systems, *Dyn. Contin. Discrete Impuls. Syst.* **3**, pp. 225–244.

[82] Fu, X., Liu, X. and Sivaloganathan, S. (2001). Oscillation criteria for impulsive parabolic systems, *Appl. Anal.* **79**, pp. 239–255.

[83] Fu, X., Liu, X. and Sivaloganathan, S. (2002). Oscillation criteria for impulsive parabolic differential equations with delay, *J. Math. Anal. Appl.* **268**, pp. 647–664.

[84] Fu, X. and Zhang, L. Q. (2004). Forced oscillation for impulsive hyperbolic boundary value problems with delay, *Appl. Math. Comput.* **158**, pp. 761–780.

[85] Fu, X. and Zhuang, W. (1995). Oscillation of certain neutral delay parabolic equations, *J. Math. Anal. Appl.* **191**, pp. 473–489.

[86] Fujita, H. (1966). On the blowing up of solutions of the Cauchy problem

for $u_t = \Delta u + u^{1+\alpha}$, *J. Fac. Sci. Univ. Tokyo Sect. I* **8**, pp. 109–124.

[87] Georgiou, D. and Kreith, K. (1985). Functional characteristic initial value problems, *J. Math. Anal. Appl.* **107**, pp. 414–424.

[88] Glazman, I. M. (1965). *Direct Methods of Qualitative Spectral Analysis of Singular Differential Operators* (Israel Program for Scientific Translations, Jerusalem).

[89] Glick, I. I. (1963). On an analog of the Euler-Cauchy polygon method for the partial differential equation $u_{x_1 \cdots x_n} = f$, *Contrib. Differential Equations* **2**, pp. 1–59.

[90] Gopalsamy, K. (1992). Oscillations in parabolic neutral systems, *Mini-Conference on Free and Moving Boundary and Diffusion Problems (Canberra, 1990), Proc. Centre Math. Appl. Austral. Nat. Univ., Austral. Nat. Univ.*, Canberra, pp. 128–141.

[91] Gorbaĭchuk, V. I. and Dobrotvor, I. G. (1980). Conditions for the oscillation of solutions of a class of elliptic equations of high orders with constant coefficients, *Ukrainian Math. J.* **32**, pp. 385–391.

[92] Gorbaĭchuk, V. I. and Dobrotvor, I. G. (1982). Investigation of the oscillation of solutions of a class of equations with biharmonic operator, using the mean-value theorem, *Mat. Fiz.* **31**, pp. 71–75.

[93] Górowski, J. (1985). The mean value theorem and oscillatory behaviour for solutions of certain elliptic equations, *Math. Nachr.* **122**, pp. 259–265.

[94] Györi, I. and Ladas, G. (1991). *Oscillation Theory of Delay Differential Equations* (Clarendon Press, Oxford).

[95] Haraux, A. and Komornik, V. (1985). Oscillations of anharmonic Fourier series and the wave equation, *Rev. Mat. Iberoamericana* **1**, pp. 57–77.

[96] Haraux, A. and Zuazua, E. (1988). Super-solutions of eigenvalue problems and the oscillation properties of second order evolution equations, *J. Differential Equations* **74**, pp. 11–28.

[97] Hardy, G., Littlewood, J. E. and Pólya, G. (1988). *Inequalities, Second Edition* (Cambridge University Press).

[98] Hartman, P. and Wintner, A. (1955). On a comparison theorem for self-adjoint partial differential equations of elliptic type, *Proc. Amer. Math. Soc.* **6**, pp. 862–865.

[99] Headley, V. B. (1969). A monotonicity principle for eigenvalues, *Pacific J. Math.* **30**, pp. 663–668.

[100] Headley, V. B. (1969). Elliptic equations of order $2m$, *J. Math. Anal. Appl.* **25**, pp. 558–568.

[101] Headley, V. B. (1986). Sharp nonoscillation theorems for even-order elliptic equations, *J. Math. Anal. Appl.* **120**, pp. 709–722.

[102] Headley, V. B. (1989). Nonoscillation theorems for nonselfadjoint even-order elliptic equations, *Math. Nachr.* **141**, pp. 289–297.

[103] Hecquet, G. (1977). Etude de quelques problèmes d'existence globale concernant l'équation $\dfrac{\partial^{r+s}u}{\partial x^r \partial y^s} = f\left(x, y, u, \dfrac{\partial u}{\partial x}, \dfrac{\partial u}{\partial y}, \ldots, \dfrac{\partial^{p+q}u}{\partial x^p \partial y^q}, \ldots\right)$

$\begin{cases} 0 \le p \le r \\ 0 \le q \le s \\ p+q < r+s \end{cases}$, *Ann. Mat. Pura Appl.* **113**, pp. 173–197.

[104] Hörmander, L. (1963). *Linear Partial Differential Operators* (Springer-Verlag, Berlin).

[105] Hsiang, W. -T. and Kwong, M. K. (1980). A comparison theorem for the first nodal line of the solutions of quasilinear hyperbolic equations with non-increasing initial values, *Proc. Roy. Soc. Edinburgh Sect. A* **86**, pp. 139–151.

[106] Hsiang, W. -T. and Kwong, M. K. (1982). On the oscillation of nonlinear hyperbolic equations, *J. Math. Anal. Appl.* **85**, pp. 31–45.

[107] Hsiang, W. -T. and Kwong, M. K. (1982). Oscillation of second-order hyperbolic equations with non-integrable coefficients, *Proc. Roy. Soc. Edinburgh Sect. A* **91**, pp. 305–313.

[108] Iraniparast, N. (1987). Qualitative behaviour of solutions of the Goursat problem for hyperbolic differential equations, *Comput. Math. Appl.* **13**, pp. 889–900.

[109] Itô, S. (1957). Fundamental solutions of parabolic differential equations and boundary value problems, *Japan. J. Math.* **27**, pp. 55–102.

[110] Jaroš, J. and Kusano, T. (1997). Second-order semilinear differential equations with external forcing terms, *Sūrikaisekikenkyūsho Kōkyūroku*, no. 984, pp. 191–197.

[111] Jaroš, J. and Kusano, T. (1999). A Picone type identity for second order half-linear differential equations, *Acta Math. Univ. Comenian.* **68**, pp. 117–121.

[112] Jaroš, J., Kusano, T. and Yoshida, N. (2000). Forced superlinear oscillations via Picone's identity, *Acta Math. Univ. Comenian.* **69**, pp. 107–113.

[113] Jaroš, J., Kusano, T. and Yoshida, N. (2001). Picone-type inequalities for nonlinear elliptic equations and their applications, *J. Inequal. Appl.* **6**, pp. 387–404.

[114] Jaroš, J., Kusano, T. and Yoshida, N. (2001). Oscillatory properties of solutions of superlinear-sublinear parabolic equations via Picone-type inequalities, *Math. J. Toyama Univ.* **24**, pp. 83–91.

[115] Jaroš, J., Kusano, T. and Yoshida, N. (2002). Generalized Picone's formula and forced oscillations in quasilinear differential equations of the second order, *Arch. Math. (Brno)* **38**, pp. 53–59.

[116] Jaroš, J., Kusano, T. and Yoshida, N. (2002). Picone-type inequalities for half-linear elliptic equations and their applications, *Adv. Math. Sci. Appl.* **12**, pp. 709–724.

[117] Jaroš, J., Kusano, T. and Yoshida, N. (2002). Oscillation properties of solutions of a class of nonlinear parabolic equations, *J. Comput. Appl. Math.* **146**, pp. 277–284.

[118] Jaroš, J., Kusano, T. and Yoshida, N. (2006). Picone-type inequalities for elliptic equations with first order terms and their applications, *J. Inequal. Appl.* **2006**, pp. 1–17.

[119] Jin, M. Z., Dong, Y. and Li, C. X. (1996). Forced oscillations of boundary value problems of higher order functional partial differential equations, *Appl. Math. Mech. (English Ed.)* **17**, pp. 889–900.

[120] John, F. (1980). *Partial Differential Equations, Third Edition* (Springer-Verlag, New York).

[121] Kahane, C. (1973). Oscillation theorems for solutions of hyperbolic equations, *Proc. Amer. Math. Soc.* **41**, pp. 183–188.

[122] Kamenev, I. V. (1978). An integral criterion for oscillation of linear differential equations of second order, *Math. Zamtki* **23**, pp. 249–251: *Math. Notes* **23**, pp. 136–138.

[123] Kartsatos, A. G. (1975) On nth order differential inequalities, *J. Math. Anal. Appl.* **52**, pp. 1–9.

[124] Kiguradze, I. T. (1962). Oscillation properties of solutions of certain ordinary differential equations, *Dokl. Akad. Nauk SSSR* **144**, pp. 33–36: *Soviet Math. Dokl.* **3**, pp. 649–652.

[125] Kiguradze, I. T. (1964). On the oscillation of solutions of the equation $\dfrac{d^m u}{dt^m} + a(t)|u|^n \operatorname{sign} u = 0$, *Mat. Sb.* **65**, pp. 172–187.

[126] Kiguradze, T., Kusano, T. and Yoshida, N. (2002). Oscillation criteria for a class of partial functional-differential equations of higher order, *J. Appl. Math. Stochastic Anal.* **15**, pp. 271–282.

[127] Kiguradze, T. and Stavroulakis, I. P. (2001). On oscillatory properties of solutions of higher order linear hyperbolic equations, *Adv. Math. Sci. Appl.* **11**, pp. 645–672.

[128] Kitamura, Y. and Kusano, T. (1978). Nonlinear oscillation of a fourth order elliptic equation, *J. Differential Equations* **30**, pp. 280–286.

[129] Kitamura, Y. and Kusano, T. (1978). An oscillation theorem for a sublinear Schrödinger equation, *Utilitas Math.* **14**, pp. 171–175.

[130] Kitamura, Y. and Kusano, T. (1980). Oscillation criteria for semilinear metaharmonic equations in exterior domains, *Arch. Rational Mech. Anal.* **75**, pp. 79–90.

[131] Kitamura, Y. and Kusano, T. (1980). Oscillation of first-order nonlinear differential equations with deviating arguments, *Proc. Amer. Math. Soc.* **78**, pp. 64–68.

[132] Kobayashi, K. and Yoshida, N. (2000). Oscillations of solutions of initial value problems for parabolic equations, *Math. J. Toyama Univ.* **23**, pp. 149–155.

[133] Kobayashi, K. and Yoshida, N. (2002). Unboundedness of solutions of time-dependent differential systems of parabolic type, *Math. J. Toyama Univ.* **25**, pp. 65–75.

[134] Kong, Q. (1999). Interval criteria for oscillation of second-order linear ordinary differential equations, *J. Math. Anal. Appl.* **229**, pp. 258–270.

[135] Kopáčková, M. (1983). On periodic solution of a nonlinear beam equation, *Appl. Math.* **28**, pp. 108–115.

[136] Kreith, K. (1969). Sturmian theorems for characteristic initial value problems, *Atti Accad. Naz. Lincei Rend. Cl. Sci. Fis. Mat. Natur.* **47**, pp. 139–

144.

[137] Kreith, K. (1969). Sturmian theorems for hyperbolic equations, *Proc. Amer. Math. Soc.* **22**, pp. 277–281.

[138] Kreith, K. (1973). *Oscillation Theory, Lecture Notes in Mathematics*, Vol. 324 (Springer-Verlag, Berlin).

[139] Kreith, K. (1974). A nonselfadjoint dynamical system, *Proc. Edinburgh Math. Soc. (2)* **19**, pp. 77–87.

[140] Kreith, K. (1975). Picone's identity and generalizations, *Rend. Mat.* **8**, pp. 251–261.

[141] Kreith, K. (1981). A class of comparison theorems for non-linear hyperbolic initial value problems, *Proc. Roy. Soc. Edinburgh Sect. A* **87**, pp. 189–191.

[142] Kreith, K., Kusano, T. and Yoshida, N. (1984). Oscillation properties of nonlinear hyperbolic equations, *SIAM J. Math. Anal.* **15**, pp. 570–578.

[143] Kreith, K. and Ladas, G. (1985). Allowable delays for positive diffusion processes, *Hiroshima Math. J.* **15**, pp. 437–443.

[144] Kreith, K. and Pagan, G. (1983). Qualitative theory for hyperbolic characteristic initial value problems, *Proc. Roy. Soc. Edinburgh Sect. A* **94**, pp. 15–24.

[145] Kreith, K. and Swanson, C. A. (1985). Kiguradze classes for characteristic initial value problems, *Comput. Math. Appl.* **11**, pp. 239–247.

[146] Kreith, K. and Travis, C. C. (1972). Oscillation criteria for selfadjoint elliptic equations, *Pacific J. Math.* **41**, pp. 743–753.

[147] Kubiaczyk, I. and Saker, S. H. (2002). Oscillation of delay parabolic differential equations with several coefficients, *J. Comput. Appl. Math.* **147**, pp. 263–275.

[148] Kuks, L. M. (1962). Sturm's theorem and oscillation of solutions of strongly elliptic systems, *Soviet Math. Dokl.* **3**, pp. 24–27.

[149] Kuks, L. M. (1978). Unboundedness of solutions of high-order parabolic systems in the plane and a Sturm-type comparison theorem, *Differ. Uravn.* **14**, pp. 878–884: *Differ. Equ.* **14**, pp. 623–627.

[150] Kura, T. (1982). Oscillation criteria for a class of sublinear elliptic equations of the second order, *Utilitas Math.* **22**, pp. 335–341.

[151] Kusano, T., Jaroš, J. and Yoshida, N. (2000). A Picone-type identity and Sturmian comparison and oscillation theorems for a class of half-linear partial differential equations of second order, *Nonlinear Anal.* **40**, pp. 381–395.

[152] Kusano, T. and Naito, M. (1982). Oscillation criteria for a class of perturbed Schrödinger equations, *Canad. Math. Bull.* **25**, pp. 71–77.

[153] Kusano, T. and Naito, Y. (1997). Oscillation and nonoscillation criteria for second order quasilinear differential equations, *Acta Math. Hungar.* **76**, pp. 81–99.

[154] Kusano, T., Naito, Y. and Ogata, A. (1994). Strong oscillation and nonoscillation of quasilinear differential equations of second order, *Differential Equations Dynam. Systems* **2**, pp. 1–10.

[155] Kusano, T. and Narita, M. (1977). Unboundedness of solutions of parabolic differential inequalities, *J. Math. Anal. Appl.* **57**, pp. 68–75.

[156] Kusano, T. and Yoshida, N. (1975). Nonoscillation criteria for strongly

elliptic systems, *Boll. Un. Mat. Ital. (4)* **11**, pp. 262–265.

[157] Kusano, T. and Yoshida, N. (1980). Nonlinear oscillation criteria for singular elliptic differential operators, *Funkcial. Ekvac.* **23**, pp. 135–142.

[158] Kusano, T. and Yoshida, N. (1981). Oscillation criteria for a class of nonlinear partial differential equations, *J. Math. Anal. Appl.* **79**, pp. 236–243.

[159] Kusano, T. and Yoshida, N. (1985). Forced oscillations of Timoshenko beams, *Quart. Appl. Math.* **43**, pp. 167–177.

[160] Kusano, T. and Yoshida, N. (1994). Oscillation of parabolic equations with oscillating coefficients, *Hiroshima Math. J.* **24**, pp. 123–133.

[161] Kusano, T. and Yoshida, N. (1999). Oscillation criteria for a class of functional parabolic equations, *J. Appl. Anal.* **5**, pp. 1–16.

[162] Ladde, G. S., Lakshmikantham, V. and Zhang, B. G. (1987). *Oscillation Theory of Differential Equations with Deviating Arguments* (Marcel Dekker, Inc., New York).

[163] Lakshmikantham, V., Bainov, D. D. and Simeonov, P. S. (1989). *Theory of Impulsive Differential Equations*, *Series in Modern Applied Mathematics*, 6, (World Scientific Publishing Co., Inc., Teaneck, NJ).

[164] Lalli, B. S., Yu, Y. H. and Cui, B. T. (1992). Oscillations of certain partial differential equations with deviating arguments, *Bull. Austral. Math. Soc.* **46**, pp. 373–380.

[165] Leighton, W. (1952). On selfadjoint differential equations of second order, *J. London Math. Soc.* **27**, pp. 37–47.

[166] Levine, H. A. and Payne, L. E. (1976). On the nonexistence of entire solutions to nonlinear second order elliptic equations, *SIAM J. Math. Anal.* **7**, pp. 337–343.

[167] Li, W. N. (2000). Oscillation properties for systems of hyperbolic differential equations of neutral type, *J. Math. Anal. Appl.* **248**, pp. 369–384.

[168] Li, W. N. (2003). Forced oscillation properties for certain systems of partial functional differential equations, *Appl. Math. Comput.* **143**, pp. 223–232.

[169] Li, W. N. (2005). On the forced oscillation of solutions for systems of impulsive parabolic differential equations with several delays, *J. Comput. Appl. Math.* **181**, pp. 46–57.

[170] Li, W. N. and Cui, B. T. (1999). Oscillation for systems of parabolic equations of neutral type, *Southeast Asian Bull. Math.* **23**, pp. 447–456.

[171] Li, W. N. and Cui, B. T. (1999). Oscillation of solutions of neutral partial functional differential equations, *J. Math. Anal. Appl.* **234**, pp. 123–146.

[172] Li, W. N., Cui, B. T. and Debnath, L. (2003). Oscillation of systems of certain neutral delay parabolic differential equations, *J. Appl. Math. Stochastic Anal.* **16**, pp. 81–94.

[173] Li, W. N. and Debnath, L. (2003). Oscillation of higher-order neutral partial functional differential equations, *Appl. Math. Lett.* **16**, pp. 525–530.

[174] Li, W. N., Han, M. and Meng, F. W. (2004). *H*-oscillation of solutions of certain vector hyperbolic differential equations with deviating arguments, *Appl. Math. Comput.* **158**, pp. 637–653.

[175] Li, W. N., Han, M. and Meng, F. W. (2005). Necessary and sufficient conditions for oscillation of impulsive parabolic differential equations with

delays, *Appl. Math. Lett.* **18**, pp. 1149–1155.
[176] Li, W. N. and Meng, F. W. (2001). Oscillation for systems of neutral partial differential equations with continuous distributed deviating arguments, *Demonstratio Math.* **34**, pp. 619–633.
[177] Li, W. N. and Meng, F. W. (2003). On the forced oscillation of systems of neutral parabolic differential equations with deviating arguments, *J. Math. Anal. Appl.* **288**, pp. 20–27.
[178] Li, Y. K. (1997). Oscillation of systems of hyperbolic differential equations with deviating arguments, *Acta Math. Sinica* **40**, pp. 100–105.
[179] Ličko, I. and Švec, M. (1963). Le caractère oscillatoire des solutions de l'équation $y^{(n)} + f(x)y^{\alpha} = 0$, $n > 1$, *Czechoslovak Math. J.* **13**, pp. 481–491.
[180] Liu, A. P., Xiao, L. and Liu, T. (2004). Oscillation of nonlinear impulsive hyperbolic equations with several delays, *Electron. J. Differential Equations* **2004**, No. 24, pp. 1–6.
[181] Liu, B. (1994). Oscillatory properties of systems of nonlinear delay parabolic differential equations, *Ann. Differential Equations* **10**, pp. 290–298.
[182] Liu, X. and Fu, X. (1996). Oscillation criteria for nonlinear inhomogeneous hyperbolic equations with distributed deviating arguments, *J. Appl. Math. Stochastic Anal.* **9**, pp. 21–31.
[183] Liu, X. and Fu, X. (1998). Oscillation criteria for high order delay partial differential equations, *J. Appl. Math. Stochastic Anal.* **11**, pp. 193–208.
[184] Luo, J. W. (2002). Oscillation of hyperbolic partial differential equations with impulses, *Appl. Math. Comput.* **133**, pp. 309–318.
[185] Mangeron, D. (1968). Problèmes à la frontière concernant les équations polyvibrantes, *C. R. Acad. Sci. Paris Sér. A-B* **266**, pp. 870–873; pp. 976–979; pp. 1050–1052; pp. 1103–1106; pp. 1121–1124.
[186] Mařík, R. (2000). Oscillation criteria for the Schrödinger PDE, *Adv. Math. Sci. Appl.* **10**, pp. 495–511.
[187] Mařík, R. (2000). Hartman-Wintner type theorem for PDE with p-Laplacian, *Electron. J. Qual. Theory Differ. Equ.* No. **18**, pp. 1–7.
[188] Mařík, R. (2000). Oscillation criteria for PDE with p-Laplacian via the Riccati technique, *J. Math. Anal. Appl.* **248**, pp. 290–308.
[189] Mařík, R. (2002). Positive solutions of inequality with p-Laplacian in exterior domains, *Math. Bohem.* **127**, pp. 597–604.
[190] Mařík, R. (2002). Oscillation criteria for a class of nonlinear partial differential equations, *Electron. J. Differential Equations* **2002**, No. 28, pp. 1–10.
[191] Mařík, R. (2004). Integral averages and oscillation criteria for half-linear partial differential equation, *Appl. Math. Comput.* **150**, pp. 69–87.
[192] Mařík, R. (2004). Riccati-type inequality and oscillation criteria for a half-linear PDE with damping, *Electron. J. Differential Equations* **2004**, No. 11, pp. 1–17.
[193] Mařík, R. (2005). Asymptotic estimates for PDE with p-Laplacian and damping, *Electron. J. Qual. Theory Differ. Equ.* No. **5**, pp. 1–6.

[194] Mařík, R. (2005). Oscillation of the half-linear PDE in general exterior domains — the variational approach, *Nonlinear Anal.* **60**, pp. 485–489.

[195] Mařík, R. (2006). Interval-type oscillation criteria for half-linear PDE with damping, *Appl. Appl. Math.* **1**, pp. 1–10.

[196] McNabb, A. (1962). A note on the boundedness of solutions of linear parabolic equations, *Proc. Amer. Math. Soc.* **13**, pp. 262–265.

[197] Medeiros, L. A. (1979). On a new class of nonlinear wave equations, *J. Math. Anal. Appl.* **69**, pp. 252–262.

[198] Mikusiński, J. (1951). On Fite's oscillation theorems, *Colloq. Math.* **2**, pp. 34–39.

[199] Minchev, E. and Yoshida, N. (2003). Oscillations of solutions of vector differential equations of parabolic type with functional arguments, *J. Comput. Appl. Math.* **151**, pp. 107–117.

[200] Minchev, E. and Yoshida, N. (2003). Oscillations of vector differential equations of hyperbolic type with functional arguments, *Math. J. Toyama Univ.* **26**, pp. 75–84.

[201] Mishev, D. P. (1986). Oscillatory properties of the solutions of hyperbolic differential equations with "maximum", *Hiroshima Math. J.* **16**, pp. 77–83.

[202] Mishev, D. P. (1989). Oscillation of the solutions of hyperbolic differential equations of neutral type with "maxima", *Godishnik Vissh. Uchebn. Zaved. Prilozhna Mat.* **25**, pp. 9–18.

[203] Mishev, D. P. (1989). Oscillation of the solutions of parabolic differential equations of neutral type with "maxima", *Godishnik Vissh. Uchebn. Zaved. Prilozhna Mat.* **25**, pp. 19–28.

[204] Mishev, D. P. (1991). Necessary and sufficient conditions for oscillation of neutral type of parabolic differential equations, *C. R. Acad. Bulgare Sci.* **44**, pp. 11–14.

[205] Mishev, D. P. (1991). Oscillation of the solutions of neutral type hyperbolic differential equations, *Math. Balkanica* **5**, pp. 121–128.

[206] Mishev, D. P. (1992). Oscillation of the solutions of non-linear parabolic equations of neutral type, *Houston J. Math.* **18**, pp. 259–269.

[207] Mishev, D. P. and Bainov, D. D. (1984). Oscillation properties of the solutions of hyperbolic equations of neutral type, *Proceedings of the Colloquium on Qualitative Theory of Differential Equations*, Szeged, pp.771–780.

[208] Mishev, D. P. and Bainov, D. D. (1986). Oscillation properties of the solutions of a class of hyperbolic equations of neutral type, *Funkcial. Ekvac.* **29**, pp. 213–218.

[209] Mishev, D. P. and Bainov, D. D. (1988). Oscillation of the solutions of parabolic differential equations of neutral type, *Appl. Math. Comput.* **28**, pp. 97–111.

[210] Mishev, D. P. and Bainov, D. D. (1992). A necessary and sufficient condition for oscillation of neutral type hyperbolic equations, *Rend. Mat.* **12**, pp. 545–553.

[211] Naito, M., Naito, Y. and Usami, H. (1997). Oscillation theory for semilinear elliptic equations with arbitrary nonlinearities, *Funkcial. Ekvac.* **40**, pp. 41–55.

[212] Naito, M. and Yoshida, N. (1978). Oscillation theorems for semilinear elliptic differential operators, *Proc. Roy. Soc. Edinburgh Sect. A* **82**, pp. 135–151.

[213] Naito, M. and Yoshida, N. (1989). Oscillation criteria for a class of higher order elliptic equations, *Math. Rep. Toyama Univ.* **12**, pp. 29–40.

[214] Naito, Y. and Usami, H. (2001). Oscillation criteria for quasilinear elliptic equations, *Nonlinear Anal.* **46**, pp. 629–652.

[215] Narazaki, T. (1981). On the time global solutions of perturbed beam equations, *Proc. Fac. Sci. Tokai Univ.* **16**, pp. 51–71.

[216] Narita, M. (1978). Oscillation theorems for semilinear hyperbolic and ultrahyperbolic operators, *Bull. Austral. Math. Soc.* **18**, pp. 55–64.

[217] Narita, M. and Yoshida, N. (1979). Oscillation theorems for linear ultrahyperbolic operators, *Utilitas Math.* **15**, pp. 143–159.

[218] Narita, M. and Yoshida, N. (1979). Oscillation of semilinear ultrahyperbolic differential operators, *Ann. Mat. Pura Appl.* **122**, pp. 289–300.

[219] Nasr, A. H. (1998). Sufficient conditions for the oscillation of forced superlinear second order differential equations with oscillatory potential, *Proc. Amer. Math. Soc.* **126**, pp. 123–125.

[220] Noussair, E. S. (1971). Oscillation theory of elliptic equations of order $2m$, *J. Differential Equations* **10**, pp. 100–111.

[221] Noussair, E. S. (1975). Oscillation of elliptic equations in general domains, *Canad. J. Math.* **27**, pp. 1239–1245.

[222] Noussair, E. S. and Swanson, C. A. (1972). Oscillation theorems for vector differential equations, *Utilitas Math.* **1**, pp. 97–109.

[223] Noussair, E. S. and Swanson, C. A. (1976). Oscillation theory for semilinear Schrödinger equations and inequalities, *Proc. Roy. Soc. Edinburgh Sect. A* **75**, pp. 67–81.

[224] Noussair, E. S. and Swanson, C. A. (1976). Oscillation of nonlinear vector differential equations, *Ann. Mat. Pura Appl.* **109**, pp. 305–315.

[225] Noussair, E. S. and Swanson, C. A. (1980). Oscillation of semilinear elliptic inequalities by Riccati transformations, *Canad. J. Math.* **32**, pp. 908–923.

[226] Noussair, E. S. and Yoshida, N. (1975). Nonoscillation criteria for elliptic equations of order $2m$, *Atti Accad. Naz. Lincei Rend. Cl. Sci. Fis. Mat. Natur.* **59**, pp. 57–64.

[227] Oğuztöreli, M. N. and Easwaran, S. (1971). A Goursat problem for a high order Mangeron equation, *Atti Accad. Naz. Lincei Rend. Cl. Sci. Fis. Mat. Natur.* **50**, pp. 650–653.

[228] Okikiolu, G. O. (1971). *Aspects of the Theory of Bounded Integral Operators in L^p-spaces* (Academic Press, New York).

[229] Onose, H. (1975). A comparison theorem and the forced oscillation, *Bull. Austral. Math. Soc.* **13**, pp. 13–19.

[230] Onose, H. (1982). Oscillation criteria for the sublinear Schrödinger equation, *Proc. Amer. Math. Soc.* **85**, pp. 69–72.

[231] H. Onose, (1993). Oscillation of nonlinear functional differential equations, *GAKUTO Internat. Ser. Math. Sci. Appl.* **2**, pp. 637–642.

[232] Onose, H. and Yokoyama, E. (1976). Oscillation of partial differential in-

equalities, *Tamkang J. Math.* **7**, pp. 67–70.

[233] Pagan, G. (1973). Oscillation theorems for characteristic initial value problems for linear hyperbolic equations, *Atti Accad. Naz. Lincei Rend. Cl. Sci. Fis. Mat. Natur.* **55**, pp. 301–313.

[234] Pagan, G. (1976). Existence of nodal lines for solutions of hyperbolic equations, *Amer. Math. Monthly* **83**, pp. 358–359.

[235] Pagan, G. (1977). An oscillation theorem for characteristic initial value problems in linear hyperbolic equations, *Proc. Roy. Soc. Edinburgh Sect. A* **77**, pp. 265–271.

[236] Pagan, G. and Stocks, D. (1979). Oscillation criteria for second order hyperbolic initial value problems, *Proc. Roy. Soc. Edinburgh Sect. A* **83**, pp. 239–244.

[237] Parhi, N. and Kirane, M. (1994). Oscillatory behaviour of solutions of coupled hyperbolic differential equations, *Analysis* **14**, pp. 43–56.

[238] Petrova, Z. A. (2005). Oscillations of a class of sublinear and superlinear hyperbolic equations, *C. R. Acad. Bulgare Sci.* **58**, pp. 251–256.

[239] Philos, Ch. G. (1989). Oscillation theorems for linear differential equations of second order, *Arch. Math. (Basel)* **53**, pp. 482–492.

[240] Picone, M. (1909). Sui valori eccezionali di un parametro da cui dipende un'equazione differenziale lineare ordinaria del second'ordine, *Ann. Scuola Norm. Sup. Pisa* **11**, pp. 1–141.

[241] Picone, M. (1911). Un teorema sulle soluzioni delle equazioni lineari ellittiche autoaggiunte alle derivate parziali del secondo-ordine, *Atti Accad. Naz. Lincei Rend. Cl. Sci. Fis. Mat. Natur.* **20**, pp. 213–219.

[242] Protter, M. H. and Weinberger, H. F. (1984). *Maximum Principles in Differential Equations* (Springer-Verlag, New York).

[243] Shoukaku, Y. (2003). On the oscillatory properties of parabolic equations with continuous distributed arguments, *Math. J. Toyama Univ.* **26**, pp. 93–107.

[244] Shoukaku, Y. and Yoshida, N. (2003). Oscillatory properties of solutions of nonlinear parabolic equations with functional arguments, *Indian J. Pure Appl. Math.* **34**, pp. 1469–1478.

[245] Shoukaku, Y. and Yoshida, N. (2005). Oscillations of parabolic systems with functional arguments, *Toyama Math. J.* **28**, pp. 105–131.

[246] Shoukaku, Y. and Yoshida, N. (2006). Oscillations of hyperbolic systems with functional arguments, *Appl. Appl. Math.* **1**, pp. 83–95.

[247] Śliwiński, E. (1964). On some oscillation problems for the equation $\Delta^{(n)}u - f(r)u = 0$ in a three-dimensional space, *Prace Mat.* **8**, pp. 119–120.

[248] Suleĭmanov, N. M. (1979). On the behavior of solutions of second order nonlinear elliptic equations with linear principal part, *Soviet Math. Dokl.* **20**, pp. 815–819.

[249] Sturm, C. (1836). Sur les équations différentielles linéaires du second ordre, *J. Math. Pures Appl.* **1**, pp. 106–186.

[250] Švaňa, P. (1988). Oscillation criteria for forced nonlinear elliptic equations of arbitrary order, *Časopis Pěst. Mat.* **113**, pp. 169–178.

[251] Swanson, C. A. (1968). *Comparison and Oscillation Theory of Linear Dif-*

ferential Equations (Academic Press, New York).

[252] Swanson, C. A. (1973). Strong oscillation of elliptic equations in general domains, *Canad. Math. Bull.* **16**, pp. 105–110.

[253] Swanson, C. A. (1975). Picone's identity, *Rend. Mat.* **8**, pp. 373–397.

[254] Swanson, C. A. (1979). Semilinear second order elliptic oscillation, *Canad. Math. Bull.* **22**, pp. 139–157.

[255] Swanson, C. A. (1983). Criteria for oscillatory sublinear Schrödinger equations, *Pacific J. Math.* **104**, pp. 483–493.

[256] Tanaka, S. (1997). Oscillation properties of solutions of second order neutral differential equations with deviating arguments, *Analysis* **17**, pp. 99–111.

[257] Tanaka, S. (1997). Forced oscillations of first order nonlinear neutral differential equations, *J. Appl. Anal.* **3**, pp. 23–41.

[258] Tanaka, S. and Yoshida, N. (1997). Oscillations of solutions to parabolic equations with deviating arguments, *Tamkang J. Math.* **28**, pp. 169–181.

[259] Tao, Y. and Yoshida, N. (2005). Oscillation of nonlinear hyperbolic equations, *Toyama Math. J.* **28**, pp. 27–40.

[260] Tao, Y. and Yoshida, N. (2006). Oscillation criteria for hyperbolic equations with distributed deviating arguments, *Indian J. Pure Appl. Math.* **37**, pp. 291–305.

[261] Timoshenko, S., Young, D. H. and Weaver, Jr., W. (1974). *Vibration Problems in Engineering, Fourth Edition* (Wiley, New York).

[262] Toraev, A. (1981). On the oscillation of solutions of higher order equations of elliptic type, *Soviet Math. Dokl.* **24**, pp. 195–198.

[263] Tramov, M. I. (1984). Oscillation of solutions of partial differential equations with deviating argument, *Differ. Uravn.* **20**, pp. 721–723.

[264] Travis, C. C. (1974). Comparison and oscillation theorems for hyperbolic equations, *Utilitas Math.* **6**, pp. 139–151.

[265] Travis, C. C. and Yoshida, N. (1981). Oscillation criteria for third order hyperbolic characteristic initial value problems, *Proc. Roy. Soc. Edinburgh Sect. A* **88**, pp. 135–140.

[266] Travis, C. C. and Yoshida, N. (1982). Oscillation criteria for nonlinear Bianchi equations, *Nonlinear Anal.* **6**, pp. 625–636.

[267] Travis, C. C. and Young, E. C. (1978). Comparison theorems for ultrahyperbolic equations, *Internat. J. Math. Math. Sci.* **1**, pp. 31–40.

[268] Trench, W. F. (1974). Canonical forms and principal systems for general disconjugate equations, *Trans. Amer. Math. Soc.* **189**, pp. 319–327.

[269] Uesaka, H. (1996). Oscillation of solutions of nonlinear wave equations, *Proc. Japan Acad. Ser. A* **72**, pp. 148–151.

[270] Uesaka, H. (1997). Oscillatory behavior of solutions of nonlinear wave equations, *Nonlinear Anal.* **30**, pp. 4655–4661.

[271] Uesaka, H. (2001). A pointwise oscillation property of semilinear wave equations with time-dependent coefficients II, *Nonlinear Anal.* **47**, pp. 2563–2571.

[272] Uesaka, H. (2003). A pointwise oscillation property of semilinear wave equations with time-dependent coefficients, *Nonlinear Anal.* **54**, pp. 1271–1283.

[273] Uesaka, H. (2004). Oscillation or nonoscillation property for semilinear

wave equations, *J. Comput. Appl. Math.* **164–165**, pp. 723–730.

[274] Usami, H. (1995). Nonexistence properties of positive solutions and oscillation criteria for second order semilinear elliptic inequalities, *Arch. Rational Mech. Anal.* **130**, pp. 277–302.

[275] Usami, H. (1998). Some oscillation theorems for a class of quasilinear elliptic equations, *Ann. Mat. Pura Appl.* **175**, pp. 277–283.

[276] van der Waerden, B. L. (1960). *Algebra I* (Springer-Verlag, Berlin).

[277] Wachnicki, E. (1976). On the oscillatory properties of solutions of certain elliptic equations, *Comment. Math. Prace Mat.* **9**, pp. 165–169.

[278] Wang, P. G. (1999). Forced oscillation of a class of delay hyperbolic equation boundary value problem, *Appl. Math. Comput.* **103**, pp. 15–25.

[279] Wang, P. G. (2000). Oscillation of certain neutral hyperbolic equations, *Indian J. Pure Appl. Math.* **31**, pp. 949–956.

[280] Wang, P. G. and Feng, C. H. (2000). Oscillation of parabolic equations of neutral type, *J. Comput. Appl. Math.* **126**, pp. 111–120.

[281] Wang, P. G. and Yu, Y. H. (1999). Oscillation criteria for a nonlinear hyperbolic equation boundary value problem, *Appl. Math. Lett.* **12**, pp. 91–98.

[282] Wang, T. M. and Gagnon, L. W. (1978). Vibrations of continuous Timoshenko beams on Winkler-Pasternak foundations, *J. Sound Vibration* **59**, pp. 211–220.

[283] Wintner, A. (1949). A criterion of oscillatory stability, *Quart. J. Appl. Math.* **7**, pp. 115–117.

[284] Wong, J. S. W. (2001). On Kamenev-type oscillation theorems for second-order differential equations with damping, *J. Math. Anal. Appl.* **258**, pp. 244–257.

[285] Wu, J. (1996). *Theory and Applications of Partial Functional Differential Equations* (Springer-Verlag, New York).

[286] Xie, S. L. and Cheng, S. S. (1995). Oscillation of a logistic equation with delay and diffusion, *Ann. Polon. Math.* **62**, pp. 219–230.

[287] Xu, Z. T. (2005). The oscillatory behavior of second order nonlinear elliptic equations, *Electron. J. Qual. Theory Differ. Equ.* No. **8**, pp. 1–11.

[288] Xu, Z. T. (2006). Oscillation of second-order elliptic equations with damping terms, *J. Math. Anal. Appl.* **317**, pp. 349–363.

[289] Xu, Z. T. (2006). Riccati inequality and oscillation criteria for PDE with p-Laplacian, *J. Inequal. Appl.* **2006**, pp. 1–10.

[290] Xu, Z. T. (2006). On the oscillation of second order quasilinear elliptic equations, *Math. Comput. Modelling* **43**, pp. 30–41.

[291] Xu, Z. T. and Xing, H. Y. (2003). Oscillation criteria of Kamenev-type for PDE with p-Laplacian, *Appl. Math. Comput.* **145**, pp. 735–745.

[292] Yan, L. and Yoshida, N. (2006). Oscillations of characteristic initial value problems for hyperbolic equations with delays, *Indian J. Pure Appl. Math.* **37**, pp. 357–367.

[293] Ye, Q. X. and Li, Z. Y. (1990). *Introduction to Reaction-Diffusion Equations* (Beijing Science Press, Beijing).

[294] Yoshida, N. (1976). Nonoscillation and comparison theorems for a class of

higher order elliptic systems, *Japan. J. Math. (N.S.)* **2**, pp. 419–434.
[295] Yoshida, N. (1979). An oscillation theorem for characteristic initial value problems for nonlinear hyperbolic equations, *Proc. Amer. Math. Soc.* **76**, pp. 95–100.
[296] Yoshida, N. (1983). Oscillation properties of solutions of second order elliptic equations, *SIAM J. Math. Anal.* **14**, pp. 709–718.
[297] Yoshida, N. (1984). An oscillation theorem for sublinear elliptic differential inequalities, *Bull. Austral. Math. Soc.* **30**, pp. 387–394.
[298] Yoshida, N. (1985). Forced oscillations of extensible beams, *SIAM J. Math. Anal.* **16**, pp. 211–220.
[299] Yoshida, N. (1985). Forced oscillations of nonlinear extensible beams, *Proceedings of the Tenth International Conference on Nonlinear Oscillations (Varna, 1984)*, Sofia, pp. 814–817.
[300] Yoshida, N. (1986). Oscillation of nonlinear parabolic equations with functional arguments, *Hiroshima Math. J.* **16**, pp. 305–314.
[301] Yoshida, N. (1987). On the zeros of solutions to nonlinear hyperbolic equations, *Proc. Roy. Soc. Edinburgh Sect. A* **106**, pp. 121–129.
[302] Yoshida, N. (1987). Forced oscillations of solutions of parabolic equations, *Bull. Austral. Math. Soc.* **36**, pp. 289–294.
[303] Yoshida, N. (1987). Oscillation properties of solutions of characteristic initial value problems, *Proceedings of the Eleventh International Conference on Nonlinear Oscillations (Budapest, 1987)*, Budapest, pp. 527–530.
[304] Yoshida, N. (1987). Forced oscillations of solutions of second order elliptic equations, *Appl. Anal.* **25**, pp. 149–155.
[305] Yoshida, N. (1988). On the zeros of solutions of beam equations, *Ann. Mat. Pura Appl.* **151**, pp. 389–398.
[306] Yoshida, N. (1990). On the zeros of solutions of hyperbolic equations of neutral type, *Differential Integral Equations* **3**, pp. 155–160.
[307] Yoshida, N. (1992). Oscillatory properties of solutions of certain elliptic equations, *Bull. Austral. Math. Soc.* **45**, pp. 297–303.
[308] Yoshida, N. (1992). Forced oscillations of parabolic equations with deviating arguments, *Math. J. Toyama Univ.* **15**, pp. 131–142.
[309] Yoshida, N. (1993). On the zeros of solutions of hyperbolic equations with deviating arguments, *Math. J. Toyama Univ.* **16**, pp. 125–133.
[310] Yoshida, N. (1994). On the zeros of solutions of elliptic equations with deviating arguments, *J. Math. Anal. Appl.* **185**, pp. 570–578.
[311] Yoshida, N. (1995). Forced oscillations of nonlinear parabolic equations with functional arguments, *Analysis* **15**, pp. 71–84.
[312] Yoshida, N. (1995). On the oscillation of solutions to parabolic equations with functional arguments, *Math. J. Toyama Univ.* **18**, pp. 65–78.
[313] Yoshida, N. (1995). Oscillation of partial functional-differential equations with deviating arguments, *Tamkang J. Math.* **26**, pp. 131–139.
[314] Yoshida, N. (1996). Nonlinear oscillation of first order delay differential equations, *Rocky Mountain J. Math.* **26**, pp. 361–373.
[315] Yoshida, N. (1996). On the zeros of solutions of initial value problems for hyperbolic equations in one-dimensional space, *Math. J. Toyama Univ.* **19**,

pp. 179–189.

[316] Yoshida, N. (1997). Zeros of solutions of hyperbolic equations with functional arguments, *AMS/IP Stud. Adv. Math.* **3**, pp. 613–617.

[317] Yoshida, N. (1997). Oscillation of partial functional differential equations, *Math. J. Toyama Univ.* **20**, pp. 107–139.

[318] Yoshida, N. (1999). Forced oscillations of a class of parabolic equations with functional arguments, *Math. J. Toyama Univ.* **22**, pp. 187–204.

[319] Yoshida, N. (2001). Oscillation criteria for a class of hyperbolic equations with functional arguments, *Kyungpook Math. J.* **41**, pp. 75–85.

[320] Yoshida, N. (2003). Oscillation of half-linear partial differential equations with first order terms, *Stud. Univ. Žilina* **17**, pp. 177–184.

[321] Yoshida, N. (2006). Picone-type inequalities for a class of quasilinear elliptic equations and their applications, *Proceedings of the Conference on Differential & Difference Equations and Applications (Florida, 2005)*, New York, pp. 1177–1185.

[322] Yoshida, N. (2007). Oscillation criteria for half-linear partial differential equations via Picone's identity, *Proceedings of Equadiff 11 (Bratislava, 2005)*, Bratislava, pp. 589–598.

[323] Yoshida, N. (2007). Picone-type inequality and oscillation theorems for a class of quasilinear elliptic equations, *Proceedings of the Colloquium on Differential and Difference Equations (Brno, 2006): Folia Fac. Sci. Natur. Univ. Masaryk. Brun. Math.* **16**, pp. 193–200.

[324] Yoshida, N. (2007). Oscillations and unboundedness of solutions of superlinear-sublinear parabolic equations via Picone-type inequality, *Toyama Math. J.* **30**, pp. 77–87.

[325] Zhang, B. G. and Zhou, Y. (2007). *Qualitative Analysis of Delay Partial Difference Equations* (Hindawi Publishing Corporation, New York).

[326] Zhang, L. Q. (2000). Oscillation criteria for hyperbolic partial differential equations with fixed moments of impulse effects, *Acta Math. Sinica* **43**, pp. 17–26.

[327] Zuazua, E. (1990). Oscillation properties for some damped hyperbolic equations, *Houston J. Math.* **16**, pp. 25–52.

Index